Collins

INTERNATIONAL PRIMARY MATHS

T0173346

Teacher's Guide 3

William Collins' dream of knowledge for all began with the publication of his first book in 1819. A self-educated mill worker, he not only enriched millions of lives, but also founded a flourishing publishing house. Today, staying true to this spirit, Collins books are packed with inspiration, innovation and practical expertise. They place you at the centre of a world of possibility and give you exactly what you need to explore it.

Collins. Freedom to teach.

Published by Collins
An imprint of HarperCollinsPublishers
The News Building
1 London Bridge Street
London
SE1 9GF

HarperCollins Publishers
HarperCollins
Macken House, 39/40 Mayor Street Upper,
Dublin 1, C9W8, D01
Ireland

Browse the complete Collins catalogue at
www.collins.co.uk

© HarperCollins*Publishers* Limited 2021

10 9

ISBN 978-0-00-836953-8

All rights reserved. No part of this publication may be reproduced, stored in a retrieval system, or transmitted in any form by any means, electronic, mechanical, photocopying, recording or otherwise, without the prior written permission of the Publisher or a licence permitting restricted copying in the United Kingdom issued by the Copyright Licensing Agency Ltd, 5th Floor, Shackleton House, 4 Battle Bridge Lane, London SE1 2HX.

British Library Cataloguing-in-Publication Data
A catalogue record for this publication is available from the British Library.

Author: Caroline Clissold
Series editor: Peter Clarke
Publisher: Elaine Higgleton
Product developer: Holly Woolnough
Project manager: Mike Harman (Life Lines Editorial Services)
Development editor: Joan Miller
Copyeditor: Tanya Solomons
Proofreader: Catherine Dakin
Answer checker: Steven Matchett
Cover designer: Gordon MacGilp
Cover illustrator: Ann Paganuzzi
Typesetter: QBS Learning
Illustrators: Ann Paganuzzi and QBS Learning
Production controller: Lyndsey Rogers
Printed and Bound in the UK by Ashford Colour Press Ltd

With thanks to the following teachers and schools for reviewing materials in development: Antara Banerjee, Calcutta International School; Hawar International School; Melissa Brobst, International School of Budapest; Rafaella Alexandrou, Pascal Primary Lefkosia; Maria Biglikoudi, Georgia Keravnou, Sotiria Leonidou and Niki Tzorzis, Pascal Primary School Lemessos; Taman Rama Intercultural School, Bali.

MIX
Paper | Supporting responsible forestry
FSC www.fsc.org FSC™ C007454

This book contains FSC™ certified paper and other controlled sources to ensure responsible forest management.

For more information visit: www.harpercollins.co.uk/green

The publishers gratefully acknowledge the permission granted to reproduce the copyright material in this book. Every effort has been made to trace copyright holders and to obtain their permission for the use of copyright material. The publishers will gladly receive any information enabling them to rectify any error or omission at the first opportunity.

Cambridge International copyright material in this publication is reproduced under licence and remains the intellectual property of Cambridge Assessment International Education.

Photo acknowledgements
Every effort has been made to trace copyright holders. Any omission will be rectified at the first opportunity.
p6t Tatiana Popova/Shutterstock; p6b Studio KIWI/Shutterstock; p275 Galina Petrova/Shutterstock; p283 StockAppeal/Shutterstock

Contents

Introduction

Revise activities

Collins International Primary Maths Stage 3 units

Introduction

① Key features of Collins International Primary Maths

Collins International Primary Maths places the learner at the centre of the teaching and learning of mathematics. To this end, all of the components and features of the course are aimed at helping teachers to address the distinct learning needs of *all* learners.

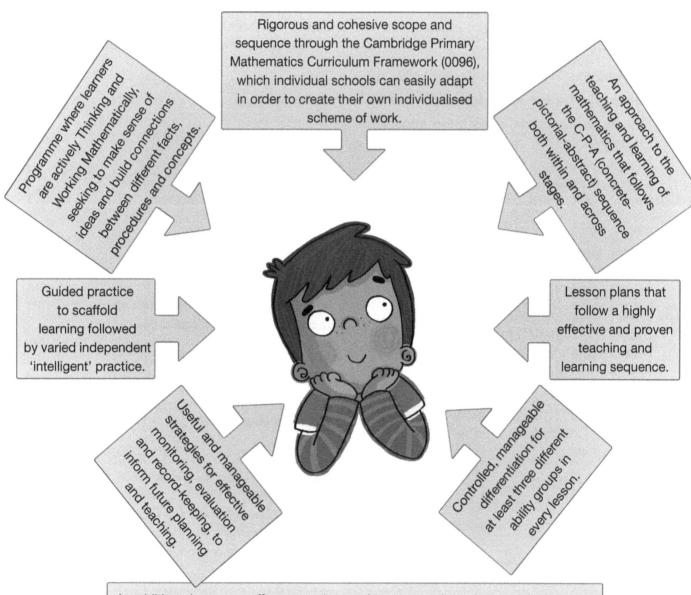

Rigorous and cohesive scope and sequence through the Cambridge Primary Mathematics Curriculum Framework (0096), which individual schools can easily adapt in order to create their own individualised scheme of work.

Programme where learners are actively Thinking and Working Mathematically, seeking to make sense of ideas and build connections between different facts, procedures and concepts.

An approach to the teaching and learning of mathematics that follows the C-P-A (concrete-pictorial-abstract) sequence both within and across stages.

Guided practice to scaffold learning followed by varied independent 'intelligent' practice.

Lesson plans that follow a highly effective and proven teaching and learning sequence.

Useful and manageable strategies for effective monitoring, evaluation and record-keeping, to inform future planning and teaching.

Controlled, manageable differentiation for at least three different ability groups in every lesson.

In addition, the course offers extensive teacher support through materials that:

- promote the most effective pedagogical methods in the teaching of mathematics
- are sufficiently detailed to aid confidence
- are rich enough to be varied and developed
- take into account issues of pace and classroom management
- give careful consideration to the key skill of appropriate and effective questioning
- provide a careful balance of teacher intervention and learner participation
- encourage communication of methods and foster mathematical rigour
- are aimed at raising levels of attainment for *every* learner.

② Collins International Primary Maths models of teaching and learning

Collins International Primary Maths is based on the constructivist model for teaching and learning developed by Piaget, Vygotsky, and later Bruner, and the importance of:

- starting from what learners know already and providing them with guidance that moves their thinking forwards
- students learning the fundamental principles of a subject, as well as the connections between ideas within the subject and across other subjects
- focusing on the process of learning, rather than the end product of it
- developing learners' intuitive thinking, by asking questions and providing opportunities for learners to ask questions

- learning through discovery and problem solving, which requires learners to make predictions, hypothesise, make generalisations, ask questions and discuss lines of enquiry
- using active methods that require rediscovering or reconstructing norms and truths
- using collaborative as well as individual activities, so that learners can learn from each other
- evaluating the level of each learner's development so that suitable tasks can be set.

At the core of the course are four of the major principles of Bruner's 'Theory of Instruction'* that characterise the organisation and content of Collins International Primary Maths.

Predisposition to learn

The concept of 'readiness for learning'.

A belief that any subject can be taught at any stage of development in a way that fits the learner's cognitive abilities.

Structure of knowledge

A body of knowledge can be structured so that it can be most readily grasped by the learner.

Effective sequencing

No one sequencing will fit every learner, but in general, curriculum content should be taught in increasing difficulty.

Modes of representation

Learning occurs through three modes of representation: *enactive* (action-based), *iconic* (image-based), and *symbolic* (language-based).

In particular, the teaching and learning opportunities in Collins International Primary Maths reflect Bruner's three modes of representation whereby learners develop an understanding of a concept through the three progressive steps (or representations) of concrete-pictorial-abstract, and that reinforcement of an idea or concept is achieved by going back and forth between these representations.

* Bruner, J. S. (1966) *Toward a Theory of Instruction*, Cambridge, MA: Belknap Press

Concrete Representation

The *enactive* stage

The learner is first introduced to a concept using physical objects. This 'hands on' approach is the foundation for conceptual understanding.

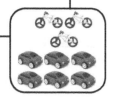

Pictorial Representation

The *iconic* stage

The learner has sufficiently understood the hands-on experiences and can now relate them to images, such as a picture, diagram or model of the concept.

Abstract Representation

The *symbolic* stage

The learner is now capable of using numbers, notation and mathematical symbols to represent the concept.

$3 + 6 = 9$

The model below illustrates the six proficiencies in mathematics that Collins International Primary Maths believe learners need to command in order to become mathematically literate and achieve mastery of the subject at each stage of learning.

All of the teaching and learning units throughout the Collins International Primary Maths course aim to achieve each of these six proficiencies, in order to help teachers to establish successful mathematics learning.

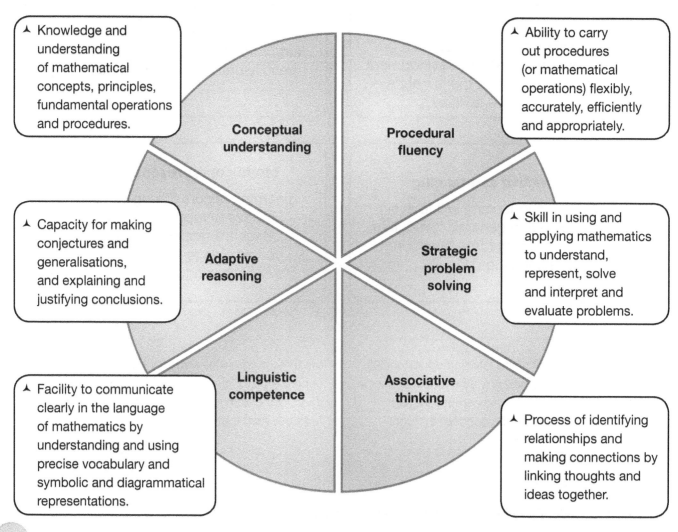

⅄ Knowledge and understanding of mathematical concepts, principles, fundamental operations and procedures.

⅄ Ability to carry out procedures (or mathematical operations) flexibly, accurately, efficiently and appropriately.

⅄ Capacity for making conjectures and generalisations, and explaining and justifying conclusions.

⅄ Skill in using and applying mathematics to understand, represent, solve and interpret and evaluate problems.

⅄ Facility to communicate clearly in the language of mathematics by understanding and using precise vocabulary and symbolic and diagrammatical representations.

⅄ Process of identifying relationships and making connections by linking thoughts and ideas together.

Conceptual understanding

Procedural fluency

Adaptive reasoning

Strategic problem solving

Linguistic competence

Associative thinking

③ How Collins International Primary Maths supports Cambridge Primary Mathematics

Cambridge Primary is typically for learners aged 5 to 11 years. It develops learner skills and understanding through the primary years in English, Mathematics and Science. It provides a flexible framework that can be used to tailor the curriculum to the needs of individual schools.

In Cambridge Primary Mathematics, learners:

- engage in creative mathematical thinking to generate elegant solutions
- improve numerical fluency and knowledge of key mathematical concepts to make sense of numbers, patterns, shapes, measurements and data

- develop a variety of mathematical skills, strategies and a way of thinking that will enable them to describe the world around them and play an active role in modern society
- communicate solutions and ideas logically in spoken and written language, using appropriate mathematical symbols, diagrams and representations
- understand that technology provides a powerful way of communicating mathematics, one that is particularly important in an increasingly technological and digital world.

Cambridge Primary Mathematics supports learners to become:

RESPONSIBLE	INNOVATIVE	ENGAGED
• Learners understand how principles of mathematics can be applied to real-life problems in a responsible way.	• Learners solve new and unfamiliar problems, using innovative mathematical thinking. They can select their own preferred mathematical strategies and can suggest alternative routes to develop efficient solutions.	• Learners are curious and engage intellectually to deepen their mathematical understanding. They are able to use mathematics to participate constructively in society and the economy by making informed mathematical choices.

CONFIDENT	REFLECTIVE
• Learners are confident and enthusiastic mathematical practitioners, able to use appropriate techniques without hesitation, uncertainty or fear. They are keen to ask mathematical questions in a structured, systematic, critical and analytical way. They are able to present their findings and defend their strategies and solutions as well as critique and improve the solutions of others.	• Learners reflect on the process of thinking and working mathematically as well as mastering mathematics concepts. They are keen to make conjectures by asking sophisticated questions and thus develop higher-order thinking skills.

Cambridge Primary is organised into six stages. Each stage reflects the teaching targets for a year group. Broadly speaking, Stage 1 covers the first year of primary teaching, when learners are approximately five years old. Stage 6 covers the final year of primary teaching, when learners are approximately 11 years old.

The Cambridge Primary Mathematics Curriculum Framework (0096) replaces the previous curriculum framework (0845), and is presented in three content areas (strands), with each strand divided into sub-strands. Thinking and Working Mathematically underpins all strands and sub-strands, while mental strategies are a key part of the Number content.

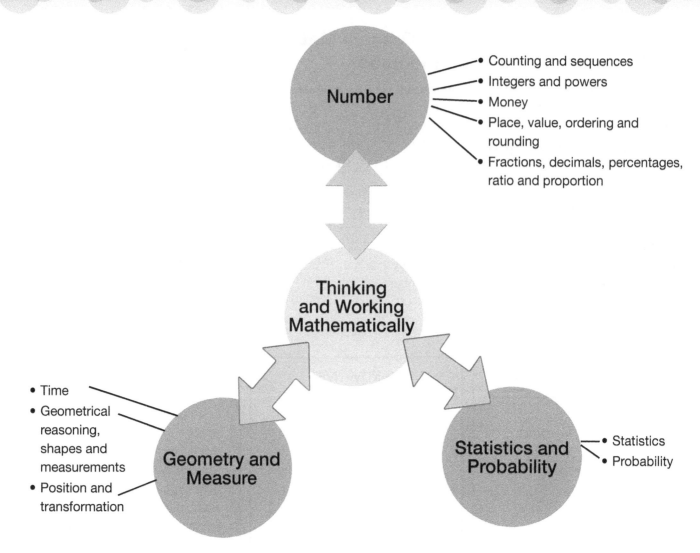

- Counting and sequences
- Integers and powers
- Money
- Place, value, ordering and rounding
- Fractions, decimals, percentages, ratio and proportion

Number

Thinking and Working Mathematically

- Time
- Geometrical reasoning, shapes and measurements
- Position and transformation

Geometry and Measure

Statistics and Probability

- Statistics
- Probability

Planning, teaching and assessment

Effective planning, teaching and assessment are the three interconnected elements that contribute to promoting learning, raising learners' attainment and achieving end-of-stage expectations (mastery). All of the components of Collins International Primary Maths emphasise, and provide guidance on, the importance of this cyclical nature of teaching in order to ensure that learners reach the end-of-stage expectations of the Cambridge Primary Mathematics Curriculum Framework (0096). This teaching and learning cycle, and the important role that the teacher plays in this cycle, are at the heart of Collins International Primary Maths.

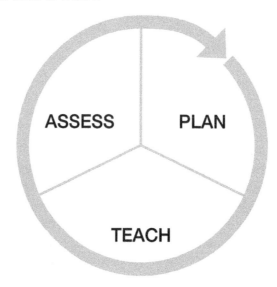

ASSESS

PLAN

TEACH

Collins International Primary Maths supports teachers in planning a successful mathematics programme for their unique teaching context and ensures:

- a clear understanding of learners' pre-requisite skills before they undertake particular tasks and learning new concepts
- considered progression from one lesson to another
- regular revisiting and extension of previous learning
- a judicious balance of objectives, and the time dedicated to each one
- the use of a consistent format and structure.

The elements of Collins International Primary Maths that form the basis for planning can be summarised as follows:

Long-term plans

The Cambridge Primary Mathematics Curriculum Framework (0096) constitutes the long-term plan for schools to follow at each stage across the school year. By closely reflecting the Curriculum Framework and the Cambridge Primary Mathematics Scheme of Work, the Collins International Primary Maths course embodies this long-term plan.

Medium-term plans

The Collins International Primary Maths Units and Recommended Teaching and Learning Sequence in Sections ⑦ and ⑧ (see pages 38–49) show termly/semester outlines of units of work with Cambridge Primary Mathematics Curriculum Framework (0096) references (including the Curriculum Framework codes). By using the Collins International Primary Maths Extended Teacher's Guide, these plans including curriculum coverage, delivery and timing, can be easily adapted to meet the specific needs of individual schools and teachers as well as learners' needs.

Short-term plans

Individual lesson plans and accompanying Additional practice activities represent the majority of each Teacher's Guide. The lessons provide short-term plans that can easily be followed closely, or used as a 'springboard' and varied to suit specific needs of particular classes. An editable 'Weekly Planning Grid' is also provided as part of the Digital content, which individual teachers can fully adapt.

This includes modifying short-term planning in order to build on learners' responses to previous lessons, thereby enabling them to make greater progress in their learning.

The most important role of teaching is to promote learning and to raise learners' attainment. To best achieve these goals, Collins International Primary Maths believes in the importance of teachers:

- promoting a 'can do' attitude, where all learners can achieve success in, and enjoy, mathematics
- having high, and ambitious, expectations for *all* learners
- adopting a philosophy of equal opportunity that means *all* learners have full access to the same curriculum content
- generating high levels of engagement (*Active learning*) and commitment to learning
- offering sharply focused and timely support and intervention that matches learners' individual needs
- being *language aware* in order to understand the possible challenges and opportunities that language presents to learning
- systematically and effectively checking learners' understanding throughout lessons, anticipating where they may need to intervene, and doing so with notable impact on the quality of learning
- consistently providing high-quality marking and constructive feedback to ensure that learners make rapid gains.

To help teachers achieve these goals, Collins International Primary Maths provides:

- highly focused and clearly defined learning objectives
- examples of targeted questioning, using appropriate mathematical vocabulary, that is aimed at both encouraging and checking learner progress
- a proven lesson structure that provides clear and accurate directions, instructions and explanations
- meaningful and well-matched activities for learners, at all levels of understanding, to practise and consolidate their learning
- a balance of individual, pair, group and whole-class activities to develop both independence and collaboration and to enable learners to develop their own thinking and learn from one another
- highly effective models and images (representations) that clearly illustrate mathematical concepts, including interactive digital resources.

The lesson sequence in Collins International Primary Maths focuses on supporting learners' understanding of the learning objectives of the Cambridge Primary Mathematics Curriculum Framework (0096), as well as building their mathematical proficiency and confidence.

Based on a highly effective and proven teaching and learning sequence, each lesson is divided into six key teaching strategies that take learners on a journey of discovery. This approach is shown on the outer ring of the diagram below.

The inner ring shows the link between the six key teaching strategies and the five phases of a Collins International Primary Maths lesson plan, as well as when to use the Student's Book and Workbook.

Pedagogical cycle

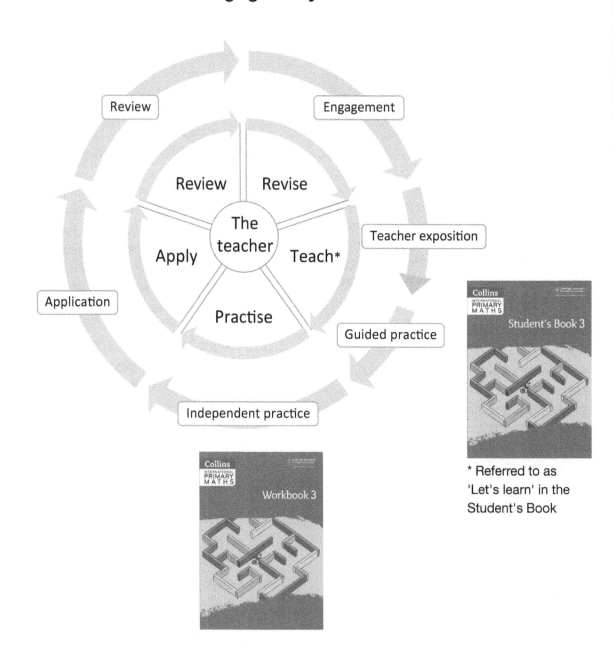

* Referred to as 'Let's learn' in the Student's Book

The chart below outlines the purposes of each phase in the Collins International Primary Maths lesson sequence, as well as the learner groupings and approximate recommended timings.

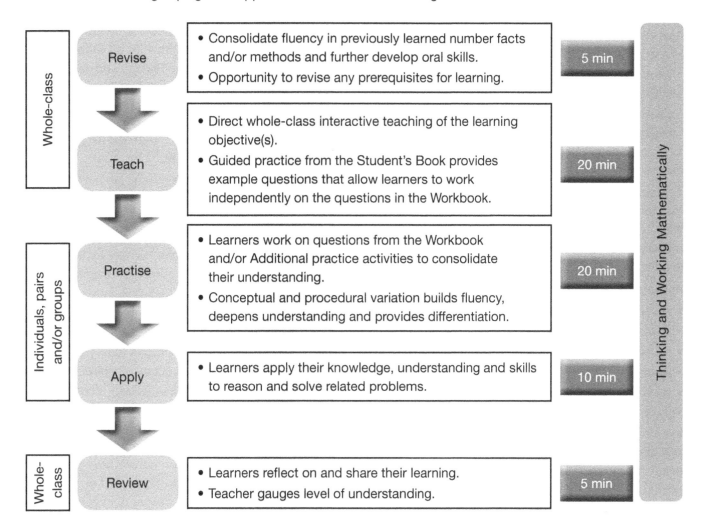

Whole-class	**Revise**	• Consolidate fluency in previously learned number facts and/or methods and further develop oral skills. • Opportunity to revise any prerequisites for learning.	5 min
	Teach	• Direct whole-class interactive teaching of the learning objective(s). • Guided practice from the Student's Book provides example questions that allow learners to work independently on the questions in the Workbook.	20 min
Individuals, pairs and/or groups	**Practise**	• Learners work on questions from the Workbook and/or Additional practice activities to consolidate their understanding. • Conceptual and procedural variation builds fluency, deepens understanding and provides differentiation.	20 min
	Apply	• Learners apply their knowledge, understanding and skills to reason and solve related problems.	10 min
Whole-class	**Review**	• Learners reflect on and share their learning. • Teacher gauges level of understanding.	5 min

Thinking and Working Mathematically

Monitoring, evaluation and feedback continue the teaching and learning cycle and are used to form the basis for adjustments to the teaching programme. Collins International Primary Maths offers meaningful, manageable and useful assessment on two of the following three levels:

Short-term 'on-going' assessment

Short-term assessments are an informal part of every lesson. A combination of carefully crafted recall, observation and thought questions is provided in each lesson of Collins International Primary Maths and these are linked to specific learning objectives.

They are designed to monitor learning and provide immediate feedback to learners and to gauge learners' progress in order to enable teachers to adapt their teaching.

Success Criteria are also provided in each unit to assist learners in identifying the steps required to achieve the unit's learning objectives.

Each unit in Collins International Primary Maths begins with a Unit introduction. One of the features of the Unit introduction is 'Common difficulties and remediation'. This feature can be used to help identify why learners do not understand, or have difficulty with, a topic or concepts and to use this information to take appropriate action to correct mistakes or misconceptions.

Medium-term 'formative' assessment

Medium-term assessments are used to review and record the progress learners make, over time, in relation to the learning objectives of the Cambridge Primary Mathematics Curriculum Framework (0096). They are used to establish whether learners have met the learning objectives or are on track to do so.

'Assessment *for* learning' is the term generally used to describe the conceptual approach to both short-term 'ongoing' assessment and medium-term 'formative' assessment.

Assessment *for* learning involves both learners and teachers finding out about the specific strengths and weaknesses of individual learners, and the class as a whole, and using this to inform future teaching and learning.

Assessment *for* learning:

- is part of the planning process
- is informed by learning objectives
- engages learners in the assessment process
- recognises the achievements of all learners
- takes account of how learners learn
- motivates learners.

In order to assist teachers with monitoring both short- and medium-term assessments, and to ensure that evidence collected is meaningful, manageable and useful, Collins International Primary Maths includes a class record-keeping document on pages 312–317. The document helps teachers:

- identify whether learners are on track to meet end-of-stage expectations
- identify those learners working *above* and *below* end-of-stage expectations
- make long-term 'summative' assessments
- report to parents and guardians
- inform the next year's teacher about which sub-strands of the Cambridge Primary Mathematics Curriculum Framework (0096) individual learners, and the class as a whole, are exceeding, meeting or are below in expectations.

For further details on how to use the class record-keeping document, please refer to pages 30–31.

Long-term 'summative' assessment

Long-term assessment is the third level of assessment. It is used at the end of the school year in order to track progress and attainment against school and external targets, and to report to other establishments and to parents on the actual attainments of learners. By ensuring complete and thorough coverage of the Cambridge Primary Mathematics Curriculum Framework (0096), Collins International Primary Maths provides an excellent foundation for the Cambridge Primary end-of-stage tests (Cambridge Primary Progression Tests) as well as the end of primary Cambridge Primary Checkpoint.

Mental strategies and Cambridge Primary Mathematics

Mental strategies learning objectives are not included in the Cambridge Primary Mathematics Curriculum Framework (0096). However, working mentally is an important feature in the curriculum framework and is embedded not just within the Number strand but throughout all strands in the curriculum framework.

Mental strategies should be applied across all of the Cambridge Primary stages (1 to 6) and to all mathematical strands. The Cambridge Primary Mathematics Curriculum Framework (0096) is, however, less prescriptive about the specific strategies that should be learned and practised by learners at each stage. Allowing teachers greater flexibility in teaching mental strategies, and allowing learners to view mental strategies as a more personal and less formal choice, means that learners will have greater ownership over the mental strategies that they choose to use, thereby developing a deeper conceptual understanding of the number system.

In keeping with the changes that have been introduced in the Cambridge Primary Mathematics Curriculum Framework (0096), mental strategies are embedded throughout Collins International Primary Maths. Learners are given opportunities to develop and practise mental strategies, using carefully chosen numbers, and are continually encouraged to articulate their strategies verbally. This is of particular importance for the Cambridge Primary Mathematics Curriculum Framework (0096), where there are no specific learning objectives relating to mental strategies.

It is not possible to exhaustively list all of the mental strategies that can be used, and there will not be one correct strategy for any particular calculation. The most appropriate mental strategy will depend on individual learners' knowledge of mathematical facts, their working memory and their conceptual understanding of different parts of the number system. Mental strategies can be explicitly learned and practised, and doing so will enable a learner to add that strategy to their 'repertoire'. It is important therefore that learners are exposed to a wide range of strategies.

Below are some of the different mental strategies that learners may employ and that are featured throughout Collins International Primary Maths.

Addition and subtraction

- counting on and back in steps
- using known addition and subtraction number facts/number bonds/complements
- applying knowledge of place value and partitioning (i.e. compose, decompose and regroup numbers)
- compensation
- putting the larger number first and counting on (addition)
- counting back from the larger number (take away)
- counting on from the smaller number (find the difference)
- recognising that when two numbers are close together it's easier to find the difference by counting on, not counting back
- using the commutative and associative properties
- using the inverse relationship between addition and subtraction

Multiplication and division

- counting on and back in steps of constant size
- using known multiplication and division facts and related facts involving multiples of 10 and 100
- applying knowledge of place value and partitioning (i.e. compose, decompose and regroup numbers), including multiplying and dividing whole numbers and decimals by 10, 100 and 1000
- using doubling
- recognising and using factor pairs
- using the commutative, distributive and associative properties
- using the inverse relationship between multiplication and division

The use of calculators and Cambridge Primary Mathematics

When used well, calculators can assist learners in their understanding of numbers and the number system. Calculators should be used as a teaching aid to promote mental calculation and mental strategies and to explore mathematical patterns. Learners should understand when it is best to use calculators to assist calculations and when to calculate mentally or use written methods.

As Cambridge International includes calculator-based assessments at Stages 5 to 9, it is recommended that learners begin to use calculators for checking calculations from the end of Stage 3, and for performing and checking calculations from Stage 4. At Stages 5 and 6, learners should be developing effective use of calculators so that they are familiar with the functionality of a basic calculator in readiness for Stage 7 onwards.

④ Thinking and Working Mathematically

In the Cambridge Primary (0096) and Lower Secondary (0862) Mathematics Curriculum Frameworks, the problem-solving strand and associated learning objectives have been replaced with four pairs of Thinking and Working Mathematically (TWM) characteristics.

The TWM characteristics represent one of the most significant changes to the Cambridge Primary Mathematics Framework (0096). In response to this, this edition of Collins International Primary Maths has incorporated and interwoven TWM throughout all of the components; this reflects the course's most substantial change to the teaching and learning of mathematics.

Thinking and Working Mathematically is based on work by Mason, Burton and Stacey*; it places an emphasis on learners:

- actively engaging with their learning of mathematics
- talking with others, challenging ideas and providing evidence that validates conjectures and solutions
- seeking to make sense of ideas
- building connections between different facts, procedures and concepts
- developing higher-order thinking skills that assists them in viewing the world in a mathematical way.

* Mason, J., Burton, L. and Stacey, K. (2010) *Thinking Mathematically*, 2nd edition, Harlow: Pearson

This contrasts with learners simply following instructions and carrying out processes that they have been shown how to do, without appreciating why such processes work or what the results mean. Through the development of each of the TWM characteristics, learners are able to see the application of mathematics in the real world more clearly and also, crucially, to develop the skills necessary to function as citizens who are autonomous problem solvers.

If learners at any of the Cambridge International stages are to gain meaning and satisfaction from their study of mathematics, then it is vital that TWM underpins their experience of learning the subject.

The four pairs of TWM characteristics that Cambridge International identifies as fundamental to a meaningful experience of learning mathematics are represented diagrammatically and referred to as 'The Thinking and Working Mathematically Star'.

The Thinking and Working Mathematically Star

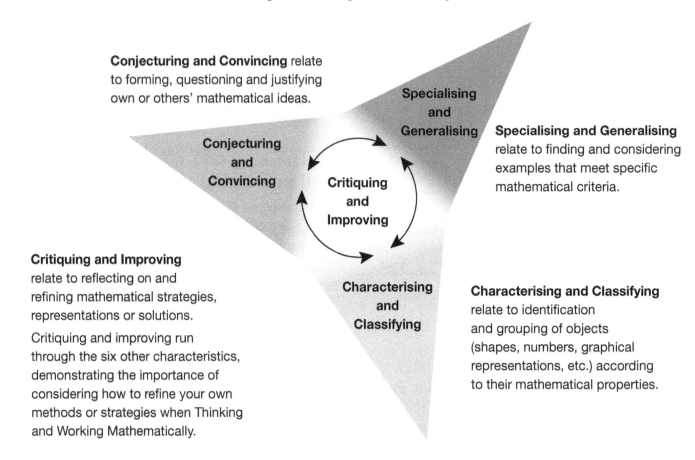

Conjecturing and Convincing relate to forming, questioning and justifying own or others' mathematical ideas.

Specialising and Generalising relate to finding and considering examples that meet specific mathematical criteria.

Critiquing and Improving relate to reflecting on and refining mathematical strategies, representations or solutions.

Critiquing and improving run through the six other characteristics, demonstrating the importance of considering how to refine your own methods or strategies when Thinking and Working Mathematically.

Characterising and Classifying relate to identification and grouping of objects (shapes, numbers, graphical representations, etc.) according to their mathematical properties.

The Thinking and Working Mathematically star, © Cambridge International, 2018

The eight characteristics of Thinking and Working Mathematically

Characteristic	Definition
TWM.01 Specialising	Choosing *an example* and checking to see if it satisfies or does not satisfy specific mathematical criteria.
TWM.02 Generalising	Recognising an underlying pattern by identifying *many* examples that satisfy the same mathematical criteria.
TWM.03 Conjecturing	Forming mathematical questions or ideas.
TWM.04 Convincing	Presenting evidence to *justify or challenge* a mathematical idea or solution.
TWM.05 Characterising	Identifying and describing the mathematical properties of an object.
TW.06 Classifying	Organising objects into groups according to their mathematical properties.
TWM.07 Critiquing	Comparing and evaluating mathematical ideas, representations or solutions to identify advantages and disadvantages.
TWM.08 Improving	Refining mathematical ideas or representations to develop a more effective approach or solution.

All eight TWM characteristics can be applied across all of the Cambridge Primary and Lower Secondary stages (1 to 9) and across all mathematical strands and sub-strands, although the prominence of different characteristics may change as learners move through the stages.

Any characteristic can be combined with any other characteristic; characteristics should be taught alongside content learning objectives and should **not** stand-alone.

The four pairs of characteristics intertwine and are interdependent, and a high-quality mathematics task may draw on one or more of them.

Thinking and Working Mathematically should **not** consist of a separate end-of-lesson or unit activity, but should be embedded throughout lessons in every unit of work. All of the characteristics identified above can be combined with most teaching topics so, when planning a unit of work, teachers should begin with one or more learning objectives and seek to draw on one or more TWM characteristics.

Thinking and Working Mathematically also enables learners' thinking to become visible, which is a crucial aspect of formative assessment.

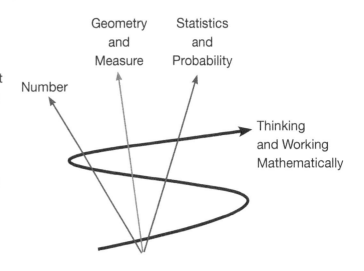

Just as TWM is at the very heart of Cambridge Primary Mathematics, so too is this approach to the teaching and learning of mathematics a core feature of Collins International Primary Maths. Opportunities are provided in each of the 27 units for learners to develop the TWM characteristics.

Specific guidance is provided at the start of each unit as part of the Unit introduction, which highlights teaching and learning opportunities, particularly in relation to the unit's learning objectives, that promote the TWM characteristics.

Unit introduction

Promoting Thinking and Working Mathematically

During this unit learners have plenty of opportunities to specialise (TWM.01) and generalise (TWM.02). For example, when counting in hundreds, they look for patterns, for example, 134, 234, 334. They should be able to choose an example in this count that will fit. For example, they can see that the tens and ones numbers are the same so any number with a different hundreds digit, such as 834, will fit the pattern. Learners form mathematical ideas (TWM.03) when they apply what they learn in number to measurement Learners discuss similarities and differences between

At each phase of the lesson (Revise, Teach, Practise, Apply and/or Review), guidance is given in the lesson plan on how to promote the TWM characteristics.

Whenever any of the eight TWM characteristics is being promoted in a lesson plan this is shown using the initials 'TWM', followed by the Cambridge Primary Mathematics Curriculum Framework (0096) code that identifies exactly which of the eight characteristics is being developed.

Teach 🅢🄑 🖥 [TWM.03/04/05/06]

- [T&T] 🄑 Display **Slide 1**. Ask: **Is it possible to know how many of each colour jelly beans there are as soon as you look at them?** Agree that there are too many to know for sure. Ask: **What could we do?** Agree learners could make a tally. Give pairs a few minutes to list the colours that they can see and make a tally of how many of each colour they think there may be.
- Tell learners that often when we make tally chart we use it to draw up a frequency table. Display **Slide 2** and referring to the table on the left inform learners that the table shows a count of the numbers of jelly beans from **Slide 1**. Tell learners that the column on the right shows the frequency. Ask: **What's the difference between tallies and frequencies?** Agree tallies are made from marks and frequencies show numbers. Ask learners to check that each tally and number match.
- Direct learners to the table on the right inform and explain that we can also show just the frequencies in a table, and that this is also called a frequency table.
- [TWM.03/04/05/06] Ask, pairs of learners to take a handful of coloured counters. They sort them

The **Teach** phase of this lesson aims to promote the Conjecturing [TWM.03], Convincing [TWM.04], Characterising [TWM.05] and Classifying [TWM.06] characteristics.

The **Apply** phase of this lesson aims to promote the Conjecturing [TWM.03] and Convincing [TWM.04] characteristics.

Apply 👥 🖥 [TWM.03/04]

- Display **Slide 2**. Learners are given part of the information to a problem and the answer.
- They work out the missing information so it gives the correct answer.
- They then make up their own similar problems before asking their partner to solve them.

Practise 🅦 [TWM.03/04]

- Workbook

Title: Rounding to the nearest 10

Pages: 122–123

- Refer to Activity 2 from the Additional practice activities.

The **Practise** phase of this lesson aims to promote the Conjecturing [TWM.03] and Convincing [TWM.04] characteristics.

In addition to the support provided in the Unit introductions and individual lesson plans, The Thinking and Working Mathematically Star below provides a list of prompting questions that teachers may find helpful when asking learners questions specifically aimed at developing each of the TWM characteristics.

The Thinking and Working Mathematically Star
Teacher prompting questions

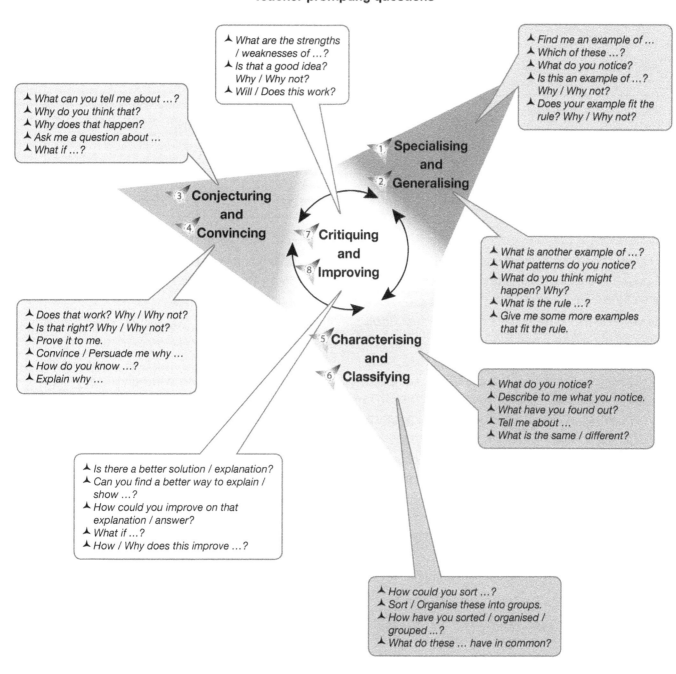

⅄ What are the strengths / weaknesses of ...?
⅄ Is that a good idea? Why / Why not?
⅄ Will / Does this work?

⅄ Find me an example of ...
⅄ Which of these ...?
⅄ What do you notice?
⅄ Is this an example of ...? Why / Why not?
⅄ Does your example fit the rule? Why / Why not?

⅄ What can you tell me about ...?
⅄ Why do you think that?
⅄ Why does that happen?
⅄ Ask me a question about ...
⅄ What if ...?

3 **Conjecturing and**
4 **Convincing**

7 **Critiquing and**
8 **Improving**

1 **Specialising and**
2 **Generalising**

⅄ What is another example of ...?
⅄ What patterns do you notice?
⅄ What do you think might happen? Why?
⅄ What is the rule ...?
⅄ Give me some more examples that fit the rule.

⅄ Does that work? Why / Why not?
⅄ Is that right? Why / Why not?
⅄ Prove it to me.
⅄ Convince / Persuade me why ...
⅄ How do you know ...?
⅄ Explain why ...

5 **Characterising and**
6 **Classifying**

⅄ What do you notice?
⅄ Describe to me what you notice.
⅄ What have you found out?
⅄ Tell me about ...
⅄ What is the same / different?

⅄ Is there a better solution / explanation?
⅄ Can you find a better way to explain / show ...?
⅄ How could you improve on that explanation / answer?
⅄ What if ...?
⅄ How / Why does this improve ...?

⅄ How could you sort ...?
⅄ Sort / Organise these into groups.
⅄ How have you sorted / organised / grouped ...?
⅄ What do these ... have in common?

As in the Teacher's Guide, learning opportunities aimed at developing TWM are also identified in the Stages 3 to 6 Student's Book and Workbook. Where an activity or question promotes TWM, this is clearly indicated using the TWM icon and the Cambridge Primary Mathematics Curriculum Framework (0096) code that identifies exactly which of the eight characteristics is being developed.

This paired activity from the **Student's Book** aims to promote the Convincing [TWM.04] characteristic.

Tom increases the time he runs by the same amount each day. On Monday he ran for 5 minutes and on Saturday he ran for 45 minutes. How many minutes did Tom run for on Tuesday, Wednesday, Thursday and Friday?

Draw a diagram that explains the problem and how to solve it.

This question from the **Workbook** aims to promote the Characterising [TWM.05] and Classifying [TWM.06] characteristics.

5 Investigate the following statement: **A triangular number can never end in 2, 4, 7 or 9.**

Is this statement true or false? Write your working in the box below.

Similar to 'The Thinking and Working Mathematically Star – Teacher prompting questions' on page 17, the star on page 19, which is located at the back of the Stages 3 to 6 Student's Books, defines in pupil-friendly language, each of the eight TWM characteristics and numbers them 1 to 8 accordingly.

The star is aimed at helping learners think specifically about what is required when they are undertaking an activity designed to develop a specific TWM characteristic.

If used in conjunction with the Teacher's Guide, Student's Book and Workbook, which uses the initials 'TWM', followed by the Cambridge Primary Mathematics Curriculum Framework (0096) code to identify exactly which of the eight characteristics an activity, discussion prompt or practise question is developing, this star will help learners better understand the meaning and purpose of each of the eight TWM characteristics.

The star also includes some sentence stems that aim to help learners to talk with others, challenge ideas and explain their reasoning. Learners should be encouraged to use the star whenever working on an activity that develops TWM. This includes whole-class discussions and activities, group and paired activities (including those located in the Student's Book as well as Additional practice activities and Apply), and individual questions from the Workbook.

In Stages 1 and 2 a similar star is provided at the back of the Student's Book, however this star does not include the sentence stems.

The Thinking and Working Mathematically Star

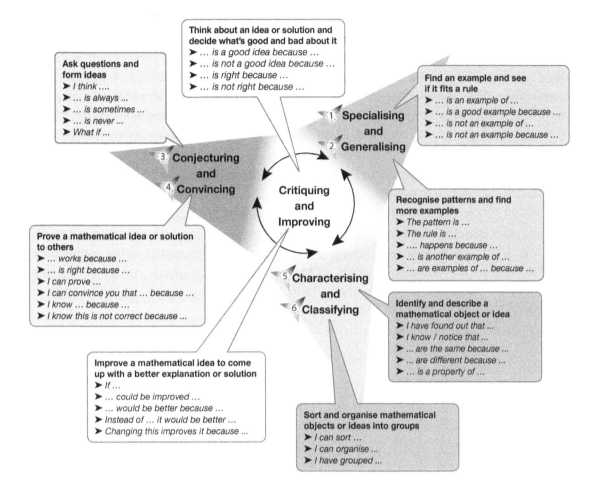

Think about an idea or solution and decide what's good and bad about it
- ➤ ... is a good idea because ...
- ➤ ... is not a good idea because ...
- ➤ ... is right because ...
- ➤ ... is not right because ...

Ask questions and form ideas
- ➤ *I think ....*
- ➤ *... is always ...*
- ➤ *... is sometimes ...*
- ➤ *... is never ...*
- ➤ *What if ...*

Find an example and see if it fits a rule
- ➤ *... is an example of ...*
- ➤ *... is a good example because ...*
- ➤ *... is not an example of ...*
- ➤ *... is not an example because ...*

1 **Specialising** and 2 **Generalising**

3 **Conjecturing** and 4 **Convincing**

Critiquing and Improving

Recognise patterns and find more examples
- ➤ *The pattern is ...*
- ➤ *The rule is ...*
- ➤ *.... happens because ...*
- ➤ *... is another example of ...*
- ➤ *... are examples of ... because ...*

Prove a mathematical idea or solution to others
- ➤ *... works because ...*
- ➤ *... is right because ...*
- ➤ *I can prove ...*
- ➤ *I can convince you that ... because ...*
- ➤ *I know ... because ...*
- ➤ *I know this is not correct because ...*

5 **Characterising** and 6 **Classifying**

Identify and describe a mathematical object or idea
- ➤ *I have found out that ...*
- ➤ *I know / notice that ...*
- ➤ *... are the same because ...*
- ➤ *... are different because ...*
- ➤ *... is a property of ...*

Improve a mathematical idea to come up with a better explanation or solution
- ➤ *If ...*
- ➤ *... could be improved ...*
- ➤ *... would be better because ...*
- ➤ *Instead of ... it would be better ...*
- ➤ *Changing this improves it because ...*

Sort and organise mathematical objects or ideas into groups
- ➤ *I can sort ...*
- ➤ *I can organise ...*
- ➤ *I have grouped ...*

⑤ Cambridge Global Perspectives™

Cambridge Global Perspectives is a unique programme that helps learners develop outstanding transferable skills, including critical thinking, research and collaboration. The programme is available for learners aged 5–19, from Cambridge Primary through to Cambridge Advanced. For Cambridge Primary and Lower Secondary learners, the programme is made up of a series of Challenges covering a wide range of topics, using a personal, local and global perspective. The programme is available to Cambridge schools but participation in the programme is voluntary. However, whether or not your school is involved with the programme, the six skills it focuses on are relevant to **all** students in the modern world. These skills are: research, analysis, evaluation, reflection, collaboration and communication.

More information about the Cambridge Global Perspectives programme can be found on the Cambridge Assessment International Education website:
www.cambridgeinternational.org/programmes-and-qualifications/cambridge-global-perspectives.

Collins supports Cambridge Global Perspectives by including activities, tasks and projects in our Cambridge Primary and Lower Secondary courses which develop and apply these skills. Note that the content of the activities is not intended to correlate with the specific topics in the Cambridge Challenges; rather, they encourage practice and development of the Cambridge Global Perspectives to support teachers in integrating and embedding them into students' learning across all school subjects.

Activities in this book that link to the Cambridge Global Perspectives are listed at the back of this book on page 318.

⑥ The components of Collins International Primary Maths

Each of the six stages in Collins International Primary Maths consists of these four components:

Teacher's Guide

The Teacher's Guide comprises:

- a bank of **Revise activities** for the first 'warm up' phase of the lesson
- **Teaching and learning units**, which consist of Unit introductions, Lesson plans and Additional practice activities
- **Resource sheets** for use with particular lessons and activities
- **Answers** to the questions in the Workbook
- **Record-keeping document** to assist teachers with both short-term 'on-going' and medium-term 'formative' assessments.

A key aim of Collins International Primary Maths is to support teachers in planning, teaching and assessing a successful mathematics programme of work, in line with the Cambridge Primary Mathematics Curriculum Framework (0096).

To ensure complete curriculum coverage and adequate revision of the learning objectives, for each stage the learning objectives from the Cambridge Primary Mathematics Curriculum Framework (0096) have been grouped into 27 topic areas or 'units'. For a more detailed explanation of the 27 units in Collins International Primary Maths Stage 3, including a recommended teaching sequence, refer to Section ⑦ (pages 38–41).

The charts in Section ⑧ (pages 42–49) provide a medium-term plan, showing each of the 27 units in Collins International Primary Maths Stage 3 and which Stage 3 Cambridge Primary Mathematics Curriculum Framework (0096) strand, sub-strand and learning objectives (and codes) each of the units is teaching.

Similarly, the charts in Section ⑨ (pages 50–53) show when each of the Stage 3 learning objectives in the Cambridge Primary Mathematics Curriculum Framework (0096) are taught in the 27 units in Collins International Primary Maths Stage 3.

Icons used in Collins International Primary Maths

 work individually

 work in pairs

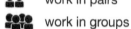 work in groups

 work as a whole class

[T&T] turn and talk (*Talk Partners*)

 progress check question

 SB refer to the Student's Book

 WB refer to the Workbook

interactive digital resource

slide

 activity that promotes Thinking and Working Mathematically

 1 question/activity number typeset on a circle indicates that the question/activity is suitable for learners who require additional support with either easier questions/activities or revising pre-requisite knowledge

 2 question/activity number typeset on a triangle indicates that the question/activity is suitable for the majority of learners to practise and consolidate the lesson's learning objective(s)

 3 question/activity number typeset on a square indicates that the question/activity is suitable for learners who require enrichment and/or extension

---

A note on the use of dice

Although some activities in Collins International Primary Maths suggest the use of dice, these are not always readily available in some countries. Where dice are unavailable or the use of dice is not appropriate, we suggest using a spinner, and have provided spinners on several Resource sheets along with instructions on how to use them using a pencil and paper clip.

Revise activities

A bank of 5-minute 'warm-up' or 'starter' activities is provided for teachers to use at this first phase of the mathematics lesson. Reference is given in each lesson plan to appropriate Revise activities.

The majority of activities are for whole-class work. However, some activities may involve individual learners demonstrating something to the rest of the class, or pairs or groups working together on an activity or a game.

Unit number and title

2 The majority of Revise activities are suitable for the majority of learners to practise and consolidate the activity's learning objective(s).

Strand and sub-strand

The relevant Cambridge Primary Mathematics Curriculum Framework (0096) strand and sub-strand covered is stated in the sidebar.

Title

Each activity is given a title. This is designed to help both teacher and learners identify a particular activity.

Learning objective(s)

Each activity has clearly defined learning objective(s) to assist teachers in choosing the most appropriate activity for the concept they want the learners to practise and consolidate.

Resources

To aid preparation, any resources required are listed, along with whether they are for the whole class, per group, per pair or per learner. Most of these resources are readily available in classrooms.

Classroom organisation

Icons are used to indicate whether the activity is designed to be used by the whole class working together, or for some activities for learners working in groups, pairs or individually.

What to do

The activity is broken down into clear steps to support teachers in achieving the objective(s) of the activity and facilitate interactive whole-class teaching.

Variations

Where appropriate, variations are included. Variations may be designed to make the activity easier **1** or more difficult **3**, or change the focus of the activity completely. Where the variation affects the challenge level of the activity, the new challenge level is given.

Unit **20** 2D shapes, symmetry and angles

Revise

Symmetry **2**

Learning objective

Code	Learning objective
3Gg.09	Identify both horizontal and vertical lines of symmetry on 2D shapes and patterns.

Resources
mini whiteboard and pen (per learner)

What to do
* Say: **All regular shapes are symmetrical.**
* Ask learners to prove, or disprove, this statement. They should sketch some regular shapes and draw the lines of symmetry onto them. Establish that this is true.

* Next, ask: **Do all irregular shapes have symmetry?** Ask learners to find examples when this statement is true and when it is false.

Variation

1 Focus on one line of symmetry only, for example vertical lines of symmetry.

Angles **2**

Learning objective

Code	Learning objective
3Gg.10	Compare angles with a right angle. Recognise that a straight line is equivalent to two right angles or a half turn.

What to do
* Display the **Pattern tool**. Make a repeating pattern using the pentagon and rectangle.
* Ask: **What are the similarities and differences**

* Repeat this with other shape patterns.
* Each time, expect the learners to explain differences and similarities and also to create right angles within the shapes you use.

ical reasoning, shapes and measurements

Collins International Primary Maths units

There are 27 units in Collins International Primary Maths, each consisting of:

- Unit introduction
- Lesson plans
- Additional practice activities.

Unit introduction

The one-page introduction to each unit in the Teacher's Guide is designed to provide background information to help teachers plan, teach and assess that unit.

Unit overview

General description of the knowledge, understanding and skills taught in the unit.

Prerequisites for learning

A list of knowledge, understanding and skills that are prerequisites for learning in the unit. This list is particularly useful for diagnostic assessment.

Vocabulary

A summary is provided of key mathematical terms particularly relevant to the unit.

Common difficulties and remediation

Common errors and misconceptions, along with useful remediation hints are offered where appropriate.

Learning objectives

The Cambridge Primary Mathematics Curriculum Framework (0096) learning objectives covered in the unit.

Collins International Primary Maths unit number and title

Collins International Primary Maths Recommended Teaching and Learning Sequence

Cambridge Primary Mathematics Curriculum Framework (0096) strand and sub-strand

Promoting Thinking and Working Mathematically (TWM)

Specific guidance on how the unit promotes the TWM characteristics.

Success criteria

Success criteria are provided to help both teachers and learners identify what learners are required to know, understand and do in order to achieve the unit's learning objective(s).

Supporting language awareness

Key strategy or idea to help learners access the mathematics of the unit and overcome any barriers that the language of mathematics may present.

Unit introduction

Unit 1: Counting and sequences (A)

Collins International Primary Maths Recommended Teaching and Learning Sequence: Term 1, Week 2

Learning objectives

Code	Learning objective
3Nc.01	Estimate the number of objects or people (up to 1000).
3Nc.02	Count on and count back in steps of constant size: [1-digit numbers,] tens or hundreds, starting from any number (from 0 to 1000).
3Nc.03	Use knowledge of even and odd numbers up to 10 to recognise and sort numbers.

Unit overview

In this unit, learners will be counting in steps of 10 and 100 using practical apparatus, so they understand that, when counting in tens, the ones digit stays the same and when counting in hundreds, the tens and ones digits stay the same. They will continue to explore odd and even numbers and develop the understanding that an odd number can be described as an even number add 1.

Learners will consolidate their understanding of estimation and how an estimate is a sensible guess. They also begin to see the application of estimating, and appreciate why it is an important skill.

Prerequisites for learning

Learners need to:
- know that numbers come in a particular order
- understand that counting can be applied to anything
- know that even numbers have 2, 4, 6, 8, or 0 in the ones position
- know that odd numbers have 1, 3, 5, 7 or 9 in the ones position
- understand that an estimate is a sensible guess.

Vocabulary

zero, one, ten, hundred, thousand, even number, odd number, shared equally, divided by 2, estimate, approximately, range

Common difficulties and remediation

Often learners rote count without understanding what is happening. It is important to encourage them to look for patterns, which will lead to them noticing things and developing a depth of understanding.

Again, with odd and even numbers, many learners simply remember the rule applied to the ones digits. Patterns are important here, to help learners develop an understanding that an even number is a multiple of 2 (the number can be equally divided by two) and an odd number is a multiple of 2 add 1 (the number cannot be equally divided by two).

Most learners are averse to estimating; they want to be correct. It is important they develop the habit of estimating from the beginning, so they are able to estimate answers to calculations automatically. The way it is introduced in this unit is helpful in encouraging learners to estimate.

Supporting language awareness

When counting in ones, tens and hundreds, display the words that learners will say, for example: one, two, three, ...; ten, twenty, thirty, ... ; hundred. Invite learners to touch the words displayed when counting.

When teaching about odd and even numbers, it is helpful to have visual representations of them on display. Showing even numbers as 'pairs' and odd numbers as 'pairs and an extra one' is helpful. There are visual representations of this in the Student's Book.

Estimating can be a tricky skill to master. Provide examples that learners can use, such as a jar of marbles. Give them possible ranges to choose from, written on card, that they can discuss with a partner.

Promoting Thinking and Working Mathematically

During this unit learners have plenty of opportunities to specialise (TWM.01) and generalise (TWM.02). For example, when counting in hundreds, they look for patterns, for example, 134, 234, 334. They should be able to choose an example in this count that will fit. For example, they can see that the tens and ones numbers are the same so any number with a different hundreds digit, such as 834, will fit the pattern. Learners form mathematical ideas (TWM.03) when they apply what they learn in number to measurement. Learners discuss similarities and differences between numbers and justify their decisions (TWM.04, TWM.05, TWM.06), for example the values of particular digits in a number. Learners are given opportunities to evaluate (TWM.07) and improve (TWM.08) estimates during the lesson on estimating.

Success criteria

Learners can:
- count on and count back in tens and hundreds
- recognise odd and even numbers
- sort odd and even numbers
- estimate numbers of up to 1000 objects.

Number – Counting and sequences

101

There are two different types of teaching and learning opportunities provided for each of the 27 units in Collins International Primary Maths:

• four lesson plans
• two Additional practice activities.

The lesson plans provide a clear, structured, step-by-step approach to teaching mathematics according to the learning objective(s) being covered throughout a unit. Each of the lessons has been written in a comprehensive way in order to give teachers maximum support for mixed-ability whole-class interactive teaching. It is intended, however, that the lessons will act as a model to be adapted to the particular needs of each class.

The Additional practice activities provide teachers with a bank of practical, hands-on activities that give learners opportunities for independent practice of the learning objective(s) being taught throughout a unit.

In most instances, the Additional practice activities are designed to be undertaken by pairs or small groups of learners as part of the 'Practise' phase of a Collins International Primary Maths lesson. Teachers choose which of the two Additional practice activities provided in the unit is most appropriate for the lesson's learning objective(s) and the needs of individual learners. Guidance as to which Additional practice activity consolidates a lesson's learning objective(s) is stated in each lesson plan.

The 'Practise' phase of a Collins International Primary Maths lesson also consists of written exercises found in the accompanying Workbook. Teachers need to decide how they wish to use these two different types of independent practice for individual learners or groups of learners. For example, depending on the lesson's learning objective(s), and the needs of individual learners, learners may:

- only complete the Additional practice activity

- start with the Additional practice activity and then move onto exercises in the Workbook

- start with exercises in the Workbook and then move onto the Additional practice activity

- only complete exercises in the Workbook.

It is important that the Additional Practice activities are used at any time throughout a unit and therefore incorporated into each lesson as and when necessary in order to supplement, or provide an alternative to, the exercises in the Workbook. They should not be seen solely as providing teaching and learning content for the 'fifth' lesson of the week.

These two different types of teaching and learning opportunities form the weekly structure for each of the 27 units and are aimed at supporting flexibility so that the course can be tailored to meet the needs of individual classes.

Experience gained from other courses similar to Collins International Primary Maths shows that individual classes take different lengths of time to learn the content of a lesson. In light of this, rather than providing five lesson plans for each week, the decision was made to provide four core lessons which cover the units learning objective(s).

The intention is that as part of their weekly short-term planning, teachers make decisions as to how they will spread out the four core lessons, over the course of five days. Alternatively, as teachers progress through the week, and as part of their ongoing monitoring and evaluation, they may decide to alter their short-term planning and spend more time on teaching a particular lesson, or provide additional teaching and learning opportunities (including the Additional practice activities) in order to ensure that the class are developing a secure understanding of the unit's learning objective(s).

Introduction

Lesson plan

Collins International Primary Maths unit number and title

Reference to accompanying Student's Book page and Workbook pages

Lesson number and title

Cambridge Primary Mathematics Curriculum Framework (0096) strand and sub-strand

Lesson objective(s)

The Cambridge Primary Mathematics Curriculum Framework (0096) learning objective(s) covered in the lesson.

Resources

To aid preparation, all the resources necessary to teach the lesson are listed. Each resource clearly states whether it is for the whole class, per group, per pair or per learner. Icons are displayed within the lesson plan to indicate any digital resources used in the lesson.

Unit **1** Counting and sequences (A)

Student's Book page 6
Workbook pages 6–7

Lesson 1: **Counting**

Number – Counting and sequences

Learning objective

Code	Learning objective
3Nc.02	Count on and count back in steps of constant size: [1-digit numbers,] tens or hundreds, starting from any number (from 0 to 1000).

Resources

Base 10 equipment: ones, tens, hundreds and a thousand cube (per pair); Resource sheet 1: 100 square (per learner) (for Same day intervention)

Revise

Use the activity *Clap counting* from Unit 1: *Counting and sequences (A)* in the Revise activities.

Teach 📖 💻 📊 [TWM.01/02]

- Count together in tens from zero to 100 and back. Count in hundreds from zero to 1000 and back.
- Display **Slide 1** and discuss the house numbers. Agree they are three-digit numbers.
- 🗣 Ask: **What are these numbers? What is the value of the 1 in 123? Which is the greatest/ smallest number? Which numbers are odd/even?**
- Together, count on, then back in tens from each door number. Then count on in hundreds, stopping before 1000, and then count back.
- **[TWM.01]** Display **Slide 2**. Ask: **What number is represented? What does each part represent?**
- Agree that there are two tens, which represents 20, and seven ones, which represents 7. Together 20 and 7 are equivalent to 27.
- Ask pairs of learners to use the Base 10 equipment to make 27. Then ask them to add a tens and another and another, until they reach 97. Ask: **What is happening to your number? Which part is changing? Which part stays the same? Can you explain why?**
- **[T&T]** When they reach 97, ask learners to discuss what will happen if another tens is added.
- 🗣 Ask: **What are 10 lots of 10? What should we exchange our tens with?** Learners make the

demonstrate how to add another 10, then count on from 305 to 355 and back.
- **[TWM.02]** Learners work through the paired activity.
- Discuss the Guided practice example in the Student's Book.

Practise 📓 [TWM.05]

- Workbook

Title: Counting

Pages: 6–7

- Refer to Activity 1 from the Additional practice activities.

Apply 👥 💻 [TWM.07]

- Display **Slide 3**.
- Learners consider whether counting in tens and hundreds applies to money.

Review

- 🗣 Ask: **What have we been learning about? When we count in tens, which digit changes? Which stay the same? Why is that?**
- 🗣 Ask: **When we count in hundreds, which digit changes? Which stay the same? Why is that?**
- 🗣 Ask: **Can we count in units of 10 centimetres? What about units of 100 grams?** Agree that counting is the same, no matter what units we use.

Assessment for learning

Revise

Recommended teaching time: 5 min

A bank of Revise activities can be found on pages 55–100. Revise activities are designed to consolidate fluency in number facts and/or provide an opportunity to revise any prerequisites for learning.

Teach

Recommended teaching time: 20 min

The main teaching activity is broken down into clear steps to support teachers in achieving the lesson objective(s) and facilitate interaction with the whole class. Suggested statements and questions are provided to support the teacher. During this phase of the lesson teachers also draw learners' attention to the 'Let's learn', Paired/TWM activity, and 'Guided practice' features in the Student's Book.

Other features of Teach include:

T&T: Turn and talk – Using *Talk Partners* helps to create a positive learning environment. Many learners feel more confident discussing with a partner before giving an answer to the whole class, and learners get opportunities to work with different students.

₽: Progress check questions – These questions are designed to obtain an overview of learners' prior experiences before introducing a new concept or topic, to provide immediate feedback to learners and to gauge learner progress in order to adapt teaching.

TWM: Teaching and learning opportunities aimed at learners developing specific Thinking and Working Mathematically characteristics.

NOTE: Timings are approximate recommendations only.

(inset panel)

Revise

Use the activity *Estimating* from Unit 1: *Counting and sequences (A)* in the Revise activities.

Teach [TWM. 02/07/08]

• Display **Slide 1**. Ask learners to estimate how many red circles there are. Allow them a few seconds and then hide the image. Establish that they didn't have time to count them all, so they had to estimate.
• Write some of their estimates on the board. Discuss the range of estimates from lowest to highest. Also write estimates of 12 and 1000.
• Show the circles briefly again.
• **[TWM.07/08] [T&T]** Ask learners to talk to each other about the estimates. Ask: **Which estimates do you think are sensible?**
• Agree that 12 and 1000 are definitely not sensible estimates.
• Display **Slide 2**. Ask: **What is different about the way the circles are shown now?**
• ₽ Ask: **How many circles are there? How do you know? Why is this arrangement easier to count?**
• **[T&T]** Ask learners, in pairs, to tip some counters onto their tables. They each make an estimate. Then they arrange them in tens, as on the slide. They compare their estimates with the actual number. They repeat several times.
• **[TWM.08]** Direct learners to the Student's Book. Ask: **How many jellybeans do you think there are? What would be a sensible range for an estimate? Why? How can you check?**
• **[TWM.02]** Introduce the paired activity in the Student's Book. Ask learners to discuss situations in everyday life when it is important and useful to make estimates. For example, estimating if you have enough money in your purse to pay for shopping, estimating the time it will take to do something, estimating lengths, heights, distances, masses and capacities. Ask pairs of learners to share their suggestions with the class.
• Discuss the Guided practice example in the Student's Book.

Practise [TWM.08]

• Workbook
Title: Estimating
Pages: 12–13
• Refer to Activity 2 from the Additional practice activities.

Apply [TWM.07]

• Display **Slide 3**.
• Learners first give a range for their estimate of how many apples are on the tree and then give a specific number for their estimate.
• Then they write an explanation of why it is impossible to know exactly how many apples there are.

Review

• Discuss what has been learned during the lesson. Ensure you talk about ranges as well as estimates.
• ₽ Display **Slide 4**. Ask: **What animal is this?** Agree that it shows a leopard.
• ₽ Ask: **How many spots can you see on this leopard? How can you give a sensible estimate?**
• Invite suggestions, such as counting the spots in one area and multiplying by the number of similar sized areas. Encourage learners to give their answer as a range.

Assessment for learning

• What is an estimate?
• Why do we estimate?
• What do we mean by range?
• What does it mean if the range is from 10 to 90?

Same day intervention

Support

• Give learners up to 30 counters to estimate and a tens frame to organise them in, so that they can easily count in tens and add any extra ones.

(side tab) Number – Counting and sequenc

105

Practise and/or and/or

Recommended teaching time: 20 min

Teach is followed by independent practice and consolidation, which provides an opportunity for all learners to focus on their newly acquired knowledge. Practice and consolidation consists of both written exercises and practical hands-on activities, with reference to the relevant Workbook pages and bank of Additional practice activities.

All of the tasks are differentiated into three different ability levels:

1 question/activity number typeset on a circle indicates that the question/activity is suitable for learners who require additional support with either easier questions/activities or revising pre-requisite knowledge

2 question/activity number typeset on a triangle indicates that the question/activity is suitable for the majority of learners to practise and consolidate the lesson's learning objective(s)

3 question/activity number typeset on a square indicates that the question/activity is suitable for learners who require enrichment and/or extension.

3Nc.05   Recognise and extend linear sequences, and describe the term-to-term rule.

Resources

mini whiteboard and pen (per pair); Resource sheet 2: Numbers in numerals and words (per pair)

Revise

Use the activity *Reciting, reading and writing odd numbers* from Unit 3: *Reading and writing numbers to 1000* in the Revise activities.

Teach [SB] 🖥 **[TWM. 04/05/06]**

- Recap the sequences from Lesson 3. Agree that the sequences they looked at were sets of numbers that increased or decreased when the same amount was added or subtracted from the previous number.
- Display **Slide 1**. Ask learners to rearrange the eight numbers so that they make an increasing sequence starting from 461. Ask them to explain what is happening each time (increasing by 2). Repeat for a decreasing sequence starting from 475.
- **[TWM.04/05/06]** ℗ Ask: **What is the same about all of these numbers?** Agree that they are all three-digit numbers, they all have 4 hundreds, and that they are all odd numbers. Accept any other observations if correct.
- ℗ Ask: **How do we know that these are odd numbers?** Agree that they end with a digit that is not a multiple of 2 and therefore cannot be divided equally by 2 – there will always be 1 left over.
- Display **Slide 2**. Ask pairs of learners to write down, as numerals, the four numbers shown by the Base 10 equipment on their whiteboard. Ask: **How is 223 similar to 323? How is 223 different from 323? What is the next number in the sequence?** Write down the next three numbers in the sequence.
- Choose two or three numbers from the sequence and ask learners to write these numbers in words.
- **[T&T]** Ask: **How is this sequence different from the previous sequence? How is it the same?** Agree that it is increasing in hundreds instead of 2 ones. The hundreds digit is increasing each time and the tens and ones are staying the same. It is the same because these are three-digit numbers and all of the numbers are odd.
- Direct learners to Let's learn in the Student's Book. Discuss what is happening in the sequence and why the numbers remain odd (each number has 7 ones). Encourage them to write the numbers in numerals and words.
- Give learners some time to work on the paired activity in the Student's Book. Ask learners who

finish quickly to try to make a decreasing sequence. Then take feedback as a class.
- Discuss the Guided practice example in the Student's Book.

Practise [WB] **[TWM.04/05]**

- Workbook

Title: Reading and writing odd numbers to 1000
Pages: 28–29

- Refer to Activity 2 from the Additional practice activities.

Apply 👥 🖥 **[TWM.05/06]**

- Display **Slide 3**. The problem requires learners to work out if it is possible to make a different odd number from the word cards shown.
- Encourage learners to discuss how this is possible. Agree that they could simply swap the nine and seven to make seven hundred and sixty-nine.

Review

- Ask pairs of learners to use the word cards from Resource sheet 2 to make some odd numbers.
- Invite individuals to write examples, in numerals, on the board and get the class to make them with their word cards and read them back.
- Recap units of a hundred, ten and one.

Assessment for learning

- How do we know if a number is odd?
- What is meant by 7 hundreds? How is this different from 7 tens?
- What is the value of 7 tens?
- What is the value of 7 ones?
- What number is created with 7 hundreds, 1 ten and 1 one? Is this an even or odd number? How do you know?

Same day intervention

Support

- Focus on reading and writing two-digit odd numbers.

(117)

Number – Integers and powers

Apply 👤 and/or 👥 and/or 👥👥

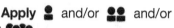

Recommended teaching time: 10 min

Apply is an investigation, problem, puzzle or cross-curricular application where the learner uses and applies knowledge, understanding and skills in an applied context. The content of Apply is located on a slide 🖥 (See Note* below).

Review 👥👥👥

Recommended teaching time: 5 min

The all-important conclusion to the lesson offers an opportunity for learners to make reflective comments about their learning, as well as to discuss misconceptions and common errors, and summarise what they have learned.

Same day intervention

Support and/or Enrichment

Offers same day intervention suggestions so that teachers can effectively provide either support or Enrichment where appropriate.

The aim is to provide guidance to teachers to ensure that:

- all learners reach a certain level of understanding by the end of the day, preventing an achievement gap from forming
- the needs of all learners are being met with respect to the lesson's learning objective(s).

Assessment for learning

Specific questions designed to assist teachers in checking learners' understanding of the lesson objective(s). These questions can be used at any time throughout the lesson.

NOTE*:

As learners will inevitably complete the 'Practise' activities at different times, it is recommended that teachers introduce the 'Apply' activity to the whole class at the end of 'Teach' and before learners work independently on the 'Practise' activities (Workbook page and/or Additional practice activities).

There should also be no expectation that *every* learner in **every** lesson should complete the 'Apply' activity. However, opportunities should be given, either during or

outside the maths lesson, for learners to work on 'Apply'. This could include (but not always) part of the teaching and learning content for the 'fifth' lesson of the week.

Finally, it is also recommended that as part of the whole class 'Review' phase of the lesson, learners are given the opportunity to share with the rest of the class the work they carried out as part of the 'Apply' activity. This includes not just providing the solution, answer or result (if there is one), but also the different methods and strategies they used, and what they learned from the activity.

Additional practice activities

As well as four lesson plans, each unit in Collins International Primary Maths provides two Additional practice activities.

Strand and Sub-strand

The relevant Cambridge Primary Mathematics Curriculum Framework (0096) strand and sub-strand covered is stated in the sidebar.

Collins International Primary Maths unit number and title

Learning objective(s)

Each activity has clearly defined learning objective(s) to assist teachers in choosing the most appropriate activity for the concept they want the learners to practise and consolidate.

Resources

To aid preparation, any resources required are listed, along with whether they are per group, per pair or per learner. Most of these resources are readily available in classrooms.

What to do

The activity is broken down into clear steps to support teachers in explaining the activity to the learners.

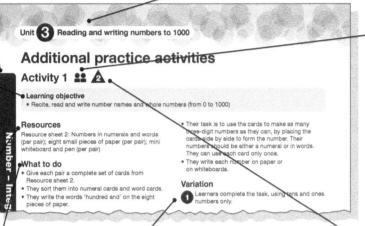

Classroom organisation

Icons are used to indicate whether the activity is designed to be used by learners working in groups , pairs or individually .

Variations

Where appropriate, variations are included. Variations may be designed to make the activity easier ❶ or more difficult ❸, or change the focus of the activity completely. Where the variation affects the challenge level of the activity, the new challenge level is given.

Challenge level

The challenge level for each activity is given:

❶ suitable for learners who require additional support

❷ suitable for the majority of learners to practise and consolidate the lesson's learning objective(s)

❸ suitable for learners who require enrichment and/or extension.

Resource sheets

Where specific paper-based resources are needed for individual lesson plans or Additional practice activities, these are provided as Resource sheets in the Teacher's Guide. Use of Resource sheets is indicated in the resources list of the relevant lesson plan or Additional practice activity.

Some Resource sheets use a spinner to generate numbers or other forms of data.

To use a spinner, hold a paper clip in the centre of the spinner using a pencil and gently flick the paper clip with your finger to make it spin.

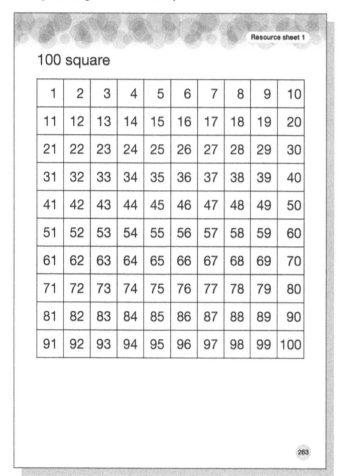

Resource sheet 1

100 square

1	2	3	4	5	6	7	8	9	10
11	12	13	14	15	16	17	18	19	20
21	22	23	24	25	26	27	28	29	30
31	32	33	34	35	36	37	38	39	40
41	42	43	44	45	46	47	48	49	50
51	52	53	54	55	56	57	58	59	60
61	62	63	64	65	66	67	68	69	70
71	72	73	74	75	76	77	78	79	80
81	82	83	84	85	86	87	88	89	90
91	92	93	94	95	96	97	98	99	100

263

Answers

Answers are provided for all the Workbook pages.

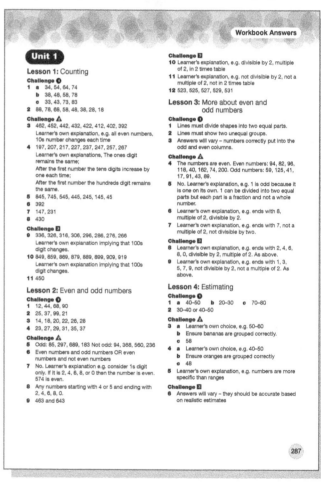

Workbook Answers

Unit 1

Lesson 1: Counting
Challenge ❶
1 a 34, 54, 64, 74
  b 38, 48, 58, 78
  c 33, 43, 73, 83
2 88, 78, 68, 58, 48, 38, 28, 18
Challenge ⚠
3 462, 452, 442, 432, 422, 412, 402, 392
  Learner's own explanation, e.g. all even numbers, 10s number changes each time
4 197, 207, 217, 227, 237, 247, 257, 267
  Learner's own explanations, The ones digit remains the same;
  After the first number the tens digits increase by one each time;
  After the first number the hundreds digit remains the same.
5 845, 745, 545, 445, 245, 145, 45
6 392
7 147, 231
8 430
Challenge ❸
9 336, 326, 316, 306, 296, 286, 276, 266
  Learner's own explanation implying that 100s digit changes.
10 849, 859, 869, 879, 889, 899, 909, 919
  Learner's own explanation implying that 100s digit changes.
11 450

Lesson 2: Even and odd numbers
Challenge ❶
1 12, 44, 68, 90
2 25, 37, 99, 21
3 14, 16, 20, 22, 26, 28
4 23, 27, 29, 31, 35, 37
Challenge ⚠
5 Odd: 85, 297, 689, 183 Not odd: 94, 368, 560, 236
6 Even numbers and odd numbers OR even numbers and not even numbers
7 No. Learner's explanation e.g. consider 1s digit only. If it is 2, 4, 6, 8, or 0 then the number is even. 574 is even.
8 Any numbers starting with 4 or 5 and ending with 2, 4, 6, 8, 0.
9 463 and 643

Challenge ❸
10 Learner's explanation, e.g. divisible by 2, multiple of 2, in 2 times table
11 Learner's explanation, e.g. not divisible by 2, not a multiple of 2, not in 2 times table
12 523, 525, 527, 529, 531

Lesson 3: More about even and odd numbers
Challenge ❶
1 Lines must divide shapes into two equal parts.
2 Lines must show two unequal groups.
3 Answers will vary – numbers correctly put into the odd and even columns.
Challenge ⚠
4 The numbers are even. Even numbers: 94, 82, 96, 118, 40, 162, 74, 200. Odd numbers: 59, 125, 41, 17, 91, 43, 89.
5 No. Learner's explanation, e.g. 1 is odd because it is one on its own. 1 can be divided into two equal parts but each part is a fraction and not a whole number.
6 Learner's own explanation, e.g. ends with 8, multiple of 2, divisible by 2.
7 Learner's own explanation, e.g. ends with 7, not a multiple of 2, not divisible by two.
Challenge ❸
8 Learner's own explanation, e.g. ends with 2, 4, 6, 8, 0, divisible by 2, multiple of 2. As above.
9 Learner's own explanation, e.g. ends with 1, 3, 5, 7, 9, not divisible by 2, not a multiple of 2. As above.

Lesson 4: Estimating
Challenge ❶
1 a 40–50   b 20–30   c 70–80
2 30–40 or 40–50
Challenge ⚠
3 a Learner's own choice, e.g. 50–60
  b Ensure bananas are grouped correctly.
  c 58
4 a Learner's own choice, e.g. 40–50
  b Ensure oranges are grouped correctly
  c 48
5 Learner's own explanation, e.g. numbers are more specific than ranges
Challenge ❸
6 Answers will vary – they should be accurate based on realistic estimates

287

Class record-keeping document

In order to assist teachers with making both short-term 'ongoing' and medium-term 'formative' assessments manageable, meaningful and useful, Collins International Primary Maths includes a class record-keeping document on pages 312–317.

It is intended to be a working document that teachers start at the beginning of the academic year and continually update and amend throughout the course of the year.

Teachers use their own professional judgement of each learner's level of mastery in each of the sub-strands, taking into account:

- mastery of the learning objectives associated with each particular sub-strand
- performance in whole-class discussions
- participation in group work
- work presented in the Workbook
- any other evidence.

Once a decision has been made regarding the degree of mastery achieved by a learner in the particular sub-strand, teachers then write the learner's name (or initials) in the appropriate column:

A: Exceeding expectations in this sub-strand

B: Meeting expectations in this sub-strand

C: Below expectations in this sub-strand.

Given that this is a working document intended to be used throughout the entire academic year, teachers may decide to write (T1), (T2) or (T3) after the learner's name to indicate in which term/semester the judgement was made. This will also help to show the progress (or regress) that learners' make during the course of a year.

Schools and/or individual teachers may decide to use a photocopy of the document at the back of this Teacher's Guide or printout the Word version from the Digital download (either enlarged to A3 if deemed appropriate). It is recommended that whichever option is taken, throughout the year teachers use pencil to fill out the document, and then at the end of the academic year complete the document in pen.

As an alternative to using a photocopy or printout, schools and/or teachers may decide to complete the document electronically using the Digital download version.

Cambridge Primary Mathematics Curriculum Framework (0096) strand and sub-strand

Cambridge Primary Mathematics Curriculum Framework (0096) learning objectives (and codes)

Overall level of mastery in the sub-strand

The degree of mastery achieved by a learner in each sub-strand is shown by writing the learner's name (or initials) in the appropriate column:

A: Exceeding expectations in this sub-strand

B: Meeting expectations in this sub-strand

C: Below expectations in this sub-strand.

An additional bonus of enlarging the document to A3 or completing it electronically, is that it will provide additional space in each of the three columns linked to each sub-strand for teachers to provide more qualitative data should they wish to do so. Teachers can write specific comments that they feel are appropriate for individual learners related not only to the entire sub-strand, but also for specific learning objectives within the sub-strand.

Finally, the class record-keeping document should be seen as an extremely useful document as it can be used to:

- identify those learners who are working *above* and *below* expectations, thereby helping teachers to better plan for the needs of individual learners
- report to parents and guardians
- inform the next year's teacher about which sub-strands of the Cambridge Primary Mathematics Curriculum Framework (0096) individual learners, and the class as a whole, are *exceeding* (A), *meeting* (B) or are *below* (C) in expectations
- assist senior managers within the school in determining whether individual learners, and the class as a whole, are on track to meet end-of-stage expectations.

Year group　　　**Class and academic year reference**

Stage 3 Record-keeping

Class: _____　　Year: _____

KEY

A: Exceeding expectations in this sub-strand	B: Meeting expectations in this sub-strand	C: Below expectations in this sub-strand

Strand: **Number**
Sub-strand: **Counting and sequences**

Code	Learning objectives
3Nc.01	Estimate the number of objects or people (up to 1000).
3Nc.02	Count on and count back in steps of constant size: 1-digit numbers, tens or hundreds, starting from any number (from 0 to 1000).
3Nc.03	Use knowledge of even and odd numbers up to 10 to recognise and sort numbers.
3Nc.04	Recognise the use of an object to represent an unknown quantity in addition and subtraction calculations.
3Nc.05	Recognise and extend linear sequences, and describe the term-to-term rule.
3Nc.06	Extend spatial patterns formed from adding and subtracting a constant.

A	B	C

312

Student's Book

There is one Student's Book for each stage in Collins International Primary Maths with one page provided for each lesson plan.

The content provided in the Student's Book is designed to be used during the 'Teach' phase of a typical Collins International Primary Maths lesson.

However, it is recommended that during the 'Practise' phase of a lesson, if appropriate, learners also use the page in the Student's Book to help them answer the questions on the accompanying Workbook pages.

On page 5 of the Student's Book is a guidance page referred to as 'How to use this book', which explains to the learners the features of the book.

The back of the Student's Book includes the TWM Star which defines, in pupil-friendly language, the eight Thinking and Working Mathematically characteristics. It also includes some sentence stems to help learners to talk with others, challenge ideas and explain their reasoning.

Reference to accompanying Workbook page

Collins International Primary Maths unit number and title

Cambridge Primary Mathematics Curriculum Framework (0096) strand

Lesson number and title

Lesson objective(s)

The Cambridge Primary Mathematics Curriculum Framework (0096) learning objective(s) covered in the lesson, written in language appropriate for learners.

Paired activity

A short paired activity or question to discuss encourages learners to explore the key mathematical idea together.

Where appropriate, indicates that the Paired activity promotes Thinking and Working Mathematically.

Key words

A list of key mathematical terms particularly relevant to the lesson.

Let's learn

Content that presents the key mathematical idea of the lesson being taught by the teacher in the 'Teach' phase.

Guided practice

Worked example(s) designed to prepare learners to work independently on the questions in the Workbook.

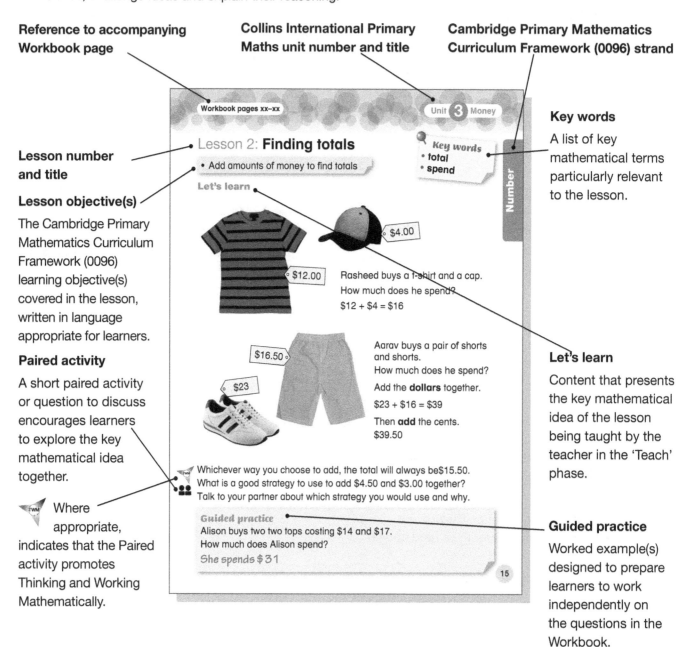

Workbook pages xx–xx

Unit **3** Money

Lesson 2: **Finding totals**

- Add amounts of money to find totals

Let's learn

Key words
- total
- spend

Number

$4.00

$12.00

Rasheed buys a t-shirt and a cap.
How much does he spend?
$12 + $4 = $16

$16.50

$23

Aarav buys a pair of shorts and shorts.
How much does he spend?
Add the **dollars** together.
$23 + $16 = $39
Then **add** the cents.
$39.50

Whichever way you choose to add, the total will always be $15.50.
What is a good strategy to use to add $4.50 and $3.00 together?
Talk to your partner about which strategy you would use and why.

Guided practice
Alison buys two two tops costing $14 and $17.
How much does Alison spend?
She spends $31

15

Workbook

All Workbook page exercises reinforce and build upon the main teaching points and learning objective(s) of a particular lesson in the Teacher's Guide. The work is intended to allow all learners in the class to practise and consolidate their newly acquired knowledge, understanding and skills.

The content provided in the Workbook is designed to be used during the 'Practise' phase of a typical Collins International Primary Maths lesson.

On pages 4 and 5 of the Workbook is a guidance page referred to as 'How to use this book', which explains to the learners the features of the book.

In Stage 3, two Workbook pages are provided for each lesson plan. There is no Workbook page for the Additional practice activities.

Each double page spread has three levels of challenge designed to cater, not only for the different abilities that occur in a mixed-ability or mixed-aged class, but also to assist those schools who 'set' or 'stream' their learners into ability groups. The three different levels of challenge are identified as follows:

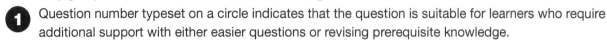 Question number typeset on a circle indicates that the question is suitable for learners who require additional support with either easier questions or revising prerequisite knowledge.

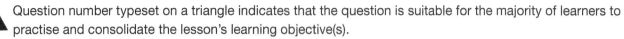 Question number typeset on a triangle indicates that the question is suitable for the majority of learners to practise and consolidate the lesson's learning objective(s).

Question number typeset on a square indicates that the question is suitable for learners who require enrichment and/or extension.

Teachers should think carefully about which of the three different levels of challenge individual learners are asked to complete. There should be no expectation that *every* learner must always answer the questions in *all* three challenge levels.

It is therefore good practice to look carefully at the questions in the Workbook and assign specific questions to specific groups of learners. An effective way of doing this is to tell learners which questions you would like them to answer, and for them to circle those question numbers on their Workbook page.

When appropriate, learners should also be encouraged to work in pairs to answer some or all of the questions. Not only does this help learners learn from each other, thereby reinforcing learners' knowledge, understandings and skills, but it also encourages discussion and mathematical talk, helps create positive self-esteem, and removes the frustrations and feelings of intellectual isolation which can so often be associated with learners working alone.

Teachers may also on occasion decide to work with a group of learners to complete some or all of the questions in the Workbook.

During the 'Review' phase of the lesson, when the whole class is back working together, teachers may decide to complete and/or mark some or all of the questions – perhaps using a different-coloured pencil to differentiate those questions that learners answered independently from those they answered with some assistance.

Finally, it is important to be aware that the Workbook is **not** designed for assessment purposes, nor as a record of what learners can or can't do, nor as proof of what has been taught / learned. Its purpose is for learners to practise and consolidate the mathematical ideas that have been taught during the lesson.

Cambridge Primary Mathematics Curriculum Framework (0096) strand

Collins International Primary Maths unit number and title

Lesson number and title

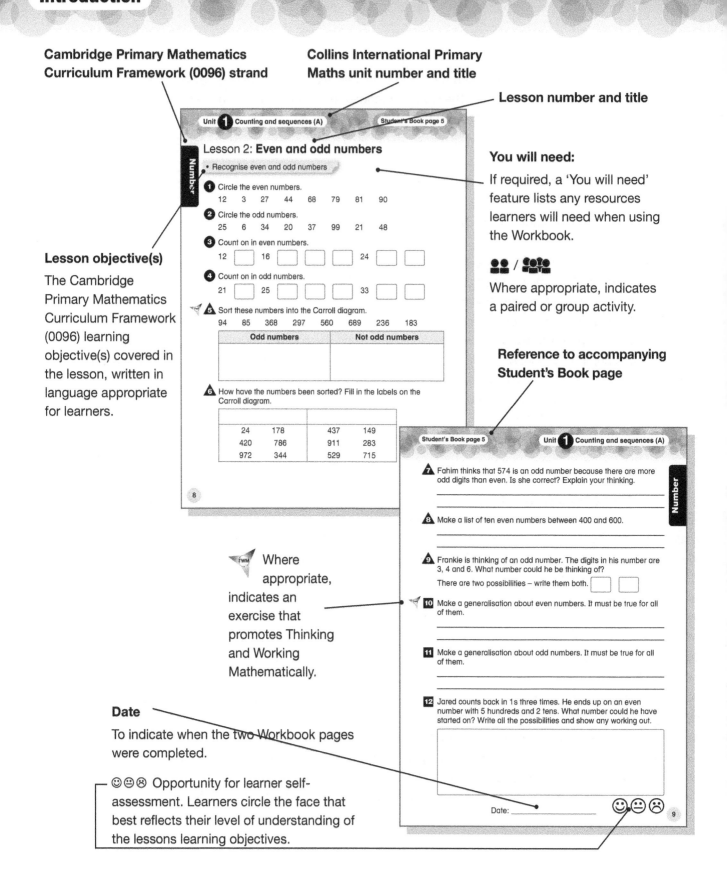

Lesson objective(s)

The Cambridge Primary Mathematics Curriculum Framework (0096) learning objective(s) covered in the lesson, written in language appropriate for learners.

You will need:

If required, a 'You will need' feature lists any resources learners will need when using the Workbook.

Where appropriate, indicates a paired or group activity.

Reference to accompanying Student's Book page

Where appropriate, indicates an exercise that promotes Thinking and Working Mathematically.

Date

To indicate when the two Workbook pages were completed.

☺☺☹ Opportunity for learner self-assessment. Learners circle the face that best reflects their level of understanding of the lessons learning objectives.

Digital content

Collins International Primary Maths also includes a comprehensive set of digital tools and resources, designed to support teachers and learners. The digital content is organised into three sections: Teach, Interact and Support, and is available as the Extended Teacher's Guide.

Teach

The Teach section contains all of the teaching content from the Teacher's Guide, organised into units. This includes:

- Unit introductions
- Lesson plans
- Additional practice activities
- Resource sheets
- Answers

In addition to the above, the Teach section also contains:

- Weekly planning grids

  Editable short-term planning grids provide a synopsis of the teaching and learning opportunities in each of the 27 weekly units. Each Weekly planning grid highlights the content of the five phases for each of the four lessons in the unit (the 5th lesson of the week is left empty for teachers to complete) as well as providing background information, assessment opportunities and teacher and learner evaluation. The intention is for teachers to adapt the grid in order to create a bespoke weekly planning overview for the specific needs of their class.

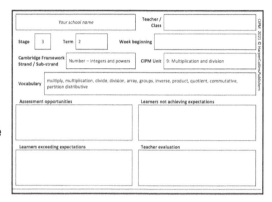

- Slideshows

  Slideshows are provided as visual aids to be shown to the whole class at various phases of a lesson (Revise, Teach, Apply and/or Review), as directed in the lesson plan.

- Interactive whiteboard mathematical tools

  Flexible interactive whiteboard (IWB) teaching tools provide additional visual representations to display to the whole class at various phases of a lesson (Revise, Teach, and/or Review), as directed in the lesson plan.

  These 41 highly adaptable teaching tools are particularly useful in generating specific examples and questions. By doing this, it enables teachers to individualise the content displayed to the class, thereby creating teaching and learning opportunities that better meet the needs of their class.

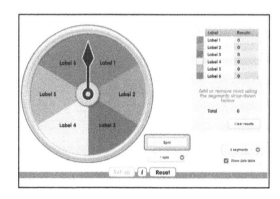

A brief description of the functionality of each of the 40 interactive whiteboard mathematical tools is provided below.

Strand	Tool name	Description of functionality
Number	Counting tool	Use the counting tool to assist with counting from 1–20.
	Place value	Explore how multiple-digit numbers are made up of millions, hundreds of thousands, tens of thousands, thousands, hundreds, tens, ones, tenths, hundredths and thousandths.
	Base ten	Demonstrate the relationship between ones, tens, hundreds and thousands.
	Place value counters	Similar in functionality to the Base ten tool above, i.e. demonstrate the relationship between ones, tens, hundreds, thousands (and millions and decimals – tenths, hundredths and thousandths). However this tool uses place value counters and not Dienes, as is used in the Base ten tool.
	Number line	Use the number line to assist with counting, calculations and exploring decimals.
	Fractions	Demonstrate fractions visually and display the accompanying numerical fraction, decimal, ratio or percentage alongside the pictorial fraction.
	Fraction wall	Demonstrate fractions visually with the fully customisable fraction wall.
	Snake fraction tool	Demonstrate fractions visually with the fully customisable fraction snake.
	Spinner	Demonstrate the concept of probability and making a calculated estimate.
	Bead sticks	Assist with counting, and show the relationship between thousands, hundreds, tens, ones, tenths, hundredths and thousandths.
	Number cards	Display sets of numbers and calculations on movable number cards.
	Number square	Demonstrate and explore counting and number patterns.
	Multiplication square	Demonstrate and explore multiplication and number patterns.
	Function machine	Demonstrate one- and two-step calculations with this animated tool.
	Tree tool	Use real-life objects to practise addition and subtraction.
	Dice tool	Demonstrate the concept of probability or use the tool in conjunction with activities which require dice.
	Tens frame	Use a tens frame template to demonstrate calculations.
Geometry and Measure	Co-ordinates	Create and interpret labelled co-ordinate grids, in one, two or four quadrants.
	Geoboard	Join the dots to depict shapes, routes between two points, to draw nets or to make patterns.
	Geometry set	Use the ruler, protractor or set square to measure and draw lines and angles.
	Rotate and reflect	Demonstrate the rotation and reflection across a horizontal, vertical and diagonal mirror line of a range of 2D shapes.
	Symmetry	Demonstrate lines of horizontal, vertical and diagonal symmetry on a range of 2D shapes.
	Pattern tool	Create patterns and sequences of shapes and design a jumper.

Strand	Tool name	Description of functionality
Geometry and Measure	Beads and laces	Create repeating patterns with the beads and laces tool.
	Shape set	Use the shape set to compare 2D and 3D shapes.
	Nets	Explore the nets of 3D shapes such as prisms, pyramids, tetrahedrons, cubes, and cuboids.
	Money	Practise counting money, or solving calculations involving money.
	Clock	Demonstrate the features of analogue and digital clocks, and explore time.
	Thermometer	Demonstrate how to measure and record temperature, using either Fahrenheit and Celsius scales.
	Capacity	Demonstrate how to measure capacity with this range of animated water containers.
	Weighing	Demonstrate mass by weighing objects of different masses on a range of animated weighing scales.
Statistics and Probability	Bar charter	Demonstrate the creation and interpretation of bar graphs.
	Pie charter	Demonstrate how to create and interpret pie charts with this dynamic data tool.
	Pictogram	Demonstrate the creation and interpretation of pictograms.
	Line grapher	Demonstrate the plotting and interpretation of line graphs with this dynamic data tool.
	Carroll diagram	Demonstrate how to classify and group data with this sorting tool.
	Venn diagram	Demonstrate how to classify and group data with this sorting tool. Import images to give a real-life context.
	Waffle diagram	Demonstrate the creation and interpretation of waffle diagrams.
	Frequency diagram	Demonstrate the creation and interpretation of frequency diagrams for continuous data.
	Scatter graph	Demonstrate the creation and interpretation of scatter graphs.
	Dot plot	Demonstrate the creation and interpretation of dot plot diagrams.

Within Teach, the planning tool allows schools and individual teachers to customise the sequence of units in Collins International Primary Maths within and across all stages. This allows schools and individual teachers to develop their own unique scheme of work.

Interact

The Interact section contains 16 interactive mathematical games. The audio glossary of terms for all stages is also located here.

Support

The Support section contains useful documents for the teacher, such as the medium-term plan, Record-keeping documents described on pages 30–31 and the Collins International Primary Maths Training Package.

Ebook

Ebooks are available for all of the components: Teacher's Guide, Student's Book and Workbook. These enable greater teacher-learner interaction during the whole-class 'Teach' phase of the lesson and also assist teachers in explaining activities and questions to learners as well as in discussing results, solutions and answers once learners have completed an activity or set of questions.

The ebooks can be used in a reader view on computer screens and are also designed to be used with interactive whiteboards (IWB) and if available, iPads and tablets.

Each ebook has standard functionality such as scrolling, zooming, an interactive Contents page and the ability to make notes and highlight sections digitally.

⑦ Collins International Primary Maths Stage 3 Units and Recommended Teaching and Learning Sequence

The Stage 3 learning objectives from the Cambridge Primary Mathematics Curriculum Framework (0096) have been grouped into the following 27 topic areas or 'units':

The Thinking and Working Mathematically characteristics are developed throughout each unit.

Cambridge Primary Mathematics Curriculum Framework (0096)		Collins International Primary Maths	
Strand	Sub-strand	Unit number	Topic
Number	Counting and sequences	1	Counting and sequences (A)
		2	Counting and sequences (B)
	Integers and powers	3	Reading and writing numbers to 1000
		4	Addition and subtraction (A)
		5	Addition and subtraction (B)
		6	Addition
		7	Subtraction
		8	Multiplication and division
		9	Times tables (A)
		10	Times tables (B)
		11	Multiplication
		12	Division
	Money	13	Money
	Place value, ordering and rounding	14	Place value and ordering
		15	Place value, ordering and rounding
	Fractions, decimals, percentages, ratio and proportion	16	Fractions (A)
		17	Fractions (B)
		18	Fractions (C)
Geometry and Measure	Time	19	Time
	Geometrical reasoning, shapes and measurements	20	2D shapes, symmetry and angles
		21	3D shapes
		22	Length, perimeter and area
		23	Mass
		24	Capacity and temperature
	Position and transformation	25	Position, direction, movement and reflection
Statistics and Probability	Statistics and Probability	26	Statistics
		27	Statistics and chance

The Cambridge Primary Mathematics Scheme of Work offers an approach to organising the learning objectives of the Stage 3 curriculum. An overview of this can be seen below.

Unit 3.1   Numbers to 1000, addition and subtraction
Unit 3.2   Time
Unit 3.3   Shapes and angles
Unit 3.4   Patterns, place value and rounding
Unit 3.5   Measurement
Unit 3.6   Multiplication and division
Unit 3.7   Fractions
Unit 3.8   Statistical methods and chance

The table below shows how the 27 units of Collins International Primary Maths Stage 3 link to the Cambridge Primary Mathematics Stage 3 Scheme of Work units. Please note that while the units in the Cambridge Primary Mathematics Scheme of Work may differ from that of Collins International Primary Maths, guidance from Cambridge states that there is no requirement for endorsed resources to follow the teaching order suggested in the Cambridge scheme of work. If a resource is endorsed, schools can be confident that all the learning objectives are covered.

Cambridge Primary Mathematics Stage 3 Scheme of Work units		Collins International Primary Maths Stage 3 units	
Unit number	Topic	Unit number	Topic
3.1	Numbers to 1000, addition and subtraction	3	Reading and writing numbers to 1000
		4	Addition and subtraction (A)
		5	Addition and subtraction (B)
		6	Addition
		7	Subtraction
		13	Money
3.2	Time	19	Time
3.3	Shapes and angles	20	2D shapes, symmetry and angles
		21	3D shapes
		25	Position, direction, movement and reflection
3.4	Patterns, place value and rounding	1	Counting and sequences (A)
		2	Counting and sequences (B)
		14	Place value and ordering
		15	Place value, ordering and rounding
3.5	Measurement	22	Length, perimeter and area
		23	Mass
		24	Capacity and temperature
3.6	Multiplication and division	8	Multiplication and division
		9	Times tables (A)
		10	Times tables (B)
		11	Multiplication
		12	Division
3.7	Fractions	16	Fractions (A)
		17	Fractions (B)
		18	Fractions (C)
3.8	Statistical methods and chance	26	Statistics
		27	Statistics and chance

STRAND: | Number | Geometry and Measure | Statistics and Probability |

The table on the next page shows a recommended teaching and learning sequence (often referred to as a 'medium-term plan') for the 27 units in Collins International Primary Maths Stage 3.

However, as with the Cambridge Primary Mathematics Scheme of Work, schools and individual teachers are free to teach the learning objectives in any order to best meet the needs of individual schools, teachers and learners.

It is important to note that in order to allow for greater flexibility, the 27 units in each stage in Collins International Primary Maths are **not** ordered according to the recommended teaching sequence. Instead, they are in numerical order: Units 1 to 27, according to how the Strands and Sub-strands are arranged in the Cambridge Primary Mathematics Curriculum Framework (0096).

In other words, progression through the components in Collins International Primary Maths does not start at the beginning of the Teacher's Guide, Student's Book and Workbook and end at the back of the Guide / Book. Rather, units are covered as and when is appropriate, according to the Recommended Teaching and Learning Sequence provided by Collins International Primary Maths, or your school's specific scheme of work.

As a note of caution, the Collins International Primary Maths Recommended Teaching and Learning Sequence has been carefully written to ensure continuity and progression both *within* the units at a particular stage and also *across* Stages 1 to 6 and onwards into Lower Secondary.

This is extremely important in ensuring that learners have the pre-requisite knowledge, understanding and skills they require in order to successfully engage with new mathematical ideas at a deeper level and in different contexts.

Learners need to develop mastery of the learning objectives of the Strands and Sub-strands *within* a stage before they are able to apply and transfer their newly acquired knowledge, understanding and skills *across* other Strands and Sub-strands and into later stages. It is for this reason that in the Collins International Primary Maths Recommended Teaching and Learning Sequence, terms/semesters begin with units from the Number Strand, so that learners develop knowledge and skills in number that they can then apply to the other strands of mathematics (Geometry and Measure, and Statistics and Probability) and to their own lives.

Therefore, schools need to think extremely carefully when altering the Collins International Primary Maths Recommended Teaching and Learning Sequence in order to ensure that new learning builds on learners' prior knowledge, understanding and skills. In order to be confident with making such amendments, it is important that teachers are extremely familiar with the lines of progression both *within* and *across* stages of the Cambridge Primary and Lower Secondary Mathematics Curriculum Frameworks.

As with the Cambridge Primary Mathematics Scheme of Work, Collins International Primary Maths has assumed an academic year of three terms/semesters, each of 10 weeks duration. This is the minimum length of a school year and thereby allows flexibility for schools to add in more teaching time as necessary to meet the needs of the learners, and also to comfortably cover the content of the curriculum into an individual schools specific term/semester times.

Collins International Primary Maths Stage 3 Recommended Teaching and Learning Sequence

	Term 1	Term 2	Term 3
Week 1	Unit 3: Reading and writing numbers to 1000	Unit 2: Counting and sequences (B)	Unit 6: Addition
Week 2	Unit 1: Counting and sequences (A)	Unit 15: Place value, ordering and rounding	Unit 7: Subtraction
Week 3	Unit 14: Place value and ordering	Unit 8: Multiplication and division	Unit 13: Money
Week 4	Unit 4: Addition and subtraction (A)	Unit 9: Times tables (A)	Unit 17: Fractions (B)
Week 5	Unit 5: Addition and subtraction (B)	Unit 10: Times tables (B)	Unit 18: Fractions (C)
Week 6	Unit 16: Fractions (A)	Unit 11: Multiplication	Unit 27: Statistics and chance
Week 7	Unit 19: Time	Unit 12: Division	Unit 22: Length, perimeter and area
Week 8	Unit 20: 2D shapes, symmetry and angles	Unit 25: Position, direction, movement and reflection	Unit 23: Mass
Week 9	Unit 21: 3D shapes	Unit 26: Statistics	Unit 24: Capacity and temperature
Week 10	Revision	Revision	Revision

STRAND:	Number	Geometry and Measure	Statistics and Probability

No material is provided in Collins International Primary Maths for the three Revision weeks each term/semester. Individual teachers will decide the content to cover during these weeks, based on monitoring and evaluation made over the course of the term/semester, and learners' levels of achievement on the topics covered throughout the term/semester.

Teachers may decide to revisit certain topics and provide further practice of various concepts that have been taught during the term/semester, or they may use this week to catch up if lessons or units have taken longer than expected.

There is also no expectation that a Revision week will only take place at the end of each term/semester. If individual teachers feel that better use can be made of this week at another time throughout the term/semester, then they should feel free to do so.

⑧ Collins International Primary Maths Stage 3 units match to Cambridge Primary Mathematics Curriculum Framework (0096) Stage 3

The recommended teaching time for each unit is 1 week.

The Thinking and Working Mathematically characteristics are developed throughout each unit. Square brackets within objectives indicate parts of the objective that are not covered in this unit, but are covered elsewhere.

These learning objectives are reproduced from the Cambridge Primary Mathematics curriculum framework (0096) from 2020. This Cambridge International copyright material is reproduced under licence and remains the intellectual property of Cambridge Assessment International Education.

Unit 1 – Counting and sequences (A)			
Cambridge Primary Mathematics Curriculum Framework (0096)			
Strand	**Sub-strand**	**Code**	**Learning objectives**
Number	Counting and sequences	3Nc.01	Estimate the number of objects or people (up to 1000).
		3Nc.02	Count on and count back in steps of constant size: [1-digit numbers], tens or hundreds, starting from any number (from 0 to 1000).
		3Nc.03	Use knowledge of even and odd numbers up to 10 to recognise and sort numbers.
		Collins International Primary Maths Recommended Teaching and Learning Sequence:	**Term 1 Week 2**

Unit 2 – Counting and sequences (B)			
Cambridge Primary Mathematics Curriculum Framework (0096)			
Strand	**Sub-strand**	**Code**	**Learning objectives**
Number	Counting and sequences	3Nc.01	Estimate the number of objects or people (up to 1000).
		3Nc.02	Count on and count back in steps of constant size: 1-digit numbers, [tens or hundreds,] starting from any number (from 0 to 1000).
		3Nc.03	Use knowledge of even and odd numbers up to 10 to recognise and sort numbers.
		3Nc.05	Recognise and extend linear sequences, and describe the term-to-term rule.
		3Nc.06	Extend spatial patterns formed from adding and subtracting a constant.
		Collins International Primary Maths Recommended Teaching and Learning Sequence:	**Term 2 Week 1**

Unit 3 – Reading and writing numbers to 1000			
Cambridge Primary Mathematics Curriculum Framework (0096)			
Strand	**Sub-strand**	**Code**	**Learning objectives**
Number	Integers and powers	3Ni.01	Recite, read and write number names and whole numbers (from 0 to 1000).
	Place value, ordering and rounding	3Np.01	Understand and explain that the value of each digit is determined by its position in that number (up to 3-digit numbers).
	Counting and sequences	3Nc.02	Count on and count back in steps of constant size: 1-digit numbers, tens or hundreds, starting from any number (from 0 to 1000).
		3Nc.03	Use knowledge of even and odd numbers up to 10 to recognise and sort numbers.
		3Nc.05	Recognise and extend linear sequences, and describe the term-to-term rule
			Collins International Primary Maths **Term 1** **Recommended Teaching and Learning Sequence:** **Week 1**

Unit 4 – Addition and subtraction (A)			
Cambridge Primary Mathematics Curriculum Framework (0096)			
Strand	**Sub-strand**	**Code**	**Learning objectives**
Number	Integers and powers	3Ni.02	Understand the commutative [and associative] properties of addition, and use these to simplify calculations.
		3Ni.03	Recognise complements of 100 and complements of multiples of [10 or] 100 (up to 1000).
		3Ni.04	Estimate, add and subtract whole numbers with up to three digits (regrouping of ones or tens).
	Counting and sequences	3Nc.04	Recognise the use of an object to represent an unknown quantity in addition and subtraction calculations.
			Collins International Primary Maths **Term 1** **Recommended Teaching and Learning Sequence:** **Week 4**

Unit 5 – Addition and subtraction (B)			
Cambridge Primary Mathematics Curriculum Framework (0096)			
Strand	**Sub-strand**	**Code**	**Learning objectives**
Number	Integers and powers	3Ni.02	Understand the [commutative and] associative properties of addition, and use these to simplify calculations.
		3Ni.03	Recognise [complements of 100 and] complements of multiples of 10 [or 100] (up to 1000).
		3Ni.04	Estimate, add and subtract whole numbers with up to three digits (regrouping of ones or tens).
	Counting and sequences	3Nc.04	Recognise the use of an object to represent an unknown quantity in addition and subtraction calculations.
			Collins International Primary Maths **Term 1** **Recommended Teaching and Learning Sequence:** **Week 5**

Unit 6 – Addition			
Cambridge Primary Mathematics Curriculum Framework (0096)			
Strand	**Sub-strand**	**Code**	**Learning objectives**
Number	Integers and powers	3Ni.04	Estimate, add [and subtract] whole numbers with up to three digits (regrouping of ones or tens).
		Collins International Primary Maths Recommended Teaching and Learning Sequence:	**Term 3 Week 1**

Unit 7 – Subtraction			
Cambridge Primary Mathematics Curriculum Framework (0096)			
Strand	**Sub-strand**	**Code**	**Learning objective**
Number	Integers and powers	3Ni.04	Estimate, [add] and subtract whole numbers with up to three digits (regrouping of ones or tens).
		Collins International Primary Maths Recommended Teaching and Learning Sequence:	**Term 3 Week 2**

Unit 8 – Multiplication and division			
Cambridge Primary Mathematics Curriculum Framework (0096)			
Strand	**Sub-strand**	**Code**	**Learning objectives**
Number	Integers and powers	3Ni.05	Understand and explain the relationship between multiplication and division.
		3Ni.06	Understand and explain the commutative and distributive properties of multiplication, and use these to simplify calculations.
		Collins International Primary Maths Recommended Teaching and Learning Sequence:	**Term 2 Week 3**

Unit 9 – Times tables (A)			
Cambridge Primary Mathematics Curriculum Framework (0096)			
Strand	**Sub-strand**	**Code**	**Learning objective**
Number	Counting and sequences	3Nc.02	Count on and count back in steps of constant size: 1-digit numbers, [tens or hundreds,] starting [from any number] (from 0 [to 1000]).
		3Nc.05	Recognise and extend linear sequences, and describe the term-to-term rule.
	Integers and powers	3Ni.07	Know [1,] 2, [3,] 4, 5, [6,] 8[, 9] and 10 times tables.
		3Ni.10	Recognise multiples of 2, 5 and 10 (up to 1000).
		Collins International Primary Maths Recommended Teaching and Learning Sequence:	**Term 2 Week 4**

Unit 10 – Times tables (B)			
Cambridge Primary Mathematics Curriculum Framework (0096)			
Strand	**Sub-strand**	**Code**	**Learning objectives**
Number	Counting and sequences	3Nc.02	Count on and count back in steps of constant size: 1-digit numbers, [tens or hundreds,] starting [from any number] (from 0 [to 1000]).
	Integers and powers	3Ni.07	Know 1, 2, 3, 4, 5, 6, 8, 9 and 10 times tables.
		Collins International Primary Maths Recommended Teaching and Learning Sequence:	**Term 2 Week 5**

Unit 11 – Multiplication

Cambridge Primary Mathematics Curriculum Framework (0096)

Strand	Sub-strand	Code	Learning objectives
Number	Integers and powers	3Ni.06	Understand and explain the [commutative and] distributive properties of multiplication, and use these to simplify calculations.
		3Ni.08	Estimate and multiply whole numbers up to 100 by 2, 3, 4 and 5.
		Collins International Primary Maths **Recommended Teaching and Learning Sequence:**	**Term 2** **Week 6**

Unit 12 – Division

Cambridge Primary Mathematics Curriculum Framework (0096)

Strand	Sub-strand	Code	Learning objectives
Number	Integers and powers	3Ni.09	Estimate and divide whole numbers up to 100 by 2, 3, 4 and 5.
	Place value, ordering and rounding	3Np.03	Compose, decompose and regroup 2-digit [3-digit] numbers, using [hundreds,] tens and ones.
		Collins International Primary Maths **Recommended Teaching and Learning Sequence:**	**Term 2** **Week 7**

Unit 13 – Money

Cambridge Primary Mathematics Curriculum Framework (0096)

Strand	Sub-strand	Code	Learning objective
Number	Money	3Nm.01	Interpret money notation for currencies that use a decimal point.
		3Nm.02	Add and subtract amounts of money to give change.
		Collins International Primary Maths **Recommended Teaching and Learning Sequence:**	**Term 3** **Week 3**

Unit 14 – Place value and ordering

Cambridge Primary Mathematics Curriculum Framework (0096)

Strand	Sub-strand	Code	Learning objectives
Number	Place value, ordering and rounding	3Np.01	Understand and explain that the value of each digit is determined by its position in that number (up to 3-digit numbers).
		3Np.03	Compose, decompose and regroup 3-digit numbers, using hundreds, tens and ones.
		3Np.04	Understand the relative size of quantities to compare [and order] 3-digit positive numbers, using the symbols =, > and <.
		Collins International Primary Maths **Recommended Teaching and Learning Sequence:**	**Term 1** **Week 3**

Unit 15 – Place value, ordering and rounding			
Cambridge Primary Mathematics Curriculum Framework (0096)			
Strand	**Sub-strand**	**Code**	**Learning objective**
Number	Place value, ordering and rounding	3Np.01	Understand and explain that the value of each digit is determined by its position in that number (up to 3-digit numbers).
		3Np.02	Use knowledge of place value to multiply whole numbers by 10.
		3Np.04	Understand the relative size of quantities to compare and order 3-digit positive numbers, using the symbols =, > and <.
		3Np.05	Round 3-digit numbers to the nearest 10 or 100.
		Collins International Primary Maths **Recommended Teaching and Learning Sequence:**	**Term 2** **Week 2**

Unit 16 – Fractions (A)			
Cambridge Primary Mathematics Curriculum Framework (0096)			
Strand	**Sub-strand**	**Code**	**Learning objectives**
Number	Fractions, decimals, percentages, ratio and proportion	3Nf.01	Understand and explain that fractions are several equal parts of an object or shape and all the parts, taken together, equal one whole.
		3Nf.02	Understand that the relationship between the whole and the parts depends on the relative size of each, regardless of their shape or orientation.
		3Nf.03	Understand and explain that fractions can describe equal parts of a quantity or set of objects.
		Collins International Primary Maths **Recommended Teaching and Learning Sequence:**	**Term 1** **Week 6**

Unit 17 – Fractions (B)			
Cambridge Primary Mathematics Curriculum Framework (0096)			
Strand	**Sub-strand**	**Code**	**Learning objectives**
Number	Fractions, decimals, percentages, ratio and proportion	3Nf.06	Recognise that two fractions can have an equivalent value (halves, quarters, fifths and tenths).
		3Nf.08	Use knowledge of equivalence to compare and order unit fractions and fractions with the same denominator, using the symbols =, > and <.
		Collins International Primary Maths **Recommended Teaching and Learning Sequence:**	**Term 3** **Week 4**

Unit 18 – Fractions (C)			
Cambridge Primary Mathematics Curriculum Framework (0096)			
Strand	**Sub-strand**	**Code**	**Learning objectives**
Number	Fractions, decimals, percentages, ratio and proportion	3Nf.04	Understand that a fraction can be represented as a division of the numerator by the denominator (half, quarter and three-quarters).
		3Nf.05	Understand that fractions (half, quarter, three-quarters, third and tenth) can act as operators.
		3Nf.07	Estimate, add and subtract fractions with the same denominator (within one whole).
		Collins International Primary Maths **Recommended Teaching and Learning Sequence:**	**Term 3** **Week 5**

Unit 19 – Time			
Cambridge Primary Mathematics Curriculum Framework (0096)			
Strand	**Sub-strand**	**Code**	**Learning objectives**
Geometry and Measure	Time	3Gt.01	Choose the appropriate unit of time for familiar activities.
		3Gt.02	Read and record time accurately in digital notation (12-hour) and on analogue clocks.
		3Gt.03	Interpret and use the information in timetables (12-hour clock).
		3Gt.04	Understand the difference between a time and a time interval. Find time intervals between the same units in days, weeks, months and years.
		Collins International Primary Maths Recommended Teaching and Learning Sequence:	**Term 1 Week 7**

Unit 20 – 2D shapes, symmetry and angles			
Cambridge Primary Mathematics Curriculum Framework (0096)			
Strand	**Sub-strand**	**Code**	**Learning objectives**
Geometry and Measure	Geometrical reasoning, shapes and measurements	3Gg.01	Identify, describe, classify, name and sketch 2D shapes by their properties. Differentiate between regular and irregular polygons.
		3Gg.09	Identify both horizontal and vertical lines of symmetry on 2D shapes and patterns.
		3Gg.10	Compare angles with a right angle. Recognise that a straight line is equivalent to two right angles or a half turn.
		Collins International Primary Maths Recommended Teaching and Learning Sequence:	**Term 1 Week 8**

Unit 21 – 3D shapes			
Cambridge Primary Mathematics Curriculum Framework (0096)			
Strand	**Sub-strand**	**Code**	**Learning objectives**
Geometry and Measure	Geometrical reasoning, shapes and measurements	3Gg.05	Identify, describe, sort, name and sketch 3D shapes by their properties.
		3Gg.08	Recognise pictures, drawings and diagrams of 3D shapes.
		Collins International Primary Maths Recommended Teaching and Learning Sequence:	**Term 1 Week 9**

Unit 22 – Length, perimeter and area			
Cambridge Primary Mathematics Curriculum Framework (0096)			
Strand	**Sub-strand**	**Code**	**Learning objectives**
Geometry and Measure	Geometrical reasoning, shapes and measurements	3Gg.02	Estimate and measure lengths in centimetres (cm), metres (m) and kilometres (km). Understand the relationship between units.
		3Gg.03	Understand that perimeter is the total distance around a 2D shape and can be calculated by adding lengths, and area is how much space a 2D shape occupies within its boundary.
		3Gg.04	Draw lines, rectangles and squares. Estimate, measure and calculate the perimeter of a shape, using appropriate metric units and area on a square grid.
		3Gg.11	Use instruments that measure length, [mass, capacity and temperature].
		Collins International Primary Maths Recommended Teaching and Learning Sequence:	**Term 3 Week 7**

Unit 23 – Mass

Cambridge Primary Mathematics Curriculum Framework (0096)

Strand	Sub-strand	Code	Learning objectives
Geometry and Measure	Geometrical reasoning, shapes and measurements	3Gg.06	Estimate and measure the mass of objects in grams (g) and kilograms (kg). Understand the relationship between units.
		3Gg.11	Use instruments that measure [length,] mass, [capacity and temperature].
		Collins International Primary Maths Recommended Teaching and Learning Sequence:	**Term 3 Week 8**

Unit 24 – Capacity and temperature

Cambridge Primary Mathematics Curriculum Framework (0096)

Strand	Sub-strand	Code	Learning objectives
Geometry and Measure	Geometrical reasoning, shapes and measurements	3Gg.07	Estimate and measure capacity in millilitres (ml) and litres (l), and understand their relationships.
		3Gg.11	Use instruments that measure [length, mass,] capacity and temperature.
		Collins International Primary Maths Recommended Teaching and Learning Sequence:	**Term 3 Week 9**

Unit 25 – Position, direction, movement and reflection

Cambridge Primary Mathematics Curriculum Framework (0096)

Strand	Sub-strand	Code	Learning objective
Geometry and Measure	Position and transformation	3Gp.01	Interpret and create descriptions of position, direction and movement, including reference to cardinal points.
		3Gp.02	Sketch the reflection of a 2D shape in a horizontal or vertical mirror line, including where the mirror line is the edge of the shape.
		Collins International Primary Maths Recommended Teaching and Learning Sequence:	**Term 2 Week 8**

Unit 26 – Statistics

Cambridge Primary Mathematics Curriculum Framework (0096)

Strand	Sub-strand	Code	Learning objectives
Statistics and Probability	Statistics	3Ss.01	Conduct an investigation to answer non-statistical and statistical questions (categorical and discrete data).
		3Ss.02	Record, organise and represent categorical and discrete data. Choose and explain which representation to use in a given situation: - Venn and Carroll diagrams - tally charts and frequency tables [- pictograms and bar charts.]
		3Ss.03	Interpret data, identifying similarities and variations, within data sets, to answer non-statistical and statistical questions and discuss conclusions.
		Collins International Primary Maths Recommended Teaching and Learning Sequence:	**Term 2 Week 9**

Unit 27 – Statistics and chance			
Cambridge Primary Mathematics Curriculum Framework (0096)			
Strand	**Sub-strand**	**Code**	**Learning objectives**
Statistics and Probability	Statistics	3Ss.01	Conduct an investigation to answer non-statistical and statistical questions (categorical and discrete data).
		3Ss.02	Record, organise and represent categorical and discrete data. Choose and explain which representation to use in a given situation: [- Venn and Carroll diagrams - tally charts and frequency tables] - pictograms and bar charts.
		3Ss.03	Interpret data, identifying similarities and variations, within data sets, to answer non-statistical and statistical questions and discuss conclusions.
	Probability	3Sp.01	Use familiar language associated with chance to describe events, including 'it will happen', 'it will not happen', 'it might happen'.
		3Sp.02	Conduct chance experiments, and present and describe the results.

Collins International Primary Maths **Recommended Teaching and Learning Sequence:**	**Term 3** **Week 6**

⑨ Cambridge Primary Mathematics Curriculum Framework (0096) Stage 3 match to Collins International Primary Maths units

The charts below show when each of the Stage 3 learning objectives in the Cambridge Primary Mathematics Curriculum Framework (0096) are taught in the 27 units in Collins International Primary Maths Stage 3.

Cambridge Primary Mathematics Curriculum Framework (0096)				Collins International Primary Maths unit(s)
Strand	Sub-strand	Code	Learning objective	
Number	Counting and sequences	3Nc.01	Estimate the number of objects or people (up to 1000).	1, 2
		3Nc.02	Count on and count back in steps of constant size: 1-digit numbers, tens or hundreds, starting from any number (from 0 to 1000).	1, 2, 3, 9, 10
		3Nc.03	Use knowledge of even and odd numbers up to 10 to recognise and sort numbers.	1, 2, 3
		3Nc.04	Recognise the use of an object to represent an unknown quantity in addition and subtraction calculations.	4, 5
		3Nc.05	Recognise and extend linear sequences, and describe the term-to-term rule.	2, 3, 9
		3Nc.06	Extend spatial patterns formed from adding and subtracting a constant.	2
	Integers and powers	3Ni.01	Recite, read and write number names and whole numbers (from 0 to 1000).	3
		3Ni.02	Understand the commutative and associative properties of addition, and use these to simplify calculations.	4, 5
		3Ni.03	Recognise complements of 100 and complements of multiples of 10 or 100 (up to 1000).	4, 5
		3Ni.04	Estimate, add and subtract whole numbers with up to three digits (regrouping of ones or tens).	4, 5, 6, 7
		3Ni.05	Understand and explain the relationship between multiplication and division.	8
		3Ni.06	Understand and explain the commutative and distributive properties of multiplication, and use these to simplify calculations.	8, 11

Cambridge Primary Mathematics Curriculum Framework (0096)				Collins International Primary Maths unit(s)
Strand	**Sub-strand**	**Code**	**Learning objective**	
Number	Integers, powers and roots	3Ni.07	Know 1, 2, 3, 4, 5, 6, 8, 9 and 10 times tables.	9, 10
		3Ni.08	Estimate and multiply whole numbers up to 100 by 2, 3, 4 and 5.	11
		3Ni.09	Estimate and divide whole numbers up to 100 by 2, 3, 4 and 5.	12
		3Ni.10	Recognise multiples of 2, 5 and 10 (up to 1000).	9
	Money	3Nm.01	Interpret money notation for currencies that use a decimal point.	13
		3Nm.02	Add and subtract amounts of money to give change.	13
	Place value, ordering and rounding	3Np.01	Understand and explain that the value of each digit is determined by its position in that number (up to 3-digit numbers).	3, 14, 15
		3Np.02	Use knowledge of place value to multiply whole numbers by 10.	15
		3Np.03	Compose, decompose and regroup 3-digit numbers, using hundreds, tens and ones.	12, 14
		3Np.04	Understand the relative size of quantities to compare and order 3-digit positive numbers, using the symbols =, > and <.	14, 15
		3Np.05	Round 3-digit numbers to the nearest 10 or 100.	15
	Fractions, decimals, percentages, ratio and proportion	3Nf.01	Understand and explain that fractions are several equal parts of an object or shape and all the parts, taken together, equal one whole.	16
		3Nf.02	Understand that the relationship between the whole and the parts depends on the relative size of each, regardless of their shape or orientation.	16
		3Nf.03	Understand and explain that fractions can describe equal parts of a quantity or set of objects.	16
		3Nf.04	Understand that a fraction can be represented as a division of the numerator by the denominator (half, quarter and three-quarters).	18

Cambridge Primary Mathematics Curriculum Framework (0096)				Collins International Primary Maths unit(s)
Strand	Sub-strand	Code	Learning objective	
Number	Fractions, decimals, percentages, ratio and proportion	3Nf.05	Understand that fractions (half, quarter, three-quarters, third and tenth) can act as operators.	18
		3Nf.06	Recognise that two fractions can have an equivalent value (halves, quarters, fifths and tenths).	17
		3Nf.07	Estimate, add and subtract fractions with the same denominator (within one whole).	18
		3Nf.08	Use knowledge of equivalence to compare and order unit fractions and fractions with the same denominator, using the symbols =, > and <.	17
Geometry and Measure	Time	3Gt.01	Choose the appropriate unit of time for familiar activities.	19
		3Gt.02	Read and record time accurately in digital notation (12-hour) and on analogue clocks.	19
		3Gt.03	Interpret and use the information in timetables (12-hour clock).	19
		3Gt.04	Understand the difference between a time and a time interval. Find time intervals between the same units in days, weeks, months and years.	19
Geometry and Measure	Geometrical reasoning, shapes and measurements	3Gg.01	Identify, describe, classify, name and sketch 2D shapes by their properties. Differentiate between regular and irregular polygons.	20
		3Gg.02	Estimate and measure lengths in centimetres (cm), metres (m) and kilometres (km). Understand the relationship between units.	22
		3Gg.03	Understand that perimeter is the total distance around a 2D shape and can be calculated by adding lengths, and area is how much space a 2D shape occupies within its boundary.	22
		3Gg.04	Draw lines, rectangles and squares. Estimate, measure and calculate the perimeter of a shape, using appropriate metric units and area on a square grid.	22
		3Gg.05	Identify, describe, sort, name and sketch 3D shapes by their properties.	21

Cambridge Primary Mathematics Curriculum Framework (0096)				Collins International Primary Maths unit(s)
Strand	**Sub-strand**	**Code**	**Learning objective**	
Geometry and Measure	Geometrical reasoning, shapes and measurements	3Gg.06	Estimate and measure the mass of objects in grams (g) and kilograms (kg). Understand the relationship between units.	23
		3Gg.07	Estimate and measure capacity in millilitres (ml) and litres (l), and understand their relationships.	24
		3Gg.08	Recognise pictures, drawings and diagrams of 3D shapes.	21
		3Gg.09	Identify both horizontal and vertical lines of symmetry on 2D shapes and patterns.	20
		3Gg.10	Compare angles with a right angle. Recognise that a straight line is equivalent to two right angles or a half turn.	20
		3Gg.11	Use instruments that measure length, mass, capacity and temperature.	22, 23, 24
Geometry and Measure	Position and transformation	3Gp.01	Interpret and create descriptions of position, direction and movement, including reference to cardinal points.	25
		3Gp.02	Sketch the reflection of a 2D shape in a horizontal or vertical mirror line, including where the mirror line is the edge of the shape.	25
Statistics and Probability	Statistics	3Ss.01	Conduct an investigation to answer non-statistical and statistical questions (categorical and discrete data).	26, 27
		3Ss.02	Record, organise and represent categorical and discrete data. Choose and explain which representation to use in a given situation: - Venn and Carroll diagrams - tally charts and frequency tables - pictograms and bar charts.	26, 27
		3Ss.03	Interpret data, identifying similarities and variations, within data sets, to answer non-statistical and statistical questions and discuss conclusions.	26, 27
	Probability	3Sp.01	Use familiar language associated with chance to describe events, including 'it will happen', 'it will not happen', 'it might happen'.	27
		3Sp.02	Conduct chance experiments, and present and describe the results.	27

Revise

Clap counting

Learning objective

Code	Learning objective
3Nc.02	Count on and count back in steps of constant size: 1-digit numbers, tens or hundreds, starting from any number (from 0 to 1000).

What to do

- Ask learners to sit in a circle.
- Starting with a given number between 100 and 200, such as 154, learners count clockwise around the circle in ones, each saying one number.
- At various points, clap. When learners hear the clap, they change counting direction and begin to count backwards. They continue to count in a clockwise direction so that learners experience saying different numbers.

Variation

 Learners count on and back in steps of 10 and 100 from multiples of 10, for example 60 and 120.

A counting wave

Learning objective

Code	Learning objective
3Nc.02	Count on and count back in steps of constant size: 1-digit numbers, tens or hundreds, starting from any number (from 0 to 1000).

What to do

- Ask learners to either stand or sit in a circle or a line.
- Starting with a given two-digit or three-digit number, such as 346, learners count around the circle (or along the line) in steps of 1, 10 or 100. As they say their number, learners jump in the air (if starting from a standing position) or stand up (if starting from a sitting position), raising their hands in the air to form a Mexican wave.
- Encourage learners to consider the digits that change when counting in ones, tens or hundreds.

Variation

 Learners count on and back in steps of 10 and 100 from multiples of 10, for example 30 and 140.

Number – Counting and sequences

Revise

Even and odd numbers

Number – Counting and sequences

Learning objective

Code	Learning objective
3Nc.03	Use knowledge of even and odd numbers up to 10 to recognise and sort numbers.

What to do

- Recap what even and odd numbers are.
- Call out different numbers in the range of 1 to 1000.
- Learners decide if the number is even or odd.
- Remind them that they only need to consider the ones digit.
- If the number you say is even, they clap.
- If the number is odd, they wave.
- Invite learners to call out some numbers for the class to clap or wave.

Variation

1 Write two- or three-digit numbers down on paper or the board. These learners can then look at the last digit and make their decision.

Estimating

Learning objective

Code	Learning objective
3Nc.01	Estimate the number of objects or people (up to 1000).

Resources

mini whiteboard and pen (per learner)

What to do 🖵

- Display **Unit 1, Slide 1** from the Revise slides, which shows a group of coloured circles.
- Ask learners to talk to a partner about how many they think are there.
- They write this down.
- Take some of their estimates and write them on the board.
- Display **Unit 1, Slide 2** from the Revise slides. Ask: **What is the same about this slide and the last? What is different?**
- Agree that a group of 10 has been looped.

- Ask learners to review their estimates. Let them change them if they want.
- Display **Unit 1, Slide 3** from the Revise slides and repeat the questions.
- Agree another group of 10 has been looped.
- Ask learners to review their estimates again.
- Repeat for the next six slides so that learners can clearly see eight groups of 10 and six individual marbles. There are 86 altogether.

Variation

 When learners estimate the number of marbles, they give a range, for example, between 80 and 90.

Revise

Estimating

Learning objective

Code	Learning objective
3Nc.01	Estimate the number of objects or people (up to 1000).

What to do 🖥

- Display **Unit 2, Slide 1** from the Revise slides and ask learners to identify the shapes that they can see.
- Agree that there are rectangles, squares, circles and triangles.
- Ask learners to estimate how many shapes there are altogether. Write some of their suggestions on the board.
- Tell them that there are between 100 and 150 shapes and ask them to review their estimates.
- Inform them that there are 130 shapes. Who had the closest estimate?
- Now tell learners that the number of rectangles, squares and circles is the same.
- Ask them to estimate how many there are of each of those three shapes.
- Count the rectangles to check. Agree that there are 28 of each.

Variation

3 Tell learners that there are more triangles than any other shape.

Expect them to use their knowledge that there are 28 of each of the other three shapes and that there are 130 altogether to estimate the number of triangles. They do not need to calculate, just estimate.

Take feedback of their estimates. Count to see how many there actually are. Agree that there are 46 triangles.

Counting in steps

Learning objective

Code	Learning objective
3Nc.02	Count on and count back in steps of constant size: 1-digit numbers, tens or hundreds, starting from any number (from 0 to 1000).

What to do

- As a class, rehearse counting in steps of 10 from zero to 100 and back to zero.
- Change the starting number, for example count in tens from 7, 16 or 23.
- Repeat for steps of 100, initially from zero to 100 and back, then starting from any one-digit number, for example, 76, 99 or 172.
- Next count in ones from zero to 10 and back to zero. Then change the starting number to, for example, 63, 179, 265 or 898.
- Learners should be familiar with counting in steps of 2 and 5, so practise counting in twos and fives from a variety of starting points, including some in the hundreds, for example counting in steps of 5 from 210 to 260 and back.

Variations

1 Learners count in steps of 2 and 5 within 100.

3 If learners can count in steps of 2, they should be able to use this knowledge to count in steps of 4 (doubling the 2 count) and 8 (doubling the 4 count).

Give them opportunities to try this out, beginning on 4 or 8 each time.

Then extend the starting points so that they begin counting from other multiples of 4 and 8, for example 16, 24, 32.

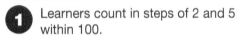

Number – Counting and sequences

Revise

Sequences in counting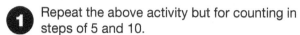

Learning objectives

Code	Learning objective
3Nc.02	Count on and count back in steps of constant size: 1-digit numbers, tens [or hundreds], starting from any number (from 0 to 1000).
3Nc.05	Recognise and extend linear sequences, and describe the term-to-term rule.

What to do

- Ask learners to count in steps of 2 from zero to 20 and back again.
- Repeat this for counting in steps of 4.
- Ask learners to tell you what they know about counting in fours.
- Establish that they are simply doubling counting in twos.
- Discuss the emerging pattern of doubling the 2 count.
- Together say: **Double two equals four, double four equals eight, double six equals 12** and so on to double 20 equals 40.
- Count in steps of 8. Discuss the pattern here.

- Agree that counting in eights is double counting in fours.
- Together say: **Double four equals eight, double eight equals 16, double 16 equals 32** and so on to double 40 equals 80.

Variation

1 Repeat the above activity but for counting in steps of 5 and 10.

Learners will easily be able to count in tens. This time stress the link that counting in tens is the same as doubling the count for 5.

Together say: **Double five equals ten, double ten equals 20** and so on to double 50 equals 100.

Patterns

Learning objective

Code	Learning objective
3Nc.06	Extend spatial patterns formed from adding and subtracting a constant.

What to do 🖥

- Display **Unit 2, Slide 2** from the Revise slides and ask learners what these shapes are. Establish that they are 2D shapes called squares. Do not accept diamonds. Recap that squares are regular shapes because they have four equal sides and four right angles. They are rectangles with equal sides.
- Ask learners to talk to their partners about what they notice about the patterns made by the groups of squares. Elicit that each group contains two more squares than the one before it.
- Ask: **How many squares will be in the next group of squares? Can you relate this to counting in steps of 2?**
- Rehearse counting in steps of 2 from zero to 40 and back.

- Ask learners to work out the numbers of squares in different groups, for example the tenth group, the 12th, 20th, and so on.
- Ask: **How can we work this out quickly?** Establish that they can use their knowledge of the 2 times table.
- Rehearse the 2 times table to 2 × 10 by reciting it together.
- Repeat for the 5 times table and then the 10 times table.

Variation

3 Repeat the above activity, but include counting in steps of 3 and 4.

Revise

Reading numbers

Learning objective

Code	Learning objective
3Ni.01	Recite, read and write number names and whole numbers (from 0 to 1000).

Resources

Resource sheet 2: Numbers in numerals and words (per class)

What to do

- Invite a learner to help you. Give them the cards for 'two', 'thirty', 'four', 'hundred' and 'and' from the word cards on Resource sheet 2.
- Ask them to arrange the cards to make a three-digit number.
- Ask the rest of the class to predict what number this learner will make.

- The learner then shows the number.
- Who predicted correctly? Discuss what numbers could be made with the words: 'two', 'hundred', 'and', 'thirty' and 'four'. (two hundred and thirty-four and four hundred and thirty-two)
- Establish that the tens and ones numbers are preceded by the word 'and'.
- Repeat with other cards.

Variation

1 Follow the instructions for the activity but focus on tens and ones cards.

Numerals and words

Learning objectives

Code	Learning objective
3Ni.01	Recite, read and write number names and whole numbers (from 0 to 1000).
3Np.01	Understand and explain that the value of each digit is determined by its position in that number (up to 3-digit numbers).

Resources

digit cards from Resource sheet 2: Numbers in numerals and words (per class)

What to do

- Draw a simple table on the board. Head the columns 'hundreds', 'tens' and 'ones', for example:

100s	10s	1s

- Invite a learner to pick a digit card from a pile face down on the table.
- The learner can choose where to place the digit and writes it in the appropriate column.

- Repeat this until the table is complete.
- Ask the class to read out the number. Invite a different learner to write the number in words.
- Repeat until all learners have had a chance to take part, either picking a card and choosing where to position the digit or writing the number in words.

Variation

1 Invite learners to choose numbers to place in the tens and ones positions.

Place the digit 1 in the hundreds section of the table.

When confident, exchange the 1 for 2, so that there is a gradual build up to larger numbers.

Number – Integers and powers

Revise

Reciting, reading and writing even numbers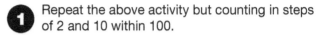

Learning objective	
Code	Learning objective
3Ni.01	Recite, read and write number names and whole numbers (from 0 to 1000).

Resources

mini whiteboard and pen (per learner)

What to do [TWM.05/06]

- Ask learners to count in steps of 10 from zero to 100 and back.
- Ask them to tell you what they notice about each number. Agree that they are all multiples of 10, all end with zero and are all even numbers.
- Ask them to talk to a partner and discuss other steps that they can count in, where the numbers are all even. Agree 2, 100 and anything else that they say that is correct, for example multiples of 10 such as 20, 50.

- Count in steps of 2 from 360.
- Stop at various places and ask learners to write the number said as a numeral and in words.
- Ask: **How many hundreds are there in the number? How many tens? How many ones?**

Variation

 Repeat the above activity but counting in steps of 2 and 10 within 100.

Ensure that learners write different numbers as a numeral and in words and that they can tell you how many tens and ones there are.

Reciting, reading and writing odd numbers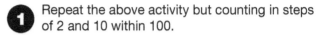

Learning objective	
Code	Learning objective
3Ni.01	Recite, read and write number names and whole numbers (from 0 to 1000).

Resources

mini whiteboard and pen (per learner)

What to do [TWM.05/06]

- Ask learners to count in steps of 5 from zero to 50 and back.
- Ask them to tell you what they notice about each number. Agree that they are all multiples of 5, and that the numbers alternate between even and odd numbers.
- Ask them to talk to a partner and discuss other steps that they can count in where the numbers alternate between even and odd numbers. Agree 3 and any odd number, for example 7, 9 and 11.
- Count in steps of 3 from 103.
- Stop at various places and ask learners to write the number said as a numeral and in words.

- Ask: **How many hundreds are there in the number? How many tens? How many ones?**

Variation

 Repeat the above activity but for counting in steps of 3 and 5, from different starting numbers within 100.

Ensure that learners write different numbers as a numeral and in words and that they can tell you how many tens and ones there are.

Number – Integers and powers

Revise

Relationships 👥👥 👤👤 **1**

Learning objective

Code	Learning objective
3Ni.02	Understand the commutative [and associative] properties of addition, and use these to simplify calculations.

Resources

Resource sheet 3: Digit cards (per pair); mini whiteboard and pen (per pair)

What to do

- Write a calculation on the board, for example: 7 + 2 =. Ask learners to add the two numbers together. Agree that the total is 9. Ask: **What other addition do you know if you know this?** Agree 2 + 7 = 9

- Ask: **What subtraction calculation do you know from the addition?** Agree 9 – 2 = 7 and 9 – 7 = 2.

- Write this calculation on the board: 7 + □ = 9. Ask learners to tell you the missing digit. Agree, that if 7 add 2 equals 9, then the missing number must be 2. Repeat for □ + 2 = 9. Agree the missing number must be 7.

- Give pairs a set of digit cards from Resource sheet 3. They shuffle them and place the pack face down on the table.

- They pick two cards and use them to make an addition or subtraction calculation. So, if they pick 8 and 3 they can make 8 – 3 = 5. They then write the other three calculations to make the set of two addition and two subtraction calculations on paper or mini whiteboards.

- Give them a few minutes to make up as many calculations as they can. Invite pairs to share those they made up.

Variation

 Tell learners that the answer to a calculation is 20. Give them one number and they recall the pair that makes 20.

They write the two addition and two subtraction calculations.

Number pairs to 10 and 20 👥👥 **1**

Learning objective

Code	Learning objective
3Ni.04	[Estimate,] add and subtract whole numbers [with up to three digits (regrouping of ones or tens).]

What to do 📊

- Call out a single digit. Learners say the number that goes with it to make 10, for example call out 3, learners respond with 7.

- Repeat, but this time to 20, for example, you call out 3 and they respond with 17.

- Ensure you include all the possible pairs for 20.

- Write one of the facts on the board and invite learners to write the commutative fact and the two related subtraction facts, for example, if the fact is 3 + 17 = 20, learners write 17 + 3 = 20, 20 – 3 = 17 and 20 – 17 = 3.

Variation

2 Apply facts to 10 to facts for 100.

Write 8 + 2 = 10 on the board. Ask learners to tell you what else they know. Include commutative and inverse facts. Ask: **How can we use this fact to work out the total of 80 + 20?** Agree that 80 is 10 times 8 and 20 is 10 times 2, so the total must be 10 times 10. Call out a multiple of 10 and learners respond by giving you the appropriate fact to 100.

You may wish to use Fishy Facts. Select Level 3. They are given a calculation and learners need to quickly add or subtract to find the answer. You click on the number they say. You could invite learners to do the clicking.

Revise

Missing numbers

Number – Integers and powers

Learning objective

Code	Learning objective
3Nc.04	Recognise the use of an object to represent an unknown quantity in addition and subtraction calculations.

Resources

mini whiteboard and pen (per learner)

What to do

- Display **Unit 4, Slide 1** from the Revise slides and on the price tag to the right of the equals sign write: '$2'. Ask learners to tell you what they see. Agree that there are two pens, each with a blank price tag and a price tag showing the total cost.

- The slide shows that the cost of two pens is $2. Ask: **If two pens cost $2, what is the price of one pen? How can we find out?** Agree the total of the two amounts is $2 and so they need to halve $2 to give $1 as the price for each pen.

- Ask: **What is halving the same as?** Agree dividing by 2.

- Change the amounts on the total price tag to different even amounts of money. Expect learners to tell you the price for each pen, for each new total, by halving.

Variation

3 Change the amounts on the total price tag to different odd amounts of money, for example, $5.

Revise

Commutativity 👥 **1**

Learning objective

Code	Learning objective
3Ni.02	Understand the commutative [and associative] properties of addition, and use these to simplify calculations.

Resources

Resource sheet 3: Digit cards (per pair); mini whiteboard and pen (per pair)

What to do

• Recap the commutative property of addition.
• Ask: **What does it mean if we say that addition is commutative?** Agree that it means that numbers can be added in any order and the total will be the same.
• Give learners a set of digit cards. Ask them to take two cards and use them as the numbers to be added in a calculation.
• They write down the two possible calculations, for example, if they pick the digit cards 4 and 7, the two calculations would be 4 + 7 = 11 and 7 + 4 = 11.

• Ask learners to consider two other ways of recording this. Agree that they could record with the total first, for example, 11 = 4 + 7 and 11 = 7 + 4. It is important that learners understand that the equals symbol means 'the same as' and is not an answer to a calculation. It is therefore good practice to show it written in different positions.
• Repeat this several times, expecting learners to write the four addition number statements for each pair of cards.

Variation

 Learners make a two-digit number with their digit cards and then pick a third card, which is the addend and add that number to their two-digit number.

They write the four appropriate addition number statements.

Number pairs 👥 **2**

Learning objective

Code	Learning objective
3Ni.03	Recognise complements of 100 and complements of multiples of 10 [or 100] (up to 1000).

What to do

• Say a one-digit number. Learners say the number that goes with it to make 10, for example call out 3, learners respond with 7.
• Repeat, but this time for multiples of 10 to 100, for example call out 30 and learners respond with 70. Discuss how knowing numbers that make 10 helps with these facts.
• Ensure you include all the possible pairs of multiples of 10 to 100.
• Repeat for multiples of 100 to 1000 and then other multiples of 100, for example facts that make 700. You call out 100 and learners respond with 600.
• Write one of the facts on the board and invite learners to write the commutative fact and the two related subtraction facts, for example, if the fact is 300 + 700 = 1000, learners write 700 + 300 = 1000, 1000 − 300 = 700 and 1000 − 700 = 300.

Variation

1 Focus on facts to 10 and then multiples of 10 to 100. Write 7 + 3 = 10 on the board. Ask learners to tell you what else they know, including commutative and inverse facts.

Ask: **How can we use this fact to work out the total of 70 + 30?** Agree that 70 is 10 times greater than 7 and 30 is 10 times greater than 3, so the total must be 10 times greater than 10.

Number – Integers and powers

Revise

Using mental calculation strategies △ 2

Learning objective

Code	Learning objective
3Ni.04	Estimate, add and subtract whole numbers with up to three digits (regrouping of ones or tens).

Resources

mini whiteboard and pen (per pair)

What to do

- Write 48 + 27 = on the board. Ask learners to discuss with a partner how they could find the total of the two numbers.
- Take feedback. Did anyone use partitioning to add 40 and 20 and 8 and 7, then recombine 60 and 15? Did anyone take 2 from the 7 and add it to 48 to make 50 + 25? Did anyone add 50 and 27 and subtract 2?
- Work through each strategy and any others suggested by the learners.
- Repeat for other examples, including adding a one-digit number to a two-digit number.
- Write 72 – 18 = on the board. Again, ask learners to discuss with a partner how they could find the difference between these two numbers.

- Take feedback. Did anyone partition 72 into 60 and 12, subtract 8 from 12, 10 from 50 and recombine the numbers? Did anyone subtract 20 and adjust? Did anyone add 2 to both numbers to make the calculation 74 subtract 20?
- Again, work through each strategy and any others that learners used.
- Repeat for other examples, including subtracting a one-digit number from a two-digit number.

Variation

1 Repeat the activity above but focus on making a multiple of 10 for addition, for example: 27 + 18 = 30 + 15 = 45.

Revise

Mental addition: partitioning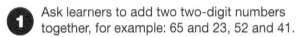

Learning objective

Code	Learning objective
3Ni.04	Estimate, add [and subtract} whole numbers with up to three digits (regrouping of ones or tens).

What to do

- On the board, write: 342 + 57 =. Ask: **What is an efficient strategy to use to add these two numbers together?** Agree that there is no regrouping involved. A sensible strategy would be to partition.
- Ask learners to partition the numbers, add them and recombine. For example, 300 + 40 + 2 + 50 + 7 = 300 + 90 + 9 = 399.
- Ask: **Is there another way?** If learners don't mention sequencing, suggest it. This means keeping the first number whole, partitioning the second number and adding the parts in stages, for example, 342 + 50 + 7.

- Give learners other 'three-digit add two-digit' calculations to add by this method.
- At this stage, ensure that your examples don't involve regrouping.

Variation

1 Ask learners to add two two-digit numbers together, for example: 65 and 23, 52 and 41.

Encourage them to make jottings, recording the steps they take.

Mental addition: compensation

Learning objective

Code	Learning objective
3Ni.04	Estimate, add [and subtract] whole numbers with up to three digits (regrouping of ones or tens).

What to do

- On the board, write: 125 + 19 =. Ask: **What is an efficient strategy to use to add these two numbers together?** Agree that they can solve this without regrouping, if they use compensation.
- Focus on the compensation strategy. Learners add 20 to 125 and subtract 1. Ask: **Why do we subtract 1?** Agree that in adding 20, they have added one too many and therefore the extra 1 needs to be subtracted.
- Give learners other three-digit numbers. They add 19 in the same way as discussed.
- Move on to adding other two-digit numbers with 9 ones, for example: 29, 39, 49, 59. Discuss what multiple of 10 needs to be added.

- Repeat for two-digit numbers that have 8 ones, for example: 28, 38, 48, 58. Ensure that you discuss the multiple of 10 that needs to be added and why, this time, they need to subtract 2.

Variation

1 Focus on adding 9 to two-digit numbers by adding 10 and subtracting 1.

When learners are confident, move on to adding 9 to three-digit numbers.

Number – Integers and powers

Revise

Mental addition: making 10s

Learning objective

Code	Learning objective
3Ni.04	Estimate, add [and subtract] whole numbers with up to three digits (regrouping of ones or tens).

Resources

mini whiteboard and pen (per pair)

What to do [TWM.07/08]

- On the board, write: 148 + 127 =. Ask learners to discuss with a partner how they could find the sum of the two numbers.
- Take feedback of their ideas. Did anyone use partitioning or sequencing? Did anyone take 2 from the 7 and add it to 148 to make 150 + 125? Did anyone add 150 and 127 and subtract 2?
- Work through each strategy and any others that learners used.
- Focus on the 'making 10s' strategy. Inform learners that they can take from one number and add to the other and the total will be the same. For example, if adding 358 and 234, 2 can be taken from 234 and added to 358 to make the calculation 360 + 232 =, or 6 can be taken from 358 and added to 234 to make the calculation 352 + 240 =.

- Learners record what they do on paper or mini whiteboards.
- On the board, write pairs of three-digit calculations that could be added using this strategy, for example: 156 + 137 = (160 + 133 or 153 + 140), 345 + 127 = (350 + 122 or 342 + 130).
- Ask learners to give you pairs of three-digit numbers to add using this strategy.

Variation

1 Repeat the activity but focus on two-digit numbers, for example: 76 + 48 =.

Revise

Mental subtraction: partitioning

Learning objective

Code	Learning objective
3Ni.04	Estimate, [add] and subtract whole numbers with up to three digits (regrouping of ones or tens).

What to do

- On the board, write: 365 – 42 =. Ask: **What is an efficient strategy to use to subtract, to complete this calculation?** Agree that there is no regrouping involved. An efficient strategy might be to partition.

- Ask learners to partition the numbers, subtract them and recombine. For example, 300, 60, 5 and 40, 2 → 60 – 40 = 20, 5 – 2 = 3, 300 + 20 + 3 = 323. Encourage them to use jottings to help them work through the process.

- Ask: **Is there another way?** Learners should know about sequencing from this strategy for addition in Unit 6. Suggest this: keep the first number whole, partition the second number and subtract the parts, for example, 365 – 40 = 325, 325 – 2 = 323.

- Give learners other three-digit and two-digit calculations to subtract using this method.

- At this stage ensure that your examples don't involve regrouping.

Variation

1 Ask learners to subtract two two-digit numbers, for example 87 – 54 =, 79 – 63 =.

Encourage them to make jottings, recording the steps they take.

Mental subtraction: compensation

Learning objective

Code	Learning objective
3Ni.04	Estimate[, add] and subtract whole numbers with up to three digits (regrouping of ones or tens).

What to do

- Write this calculation on the board: 236 – 19 =. Ask: **What is an efficient strategy to use to find the difference between these two numbers?** Agree that learners can solve this without regrouping, if they use compensation.

- Focus on the compensation strategy. Learners subtract 20 from 236 and add 1. Ask: **Why do we add 1?** Agree that in subtracting 20, they have subtracted one too many and therefore the extra 1 needs adding back.

- Give learners other three-digit numbers. They subtract 19 in the same way as discussed.

- Move on to subtracting other two-digit numbers with 9 ones, for example 29, 39, 49, 59. Discuss what multiple of 10 needs subtracting.

- Repeat for two-digit numbers that have 8 ones, for example 28, 38, 48, 58. Ensure that you discuss the multiple of 10 that needs to be subtracted and why, this time, they need to add 2.

Variation

1 Focus on subtracting 9 from two-digit numbers by subtracting 10 and adding 1.

When learners are confident, they move on to subtracting 9 from three-digit numbers.

Number – Integers and powers

Revise

Mental subtraction: counting on and back ▲2

Learning objective	
Code	**Learning objective**
3Ni.04	Estimate[, add] and subtract whole numbers with up to three digits (regrouping of ones or tens).

Number – Integers and powers

Resources

mini whiteboard and pen (per learner)

What to do [TWM.07/08]

- On the board, write: 143 – 137 =. Ask learners to discuss with a partner how they could find the difference between the two numbers.
- Take feedback of their ideas. Did anyone use partitioning, sequencing or compensation?
- Work through each strategy and any others the learners used.
- Suggest that these numbers are close together, so they could use a counting on or counting back strategy.
- Draw a number line on the board that begins with 137 and ends with 143. Ask learners to draw this on paper or mini whiteboards.

- Ask them to count back from 143 to 140 and then to 137.
- Repeat this for counting on. They count on from 137 to 140 and then to 143.
- Both give a difference of 6. Ask: **Which do you find easier, counting back or counting on? Why?**
- On the board, write pairs of three-digit numbers that are close together. Ask learners to find the difference between them, by either counting on or counting back. For each example, they draw a number line to help them.
- Ask learners to give you pairs of three-digit numbers to subtract by using this strategy.

Variation

1 Repeat the activity but focus on two-digit numbers, for example 42 – 38 =.

Revise

Counting stick (1)

Learning objective

Code	Learning objective
3Ni.05	Understand and explain the relationship between multiplication and division.

Resources

counting stick (or metre stick divided into 10 cm intervals) (per class)

What to do

- As a class, recite the multiplication table for 1.
- Show learners the counting stick.
- Tell them that 0 is at the left-hand end and 10 is at the right-hand end.
- Together count in ones from 0 to 10 and back, putting your finger on the divisions as you do.

- Repeat, but stop at various divisions and ask learners to tell you the multiplication statement and the corresponding division statement. For example, stop on the third division. Ask: **How many groups of one are here? What does this look like as a multiplication fact? How many ones are there in 3? What does this look like as a division fact?**
- Repeat for the multiplication table for 2.
- Repeat for the multiplication table for 5
- Repeat for the multiplication table for 10.

Variation

3 Repeat the above activity for the multiplication facts for 3 and 4.

Multiplication and division facts

Learning objective

Code	Learning objective
3Ni.06	Understand and explain the commutative and distributive properties of multiplication, and use these to simplify calculations.

Resources

mini whiteboard and pen (per pair)

What to do

- Write these numbers on the board: 2, 8, 16. Ask: **What do you notice about these numbers?** Expect learners to tell you, for example, that they are all even numbers – 16 is the greatest, 2 is the smallest, 8 is 6 more than 2. Accept any observations they make that are correct.
- Focus on the multiplication and division facts that they can make: $2 \times 8 = 16$, $8 \times 2 = 16$, $16 \div 2 = 8$, $16 \div 8 = 2$.
- Write $2 \times 9 = 9 \times 2$ on the board. Discuss the fact that this is correct because multiplication is commutative.

Distribution

What to do

- Write $24 \times 2 =$ on the board. Ask learners to tell you how they would find the product. Encourage learners to think of at least two different ways, for example, doubling and partitioning $(20 \times 2 + 4 \times 2)$.
- Repeat for other two-digit numbers multiplied by two. Then multiply the same numbers by three.

- Ask: **If multiplication is commutative, does that mean that division is also commutative?** Agree that it doesn't mean that division is also commutative.
- Write $20 \div 2 = 2 \div 20$ on the board. Ask: **Is this correct?** Agree that both calculations can be answered but, the quotients will be different. 20 divided by 2 is 10, but 2 divided by 20 is one half.
- Repeat for other sets of three numbers that can be made into multiplication and division calculations.
- Invite learners to give you sets of three numbers to write on the board for the class to make the calculations.

Variation

1 For each set of numbers, keep one of the factors the same, for example 2, 3, 6, and then 2, 4, 8.

Encourage learners to focus on partitioning or distribution where they partition the multiplicand into tens and ones.

Variation

1 Repeat the activity but use multipliers of 6 and 9.

Revise

Counting stick (2)

Learning objective

Code	Learning objective
3Ni.07	Know 1, 2, [3,] 4, 5, [6, 8, 9] and 10 times tables.

Resources

counting stick (or metre stick divided into 10 cm intervals) (per class)

What to do

- Show learners the counting stick.
- Tell them that zero is at the left-hand end and 10 is at the right-hand end.
- Ask: **What steps can we count in, to move from 0 to 10?** Agree steps of 1. Together count in steps of 1 from 0 to 10 and back, putting your finger on the divisions as you do.
- Repeat but stop at various divisions and ask learners to tell you the multiplication statement and

the corresponding division statement. For example, stop on the third division. Ask: **How many groups of 1 are here? What does this look like as a multiplication fact? How many ones are there in 3? What does this look like as a division fact?**

- Together recite the 1 times table.
- Repeat for the 2 times table.
- Repeat for the 5 times table.
- Repeat for the 10 times table.

Variations

1 Focus on the multiplication and corresponding division facts for 1 and 2.

3 Use the counting stick to introduction learners to the 4 times table.

Bingo

Learning objective

Code	Learning objective
3Ni.07	Know 1, 2, [3,] 4, 5, [6,] 8[, 9] and 10 times tables.

Resources

mini whiteboard and pen (per learner)

What to do

- Ask learners to draw a two by four grid on their whiteboards or paper and write a multiple of 2 in each of the eight sections. The multiples of 2 should not exceed 20.

- Write a number from 1 to 10 on the board. Learners multiply that number by 2. If they have the product on their grid, they cross it out.
- Continue to do this until one or more learners have crossed all their numbers out and called out 'Bingo!'
- Repeat for the 5 times table.
- Repeat for the 10 times table.

Cross it out

Resources

paper and pen (per learner)

What to do

- Ask learners to draw a 3 by 3 grid. Ask them to write 9 products from the multiplication tables 3, 6 and 9.
- Call out a multiplication, for example, 6 × 9. Learners cross out 54, if they have that in their grid.
- Continue until one learner has crossed out all of their numbers.
- Take one example and ask learners to write the commutative fact and the two division facts.

Variation

3 Rather than write the numbers 1 to 10 on the board use the **Number cards tool**. In Set up, set the *Number of cards* to 10, the *Type of numbers* to Serial and click 'Card values hidden as default'. Drag the 10 cards to the centre of the screen. One by one, display the number cards.

Repeat the above activity for the 4 and 8 times tables.

Repeat for other multiplication tables, for example, 2, 4 and 8.

Number – Counting and sequences, Integers and powers

Revise

Counting in steps of 2, 4 and 8

Learning objective

Code	Learning objective
3Ni.07	Know [1,] 2, [3,] 4, [5, 6,] 8[, 9 and 10] times tables.

Resources

counting stick (or metre stick divided into 10 cm intervals)

What to do

- Show learners the counting stick.
- Tell them that 0 is at the left-hand end and 20 at the right-hand end.
- Ask: **What steps do we count in to move from 0 to 20?** Agree steps of 2. Together count in twos from 0 to 20 and back, putting your finger on the divisions as you do.

- Repeat, but stop at various divisions and ask learners to tell you the multiplication statement and the corresponding division statement. For example, stop on the fifth division. Ask: **How many groups of 2 are here? What does this look like as a multiplication fact? How many twos are there in 10? What does this look like as a division fact?**
- Together recite the 2 times table.
- Repeat for the 4 times table.
- Repeat for the 8 times table.

Variation

1 Focus on multiplication and corresponding division facts for the 2 times table.

Counting in steps of 3, 6 and 9

Learning objective

Code	Learning objective
3Ni.07	Know [1, 2,] 3, [4, 5,] 6, [8,] 9 [and 10] times tables.

Resources

counting stick (or metre stick divided into 10 cm intervals)

What to do

- Show learners the counting stick.
- Tell them that 0 is at the left-hand end and 30 at the right-hand end.
- Ask: **What steps do we count in to move from 0 to 30?** Agree steps of 3. Together count in threes from 0 to 30 and back, putting your finger on the divisions as you do.

- Repeat, but stop at various divisions and ask learners to tell you the multiplication statement and the corresponding division statement. For example, stop on the fifth division. Ask: **How many groups of 3 are here? What does this look like as a multiplication fact? How many threes are there in 15? What does this look like as a division fact?**
- Together recite the 3 times table.
- Repeat for the 6 times table.
- Repeat for the 9 times table.

Variation

1 Focus on multiplication and corresponding division facts for the 3 times table.

Counting in steps of 1, 5 and 10

Learning objective

Code	Learning objective
3Ni.07	Know 1, [2, 3, 4,] 5, [6, 8, 9] and 10 times tables.

Resources

3 cubes on a piece of string to be used as a pendulum (per class); paper and pen (per learner)

What to do

- Swing your pendulum from side to side. Ask learners to count in steps of 1 from zero to 10 and back again. Repeat but stop swinging at various points

and ask learners to write the two multiplication and two division facts, for example, stop at 6, learners write $6 \times 1 = 6$, $1 \times 6 = 6$, $6 \div 1 = 6$ and $6 \div 6 = 1$.
- Repeat for counting in 5s and 10s.

Variation

1 Focus on counting in steps of 10.

Revise

Multiplying by 2, 3, 4 and 5 (A)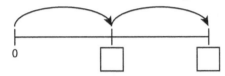

Learning objective

Code	Learning objective
3Ni.08	Estimate and multiply whole numbers up to 100 by 2, 3, 4 and 5.

What to do

- Draw a number similar to the following on the board.

- Ask: **What does this show?** Agree that it shows two jumps. Say that this could be used to model doubling.
- Ask: **What numbers could we put in the boxes?** Listen to learners' ideas. Write some of the numbers in the boxes. As a class, agree whether these will work or not. The second number should be double the first one.
- Write +12 at the top of each loop. Ask: **What numbers should we write in the boxes on the number line?** Agree 12 and 24.

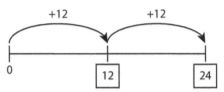

- Ask: **What have we done to 12 to get 24?** Agree that we have doubled it or added it to itself (12 + 12 = 24).
- Discuss how to double by partitioning 12 into 10 and 2, doubling each part and recombining the new numbers.
- Write another number on the loops, for example 15. Learners add 15 to itself or double. Invite a learner to write 15 and 30 in the boxes on the number line.
- Repeat for other numbers up to 50, for example 23, 28, 35, 46.

Variation

 Repeat the above activity but use different numbers for Lesson 2.

Multiplying by 2, 3, 4 and 5 (B)

Learning objective

Code	Learning objective
3Ni.08	Estimate and multiply whole numbers up to 100 by 2, 3, 4 and 5.

Resources

mini whiteboard and pen (per learner)

What to do

- Write a two-digit number, such as 16, on the board.
- Learners multiply it by 2, then by 4 as quickly as they can. Encourage them to use mental calculation strategies, such as doubling, and to make jottings on their paper or mini whiteboards.

Multiplying by 10 and halving

What to do

- Write 24 on the board. Ask learners to multiply this by 5. Ask them to share their strategies, which might include partitioning and multiplying by 10 and halving. Focus on the second strategy. Ask learners to tell you if they could halve the number first and then multiply by 10. Agree that it doesn't matter, the product will

- Once they have done this, ask them to check their products by drawing arrays of place value counters.
- Repeat for other two-digit numbers.

Variation

 Repeat the above activity but multiply by 3 and 5. For multiplying by 5, encourage learners to multiply by 10 and halve.

be the same. Write a list of two-digit even numbers on the board and ask learners to multiply each one using the multiplying by 10 and halving strategy.

Variation

 Repeat the above activity but focus on multiplying by 10.

Number – Integers and powers

Revise

Multiplication and division facts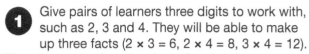

Learning objective	
Code	**Learning objective**
3Ni.09	Estimate and divide whole numbers up to 100 by 2, 3, 4 and 5.

Resources:

mini whiteboard and pen (per learner)

What to do

- On the board, write a selection of four digits, such as 2, 3, 6 and 8.
- Give pairs of learners a few minutes to write down all the different one-digit multiplication facts. They can use each digit only once in their fact, so 2 × 2 would not be permitted.
- Using 2, 3, 6 and 8, they could make 2 × 3 = 6, 2 × 6 = 12, 2 × 8 = 16, 3 × 6 = 18, 3 × 8 = 24 and 6 × 8 = 48. Encourage them to be systematic in their working.
- Once they have the six facts, they list the commutative fact and the two inverse facts, for example, 2 × 3 = 6, 3 × 2 = 6, 6 ÷ 2 = 3, 6 ÷ 3 = 2.
- Repeat for another set of four digits.

Variation

1 Give pairs of learners three digits to work with, such as 2, 3 and 4. They will be able to make up three facts (2 × 3 = 6, 2 × 4 = 8, 3 × 4 = 12). They then write the other facts they know.

Using multiplication facts to divide

Learning objective	
Code	**Learning objective**
3Ni.09	Estimate and divide whole numbers up to 100 by 2, 3, 4 and 5.

Resources:

mini whiteboard and pen (per learner)

What to do

- Write 18 ÷ 2 = on the board. Recap the fact that we can use our times tables to help us divide. To find the answer to this we need to know what number, when multiplied by 2, will have a product of 18. Agree the number is 9.
- Write all the products of the 2 times table to 20 on the board. Learners write the division statements for these, using their 2 times table facts to find the quotient each time, for example, 14 ÷ 2 = 7.
- Write 45 ÷ 5 on the board. Ask learners to tell you what the quotient is using their 5 times table. Agree 45 ÷ 5 = 9. Write all the products of the 5 times table to 50 on the board. Learners write the division statements for these, using their 5 times table facts to find the quotient.
- Repeat for facts in the 10 times table.

Variation

2 Repeat the above for the 3 and 4 times tables.

Number – Integers and powers

Revise

Partitioning in different ways

Learning objectives

Code	Learning objective
3Ni.09	Estimate and divide whole numbers up to 100 by 2, 3, 4 and 5.
3Np.03	Compose, decompose and regroup 2-digit [3-digit] numbers, using [hundreds,] tens and ones.

Resources:

mini whiteboard and pen (per learner)

What to do

- Write 67 on the board.
- Ask learners to partition the number into as many different tens and ones as possible and write them down on paper or their whiteboards.
- They compare their way of partitioning to ensure that none have been missed.
- Repeat with other two-digit numbers.

Variation

3 Give learners numbers that are multiples of 2, ask them to partition them into multiples of 2 tens and another number, for example 36, 54. They then divide each part by 2 and recombine to find the quotient.

Grouping

Learning objectives

Code	Learning objective
3Ni.09	Estimate and divide whole numbers up to 100 by 2, 3, 4 and 5.
3Np.03	Compose, decompose and regroup 2-digit [3-digit] numbers, using [hundreds,] tens and ones.

Resources:

mini whiteboard and pen (per learner) place value counters or Resource sheet 8: Place value counters (per learner) (for the variation)

What to do

- Write a number on the board, for example 48. Tell learners that they are going to divide this by 4.
- Learners partition it into 40 and 8.
- They then draw a representation of, for example, place value counters. When they have done this, they circle the group of 4 tens.
- They then make two groups of 4 ones.
- Repeat with other two-digit numbers that can be grouped simply and quickly without exchanging. Vary the divisor, for example 33, 36, 39 to be divided by 3.

Variation

1 Give learners place value counters to use instead of drawing.

Revise

Counting money (1)

Learning objectives

Code	Learning objective
3Nc.02	Count on and count back in steps of constant size: 1-digit numbers, tens [or hundreds,] starting from any number (from 0 to 1000).
3Nm.02	Add [and subtract] amounts of money [to give change].

What to do 🖵

- Display **Unit 13, Slide 1** from the Revise slides. Ask learners to identify the different coins they can see.
- Ask: **What is the difference between the coins in the top row and those in the bottom row?** Agree that they are the same coins: the top row shows images and the bottom shows values.
- Invite learners to tell you the value of each coin. If necessary, remind learners that 1 dime is equivalent to 10 cents, and that a quarter is equivalent to 25 cents.
- Together count in 1 cents from 1 cent to 20 cents and back to 1 cent.
- Repeat for counting in 5 cents, agreeing that the amounts alternate between odd and even.
- Repeat for counting in 10 cents, agreeing that the amounts will all be even.

Variation

3 Count in 50 cents. Can learners adapt their knowledge of counting in fives for this?

Count in 25 cents. Agree that two 25s are 50. Can they use this knowledge to count successfully? This is not a requirement for Stage 3, but it is interesting to see if learners can make these connections.

All about money

Learning objectives

Code	Learning objective
3Nm.01	Interpret money notation for currencies that use a decimal point.
3Nm.02	Add [and subtract] amounts of money to give change.

What to do 🖵

- Display **Unit 13, Slide 2** from the Revise slides. Ask the class to estimate how much money they can see.
- Write a few of their estimates on the board.
- Ask: **What is the most efficient way to count the coins?** Establish that it would be efficient to count the 10 cent coins first and make a note of them. Then count on the 5 cent and then 1 cent coins. Finally, add the two notes.
- Do this together, counting in tens, fives and ones as a class.
- Agree that the amount is two dollars and 107 cents. Ask learners to tell you how much that is in total, recapping the fact that there are 100c in one dollar. Invite a learner to write this amount using the correct notation: $3.07

- Call out different amounts and ask learners to tell you how they could make those amounts using only the coins and notes on the slide, for example: $1.50, $2.05, 43c, 96c. Ask them to look for different possibilities for each amount.

Variation

1 Choose a selection of ten coins to work with.

Please note that in this unit learners are interpreting dollars and cents, or local currency, as money notation where, for example, the dollars appear to the left of the decimal point and cents to the right. Learners have not yet been introduced to decimals, so cents as decimal fractions of a dollar should not be mentioned.

Number – Money

Revise

Counting money (2)

Learning objectives

Code	Learning objective
3Nc.02	Count on and count back in steps of constant size: 1-digit numbers, tens [or hundreds,] starting from any number (from 0 to 1000).
3Nm.02	Add [and subtract] amounts of money [to give change].

What to do

- Ask learners to count in steps of 5 cents from zero to 50c and back again.
- Do they think this is the same as or different from counting in steps of five?
- Establish that it is exactly the same.
- Ask learners to count in steps of 2 cents and then 4 cents. Recap that counting in fours is doubling the steps for 2.
- Ask: **How can we use what we know to count in steps of 8 cents?**
- Agree this is double counting in steps of 4 cents. Together count in steps of 8 cents.
- Count in steps of 5 and 10 cents.
- Ask: **How can we use our knowledge of counting in steps of 5 cents to count in steps of 50 cents?**
- Agree that steps of 50 are simply 10 times greater than steps of 5.
- Together count in steps of 50 cents to 500 cents and back.

Variation

3 Repeat the activity but for counting in steps of 2, 4, 8, 5 and 10 dollars.

Stress that counting in steps of numbers is the same whatever the context.

Number – Money

Revise

Place value bingo

Learning objective

Code	Learning objective
3Np.01	Understand and explain that the value of each digit is determined by its position in that number (up to 3-digit numbers).

Resources

mini whiteboard and pen (per learner)

What to do

- Learners each draw a 1 by 4 grid on their whiteboards. In each of the four sections they write a three-digit number, using any of the digits from 0 to 5.
- Learners may write the same number more than once if they choose, or two numbers made from the same digits, for example, 431, 314.

- Call out three random digits (for example, 3, 5 and 3). If a learner has written a number using these three digits (in this case: 533, 353, 335) they cross it out. They must only cross out one number each time.
- The winner is the first to have all four numbers crossed out.

Variation

1 Learners draw a 3 by 2 grid and write six two-digit numbers.

Place that digit

Learning objective

Code	Learning objective
3Np.01	Understand and explain that the value of each digit is determined by its position in that number (up to 3-digit numbers).

Resources

mini whiteboard and pen (per learner)

What to do

- Prior to the activity, inform learners of their target: to make the greatest number they can.
- Learners draw a 1 by 3 grid on a mini whiteboard. The sections of the grid represent the hundreds, tens and ones of a three-digit number respectively.
- Explain that you will call out four random digits. Learners decide where to write each digit as it is called out. Once a digit is written in a box, it cannot be erased. For example, if you call out 2, learners may decide to place that in the ones or discard it, if you say 9 they would place that in the hundreds, bearing in mind that they are trying to make the greatest possible number with the digits. They must choose to discard one of the four digits.

- Once you have called out the four digits and learners have written a three-digit number on their whiteboard, compare numbers across the class. Learners who have made the greatest number win a point.
- Repeat several times.

Variation

1 Learners draw two 1 by 3 grids with the aim of making the greatest number in one and the smallest number in the other. Call out seven random digits. Learners decide where to write each digit as it is called out and should choose to discard one – as in the original activity.

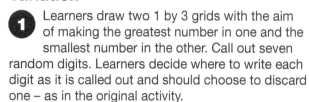

Revise

Rearrange those digits

Learning objective

Code	Learning objective
3Np.04	Understand the relative size of quantities to compare [and order] 3-digit positive numbers, using the symbols =, > and <.

What to do

- Split the class into two teams, A and B.
- Write a three-digit number on the board.
- Choose a learner from Team A to write the > or < symbol on the board and then rearrange the digits in the number so that the comparison is true.
- They get one point if they can do so correctly.
- Choose a learner from Team B. They also get one point if they are able to rearrange the digits a further time to make a number that lies between the two numbers. For example:
  - write the number 392 on the board
  - Team A learner writes the < symbol and rearranges the digits to show 392 < 932
  - Team B learner rearranges the digits and writes 923.

Variation

1 Complete the activity without using the > and < symbols. Team A learner writes a three-digit number on the board. Team B learner then says: 'I can make a number greater / smaller (or more/less) than this by rearranging the digits.' The class guesses what Team B learner's number might be.

Revise

Pat, click, tap

Learning objective

Code	Learning objective
3Np.01	Understand and explain that the value of each digit is determined by its position in that number (up to 3-digit numbers).

What to do

- Explain to learners that you will use gestures to identify digits within a number:
  - hundreds digit = patting head
  - tens digit = clicking fingers
  - ones digit = tapping knees
- Pat, click and tap out a number. For example, pat your head twice (200), click your fingers eight times (80) and tap your knees five times (5). Ask learners to identify the number (285).

- Repeat for different three-digit numbers.
- Encourage learners to pat, click and tap out numbers of their own for the class to guess. Include numbers with no tens (no clicking fingers) and/or no ones (no tapping knees).

Variation

1 Repeat the activity but focus on tens and ones.

Rearrange those digits

Learning objective

Code	Learning objective
3Np.04	Understand the relative size of quantities to compare and order 3-digit positive numbers, using the symbols =, > and <.

What to do

- Split the class into two teams, A and B.
- Write a three-digit number on the board.
- Choose a player from Team A to write the > or < symbol on the board and then rearrange the digits in the number to form a new number so that the comparison is true.
- They get one point if they can do so correctly.
- Choose a player from Team B. They also get one point if they are able to rearrange the digits a further time to make a number that lies between the two numbers.

- For example:
  - Write the number 392 on the board.
  - Team A player rearranges the digits and writes 392 < 932.
  - Team B player rearranges the digits and writes < 923 in between, as 392 < 923 < 932.

Variation

 Complete the activity without using the > and < symbols. Player 1 makes a three-digit number from a given set of digits. Player 2 then says: 'I can make a number more than/greater than/less than/smaller than this by rearranging the digits.' The class then try to guess Player 2's number.

Number – Place value, ordering and rounding

Revise

Mathletics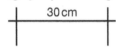

Learning objective

Code	Learning objective
3Np.05	Round 3-digit numbers to the nearest 10 or 100.

Resources

Resource sheet 3: Digit cards (at least two sets)
(per class)

What to do

- This activity is based on the idea of the high jump in an athletics competition.
- Invite four learners to come to the front and give each of them two random digit cards. They use these to make the greatest two-digit number they can.
- Draw a simple representation of a high jump on the board. Set the high jump 'bar height' at 30 cm.

$$| \qquad 30\,cm \qquad |$$

- If a learner's number is more than 30, they clear the bar. If it is less, give them two new cards and two more attempts to clear the bar.

- Repeat a few times. Increase the 'bar height' by 10 cm each time.
- The winner is the player left at the end – the one with the greatest two-digit number.

Variation

2 Give learners three digits cards to make the greatest three-digit number they can. They round their number to the nearest hundred. Go through each hundred from 100 to 1000.

If they fail to reach the new 'bar height' they must sit down. The winner is the player left at the end – the one with the greatest three-digit number.

Revise

Five-minute halving

Learning objective

Code	Learning objective
3Nf.01	Understand and explain that fractions are several equal parts of an object or shape and all the parts, taken together, equal one whole.

Resources

mini whiteboard and pen (per learner)

What to do

- Learners each write down any five even whole numbers from 2 to 20.
- Call out random even numbers from 10 to 40.
- If a learner has written down a number that is half of the number called out, they stand up and call out their number.
- They should then cross out their answer.
- The winner is the first learner to cross out all five of their numbers.

Variation

1 Learners complete the same activity but they write down the five even numbers from 2 to 10.

Fraction doodles

Learning objective

Code	Learning objective
3Nf.01	Understand and explain that fractions are several equal parts of an object or shape and all the parts, taken together, equal one whole.

What to do

- Choose a learner to come to the front and draw a 2D shape, such as a circle, triangle, square or rectangle.
- They split the shape into equal parts and shade some of the parts.
- The class then identify the fraction shaded.

- They also identify the fraction that is not shaded.
- Can they explain that the shaded parts and the unshaded parts together make the whole?

Variation

2 Repeat the activity but for unit fractions, so learners shade only one part.

Making one whole

What to do

- Write a unit fraction on the board, for example, $\frac{1}{3}$.
- Learners write the addition fact with a sum of one whole. They also write the commutative fact and the two subtraction facts, for example, $\frac{1}{3} + \frac{2}{3} = 1$, $\frac{2}{3} + \frac{1}{3} = 1$, $1 - \frac{1}{3} = \frac{2}{3}$, $1 - \frac{2}{3} = \frac{1}{3}$.
- Repeat for other unit fractions.

Variation

1 Write non-unit fractions on the board. Learners write the two addition and two subtraction facts that make one whole, for example, write $\frac{2}{5}$. Learners write: $\frac{2}{5} + \frac{3}{5} = 1$, $\frac{3}{5} + \frac{2}{5} = 1$, $1 - \frac{2}{5} = \frac{3}{5}$ and $1 - \frac{3}{5} = \frac{2}{5}$.

Number – Fractions, decimals, percentages, ratio and proportion

Revise

Fractions of numbers (1)

Learning objective

Code	Learning objective
3Nf.05	Understand that fractions (half, quarter, three-quarters, third and tenth) can act as operators.

Resources

mini whiteboard and pen (per learner)

What to do

- Write 24 on the board.
- Ask learners to work with a partner to find all the unit and non-unit fractions of 24 that they can.
- Encourage them to consider finding half, one third and two thirds, the different quarters.

Variation

3 Encourage learners to use their multiplication facts to find sixths and eighths.

Equivalent fractions

Learning objective

Code	Learning objective
3Nf.06	Recognise that two fractions can have an equivalent value (halves, quarters, fifths and tenths).

Resources

mini whiteboard and pen (per learner)

What to do

- Write 1 on the board. Ask learners to write down as many equivalent fractions as they can for 1 in a minute.
- Ask learners to tell you the different fractions they have made. If they understand that, to make one whole, the numerator and denominator must be the same, there should be no limit to the number of fractions learners make.
- Repeat for one half.
- Accept those that have been explored in Lessons 1 and 2. Accept any others that they think of if they are correct.
- Make a list of those they say on the board.

Variation

3 Repeat the activity for $\frac{1}{4}$ and $\frac{1}{5}$.

Number – Fractions, decimals, percentages, ratio and proportion

Revise

Comparing fractions

Learning objective

Code	Learning objective
3Nf.08	Use knowledge of equivalence to compare and order unit fractions and fractions with the same denominator, using the symbols =, > and <.

Resources

mini whiteboard and pen (per learner)

What to do

- Write pairs of unit fractions on the board.
- Learners compare these in two ways, using the symbols > and <.
- Encourage them to write these down on mini whiteboards or paper.
- Repeat this with fractions that have the same denominator but different numerators.
- Invite learners to explain how they know which fraction is greater/smaller.

Variation

3 Write three or four fractions on the board for learners to order: greatest to smallest.

Number – Fractions, decimals, percentages, ratio and proportion

Revise

Fractions of numbers (2)

Learning objective

Code	Learning objective
3Nf.05	Understand that fractions (half, quarter, three-quarters, third and tenth) can act as operators.

Resources

mini whiteboard and pen (per learner)

What to do

- Write 40 on the board.
- Ask learners to work with a partner to find all the unit and non-unit fractions of 40 that they can.
- Encourage them to consider finding half, quarters, fifths and tenths; for example, $\frac{3}{4}$, $\frac{2}{5}$, $\frac{3}{10}$ of 40.

Variation

 Repeat the activity but focus on finding equivalent fractions for halves, quarters, fifths and tenths. Recap equivalent fractions first.

Ask learners to find the equivalents for 40.

For example, $\frac{1}{2}$ of 40 = 20, $\frac{2}{4}$ of 40 = 20; $\frac{1}{5}$ of 40 = 8, $\frac{2}{10}$ of 40 = 8.

Adding and subtracting fractions

Learning objective

Code	Learning objective
3Nf.07	Estimate, add and subtract fractions with the same denominator (within one whole).

What to do

- Call out a variety of fractions, including half, thirds, quarters, fifths, eighths and tenths. Learners tell you what the numerators must be to make one whole.
- They explored this concept in Unit 16: *Fractions (A)* and made the generalisation that the numerator and denominator must be the same to make one whole.

Variation

 Focus on adding and subtracting different fractions but with the same denominators. Call out pairs of fractions and ask learners to add them, for example: $\frac{2}{8}$ and $\frac{3}{8}$, $\frac{1}{10}$ and $\frac{7}{10}$.

Give learners a fraction and ask them to subtract another from it, for example: $\frac{4}{5}$ subtract $\frac{2}{5}$.

Revise

Units of time

Learning objective

Code	Learning objective
3Gt.01	Choose the appropriate unit of time for familiar activities.

Resources

mini whiteboard and pen (per learner)

What to do

- Ask learners to tell you the units we use to measure time. Expect them to identify seconds, minutes, hours, days, weeks, months, years. Some may mention century, millennium. Congratulate them if they do and ask them if they know how many years these are equivalent to.
- Give examples of activities and events that learners are familiar with such as sleeping, eating a meal, walking to school. Ask learners to write what units these would be measured in on their mini whiteboards. For example, sleeping would be measured in hours, eating a meal in minutes.

- Invite learners to share what they think and, as a class, decide whether they think the unit mentioned is a sensible one.

Variation

 Ask learners to list their own activities and beside these write the units of measure. Invite learners to share the activities they have chosen and the units for measuring the length of time these take.

Do the class agree?

Reading the time

Learning objective

Code	Learning objective
3Gt.02	Read and record time accurately in digital notation (12-hour) and on analogue clocks.

Resources

clock (per learner); mini whiteboard and pen (per learner)

What to do

- Give each learner a clock.
- Say times using minutes past, for example 25 minutes past 6, 40 minutes past 11, 55 minutes past 2. Learners make these times on their clocks and then write the digital equivalent on their mini whiteboards.
- Ask them how many minutes it is until the next o'clock. Expect them to answer in a sentence, for example: 'It is 25 minutes until 7 o'clock.'

Variation

3 Say times using minutes past to the nearest minute, for example 23 minutes past 2, 48 minutes past 11, 54 minutes past 3.

Give learners problems to solve, for example:

- It is 23 minutes past 7. My watch is 15 minutes slow. Show me the time my watch says.
- Sam leaves home at 7:35. He runs to the park and arrives there 38 minutes after leaving home. Show me the time he arrives at the park.

Geometry and Measure – Time

Revise

Shapes around us

Learning objective

Code	Learning objective
3Gg.01	Identify, describe, classify, name and sketch 2D shapes by their properties. Differentiate between regular and irregular polygons.

Resources

mini whiteboard and pen (per learner)

What to do

- Say: **Tell me what 2D shapes you can see around the classroom.** Learners name and talk about the properties of any shapes they can see, for example square table top, rectangular whiteboard, circular clock face.
- Agree that the majority are squares and rectangles. Get learners' thoughts about why this might be.
- Ask them to draw the objects and label the shapes.

Variation

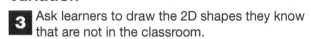 Ask learners to draw the 2D shapes they know that are not in the classroom.

Ask: **What is the same? What is different?** about the shapes and objects that learners identified from around the classroom.

Naming shapes

Learning objective

Code	Learning objective
3Gg.01	Identify, describe, classify, name and sketch 2D shapes by their properties. Differentiate between regular and irregular polygons.

Resources

mini whiteboard and pen (per learner)

What to do 📊

- Display the **Geoboard tool** showing a square. Ask learners to copy the square onto their whiteboard or paper.
- Pull one of the dots on the Geoboard to make a pentagon.
- Ask learners to draw the new 2D shape and to name it and describe the shape's properties.
- Invite learners to suggest a dot to pull and to predict the shape that you will make. They could draw this on their whiteboard or paper. Pull the dot they suggest. Ask: **Were your predictions correct?**

Variation

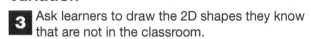 Invite a volunteer to make some 2D shapes on the **Geoboard tool**, following instructions that you give them, for example: **Make a pentagon with a line of symmetry.**

As the volunteer makes the shape, the class draws a shape to match your criterion. They then compare their shapes with the one the learner has made, discussing similarities and differences.

Revise

Symmetry

Learning objective

Code	Learning objective
3Gg.09	Identify both horizontal and vertical lines of symmetry on 2D shapes and patterns.

Resources

mini whiteboard and pen (per learner)

What to do

- Say: **All regular shapes are symmetrical.**
- Ask learners to prove, or disprove, this statement. They should sketch some regular shapes and draw the lines of symmetry onto them. Establish that this is true.

- Next, ask: **Do all irregular shapes have symmetry?** Ask learners to find examples when this statement is true and when it is false.

Variation

1 Focus on one line of symmetry only, for example vertical lines of symmetry.

Angles

Learning objective

Code	Learning objective
3Gg.10	Compare angles with a right angle. Recognise that a straight line is equivalent to two right angles or a half turn.

What to do

- Display the **Pattern tool**. Make a repeating pattern using the pentagon and rectangle.
- Ask: **What are the similarities and differences between the two shapes in this pattern?** Elicit that both the shapes have straight sides and corners. The differences include the numbers of sides and vertices. Agree that the pentagon has five sides and the rectangle has four. The pentagon doesn't have right angles and the rectangle has four right angles.
- Ask: **What could you do to the pentagon to create a right angle?** Invite learners to demonstrate their ideas. Establish that they could draw a line down the middle of the pentagon.

- Repeat this with other shape patterns.
- Each time, expect the learners to explain differences and similarities and also to create right angles within the shapes you use.

Variation

3 Ask learners to draw a pentagon with at least one right angle, then with two and then with three.

Repeat this for other polygons.

Revise

Naming shapes

Learning objective

Code	Learning objective
3Gg.05	Identify, describe, classify, name and sketch 3D shapes by their properties.

Resources

selection of 3D shapes: sphere, cylinder, triangular-based pyramid, cube, cuboid, triangular prism (per class)

What to do

- Place the 3D shapes on the table.
- Ask learners to take turns to choose a shape. They mustn't tell anyone what it is.
- The rest of the class ask questions, such as: 'Does it have square faces?'
- The learner gives yes/no answers until someone guesses the shape.
- Repeat several times. Do learners' questions improve?

Variation

1 Have fewer shapes for learners to choose from. Encourage learners to ask questions that relate to real objects, such as: 'Is it the same shape as a cereal box?'

What's my shape?

Learning objective

Code	Learning objective
3Gg.05	Identify, describe, classify, name and sketch 3D shapes by their properties.

Resources

selection of 3D shapes: sphere, cylinder, triangular-based pyramid, square-based pyramid, cube, cuboid, triangular prism, pentagonal prism, hexagonal prism (per class); bag (per class)

What to do

- Put one of the shapes in the bag, without learners seeing which shape you selected.
- Invite a learner to feel the shape inside the bag, without looking, and to tell the class a property of the shape, for example: 'It has six faces.'
- The class guess which shape the learner is describing.
- The learner keeps giving clues until the class identify which one it is.

Variation

1 Reduce the selection of 3D shapes so that learners focus on the pyramids and prisms they have been working on in class.

Geometry and Measure – Geometrical reasoning, shapes and measurements

Revise

Odd one out

Learning objective

Code	Learning objective
3Gg.05	Identify, describe, classify, name and sketch 3D shapes by their properties.

Resources

selection of 3D shapes: sphere, cylinder, triangular-based pyramid, square-based pyramid, cube, cuboid, triangular prism, pentagonal prism, hexagonal prism (per class)

What to do

- Show four of the shapes listed in the resources.
- Ask learners to tell you which is the odd one out. For example, show a cube, a square-based pyramid, a tetrahedron and a cylinder. The odd one out could be the cube, because it is the only prism. It could be the tetrahedron, because it is the only shape with four triangular faces. It could be the cylinder, because it has two circular faces and a curved surface.
- Ensure you use all of the shapes.

Variation

You could change the questions to: **What is the same? What is different?**

Ask learners to identify similarities and differences in the three 3D shapes that you show. For example, show a sphere, cylinder and cuboid. Similarities are: they are all 3D shapes; they all have at least one face. Differences could include: the cuboid has plane faces, the other two have curved surfaces.

Revise

Ruler or metre stick?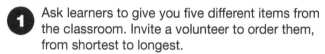

Geometry and Measure – Geometrical reasoning, shapes and measurements

Learning objectives

Code	Learning objective
3Gg.02	Estimate and measure lengths in centimetres (cm), metres (m) and kilometres (km). Understand the relationship between units.
3Gg.11	Use instruments that measure length [, mass, capacity and temperature].

Resources

metre stick (per class); ruler (per class); lengths of string: several shorter than a metre and several over a metre (per class); mini whiteboard and pen (per learner)

What to do

- Show learners the metre stick. Ask them to look at it carefully. Then, hold up each piece of string, one at a time.
- Each time, ask: **Is this shorter or longer than a metre?** Check answers against the metre stick to see if they are correct.
- Repeat with the ruler.
- Hold up each piece again and ask learners to tell you which is better for finding this length: the ruler or the metre stick.
- Ask learners to estimate the length of each piece of string. They write their estimates on their whiteboards.
- Invite learners to help you measure. How close were their estimates?

Variation

1 Ask learners to give you five different items from the classroom. Invite a volunteer to order them, from shortest to longest.

Each time, ask: **Should this be measured with a ruler or a metre stick?** Ask learners to estimate their lengths and, as before, measure to see how close their estimates were.

Units of length

Learning objective

Code	Learning objective
3Gg.02	Estimate and measure lengths in centimetres (cm), metres (m) and kilometres (km). Understand the relationship between units.

Resources

mini whiteboard and pen (per learner)

What to do

- Say: **Tell me the units we use to measure length, height width or distance.**
- Ask: **How many centimetres are there in a metre? How many metres in a kilometre?**
- Write an equivalence on the board, for example 2 m = 200 cm.
- Say: **Write as many other equivalences as you can in one minute using this fact.**

- Encourage them to double, halve, add and multiply, for example: 4 m = 400 cm, 1 m = 100 cm, 5 m = 500 cm, 10 m = 1000 cm.
- Ask learners to share some of the facts they have made and what they did with the original fact.

Variation

3 Repeat the activity above using 2 km = 2000 m.

This time, encourage learners to double, halve, add, subtract, multiply and divide, for example: 4 km = 4000 m.

Revise

Using rulers

Learning objectives

Code	Learning objective
3Gg.02	Estimate and measure lengths in centimetres (cm), metres (m) and kilometres (km). Understand the relationship between units.
3Gg.11	Use instruments that measure length[, mass, capacity and temperature].

Resources

paper, pencil and ruler (per learner)

What to do

- On the board write ten different lengths in centimetres.
- Learners quickly use their ruler to draw lines of those lengths.
- They then pass their paper to a partner who checks to see how accurate they have been.
- They label their lengths with the correct number of centimetres.
- Say: **Order your lengths, from longest to shortest.**

Variation

3 After they have drawn their lines, learners pick pairs of lines and find the difference between them. They then use a ruler to draw the difference.

If appropriate, they could also pick pairs and add their lengths together, drawing the total length on paper.

Solving length problems

Learning objective

Code	Learning objective
3Gg.02	Estimate and measure lengths in centimetres (cm), metres (m) and kilometres (km). Understand the relationship between units.

Resources

mini whiteboard and pen (per learner)

What to do

- Give learners some word problems involving length to solve.
- Make up problems that encourage scaling up and down. For example:
  o Ceris sold 3 m of material. Sian sold 5 times as much. How much did Sian sell?
  o Rasheed drew a square. Each side was 12 cm. His brother drew one with sides $\frac{1}{4}$ of that amount. What was the length of the sides of Rasheed's brother's square?
  o Sanjit went for a 5 km run. Beni went for a run that was 5 times longer. How far did Beni run?
- Learners work these out on their whiteboards.
- Invite volunteers to explain how they solved the problems. Compare their strategies with others that learners might have used.

Variation

3 Volunteers make up a problem to share with the class, who then solve the problem and explain how they did so.

Geometry and Measure – Geometrical reasoning, shapes and measurements

Revise

Kitchen scales or balance scales?

Learning objectives

Code	Learning objective
3Gg.06	Estimate and measure the mass of objects in grams (g) and kilograms (kg). [Understand the relationship between units].
3Gg.11	Use instruments that measure [length,] mass[, capacity and temperature].

What to do

- Display the **Weighing tool**. Show the kitchen scale and then the balance scale. Ask: **How are these the same?** Elicit that both let you find the mass of different objects. Next, ask: **How are these different?** Expect learners to say that the kitchen scale has a scale with intervals and that when you place something onto it, you read the mass by looking to see where the pointer stops. The balance scale requires someone to add weights to one side until both pans balance.

- Show the balance scale and demonstrate measuring a bag of sugar in one pan and the 1 kg weight in the other.

- Ask volunteers to suggest an item from the everyday items list on the **Weighing tool** and then select the mass they think it will have. Measure it with the tool on the kitchen scale, then on the balance scale, to demonstrate the different ways the masses are read on this equipment.

Variation

Ask learners to choose an object from the classroom and place it on the kitchen scales. They read the scale and identify if it is heavier or lighter than 1 kg.

Units of mass

Learning objective

Code	Learning objective
3Gg.02	Estimate and measure lengths in centimetres (cm), metres (m) and kilometres (km). Understand the relationship between units.

Resources

mini whiteboard and pen (per learner)

What to do

- Ask learners to tell you the units that are used to measure mass.
- Ask them to tell you how many grams there are in one kilogram.
- Write the equivalence on the board, for example: 2 kg = 2000 g.
- Learners write as many other equivalences as they can in one minute, using this fact.
- Encourage them to double, halve and add, for example: 4 kg = 4000 g, 1 kg = 1000 g, 5 kg = 5000 g.

- Ask learners to share some of the facts they have made and what strategy they used to make their new fact.

Revise

Using scales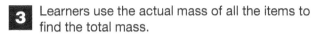

Learning objectives

Code	Learning objective
3Gg.06	Estimate and measure the mass of objects in grams (g) and kilograms (kg). Understand the relationship between units.
3Gg.11	Use instruments that measure [length,] mass[, capacity and temperature].

Resources

mini whiteboard and pen (per learner)

What to do

- Display the **Weighing tool**, set to everyday items. Ask volunteers to choose an item from those displayed.
- Learners estimate the mass of the item and write their estimates on their whiteboards.
- Once they have done this, measure the mass on the **Weighing tool** and ask learners to compare with their estimate.
- The learner with the closest estimate wins a point.
- Continue for five minutes. Who has the most points after all objects have been measured?

Variation

3 Learners use the actual mass of all the items to find the total mass.

Encourage them to use mental calculation strategies such as rounding and adjusting, sequencing and using number pairs to 10, 20, 100 and 1000.

Solving mass problems

Learning objective

Code	Learning objective
3Gg.06	Estimate and measure the mass of objects in grams (g) and kilograms (kg). Understand the relationship between units.

Resources

mini whiteboard and pen (per learner)

What to do

- Give learners some word problems to solve involving mass.
- Make up problems that encourage scaling up and down. For example:
  - Priya buys 9 kg of bananas. Sita buys peaches with mass one third of the mass of Priya's bananas. What is the mass of Sita's peaches?
  - Christos measures his dog. It has a mass of 5 kg 500 g. Jake measures his dog. Its mass is three times the mass of Christos' dog. What is the mass of Jake's dog?
  - Sharma buys 3 kg of onions. Each kilogram costs $1.50. How much does she spend?

- Learners work these out on their whiteboards.
- Invite volunteers to explain how they solved the problems. Compare their strategies with others that learners might have used.

Variation

3 Volunteers make up a problem to share with the class, who then solve the problem and explain how they did so.

Geometry and Measure – Geometrical reasoning, shapes and measurements

Revise

How many cups?

Geometry and Measure – Geometrical reasoning, shapes and measurements

Learning objective

Code	Learning objective
3Gg.11	Use instruments that measure [length, mass,] capacity [and temperature].

Resources

variety of bowls, e.g. washing-up bowl, cereal bowl (one per group); cup for measuring (per group); tray (per group); water (per group); mini whiteboard and pen (per learner); tablespoon or teaspoon, for the variation (per group)

What to do

- Each learner estimates how many cups of water will be needed to fill one of the bowls and writes their estimate on their whiteboard.
- Next, working as a group they carefully use the cup to fill the bowl, making sure that they are using a full cup each time.
- When the bowl is full, they write on their whiteboard how many cups it takes to fill their bowl. How close was their estimate to the actual number?
- They repeat for the other bowls.

Variation

2 Use teaspoons and/or tablespoons to fill the cup, making an estimate first, and then recording the number of spoons it takes to fill the cup with water.

Units of capacity

Learning objective

Code	Learning objective
3Gg.07	Estimate and measure capacity in millilitres (m*l*) and litres (*l*), and understand their relationships.

Resources

mini whiteboard and pen (per learner)

What to do

- Ask: **What units do you use to measure capacity?**
- Ask: **How many millilitres are equivalent to one litre?**
- Write an equivalence on the board, for example: 2 *l* = 2000 m*l*.
- The learners write as many other equivalences as they can in one minute using this fact.
- Encourage them to double, halve, add and multiply, for example: 4 *l* = 4000 m*l*, 1 *l* = 1000 m*l*, 5 *l* = 5000 m*l*, 6 *l* = 6000 m*l*.
- Ask learners to share some of the facts they have made and what strategy they used to make their new fact.

Variation

3 Repeat the activity, using 9000 m*l* = 9 *l*.

This time, encourage learners to double, halve, add, subtract, multiply and divide, for example: 4500 m*l* = 4 *l* 500 m*l*.

Revise

Millilitre or litre?

Learning objectives

Code	Learning objective
3Gg.07	Estimate and measure capacity in millilitres (ml) and litres (l), and understand their relationships.
3Gg.11	Use instruments that measure [length, mass,] capacity [and temperature].

Resources

mini whiteboard and pen (per learner)

What to do

- Ask: **What units do we use to measure capacity?** Agree millilitres and litres. Ask: **How can you find out the capacity of a container?** Elicit that you can use a measuring vessel.
- Display the **Capacity tool**. Show learners the different containers, one at a time.
- Demonstrate filling each container, in turn, with water. Show the container marked in both m*l* and *l*. Learners identify, from this, whether the container's capacity should be measured in millilitres or litres writing their decision on their whiteboard.

Variation

 Repeat the activity but learners must decide before the vessel is filled whether it should be measured in millilitres or litres.

Reading scales

Learning objective

Code	Learning objective
3Gg.11	Use instruments that measure length, mass, capacity and temperature.

Resources

ruler (per class); mini whiteboard and pen (per learner)

What to do

- Draw a vertical line on the board with equally spaced markers from 0 to 10.
- Use a ruler to point to somewhere between 4 and 5.
- Ask learners to tell you where your ruler is pointing. Encourage them to use the language of 'between', 'halfway', 'more than X but less than Y'.
- Repeat pointing to or between other numbers.

Variation

 Mark the scale in fives from 0 to 50/in tens from 0 to 100.

 Use the **Thermometer tool**.

Geometry and Measure – Geometrical reasoning, shapes and measurements

Geometry and Measure – Position and transformation

Revise

Where is it?

Learning objective

Code	Learning objective
3Gp.01	Interpret and create descriptions of position [, direction and movement, including reference to cardinal points].

Resources

selection of 3D shapes: sphere, cone, cylinder, tetrahedron, square-based pyramid, cube, cuboid, triangular prism (per class)

What to do

- Place the sphere beside the cube. Ask learners to describe where it is.
- Ask: **Is there another way to describe the position?**
- Place the square-based pyramid on top of the cube. Ask: **Describe, in two ways, where the cube is. For example, 'beside the sphere and below the cube'.**

- Continue to place the shapes in different positions and ask learners to describe where they are, in relation to other shapes.

Variation

1 Ask learners to describe the position of a shape without naming it: 'My shape is…'. The class needs to guess which shape the speaker is referring to.

Moving around

Learning objective

Code	Learning objective
3Gp.01	Interpret and create descriptions of [position,] direction and movement [, including reference to cardinal points].

Resources

1 cm or, ideally, 2 cm squared paper (per pair); counter (per pair); coloured pencil (per pair); dice (per pair) (for the variation)

What to do

- Ask pairs of learners to place the counter in the top left square of their squared paper.
- Say: **You must move your counter to the bottom right corner. You can only move horizontally and vertically.**
- Learners take turns to do this, drawing the path they make with their pencil. They describe the moves they make as they draw.
- Invite pairs of learners to share their routes with the class. How many different routes were created?
- Discuss which were the most interesting routes and why.

Variation

1 Ask learners to use a counter each. They place these in different squares on the top row of their squared paper. They take turns to throw a dice. If they throw a 1, 2 or 3, they move their counter that number of squares, in any direction, towards the bottom row. If they throw a 4, 5 or 6, they must stay where they are. The aim is to be the first to get to the bottom. This could be a game of strategy, where learners try to block each other's counters.

Revise

Reflection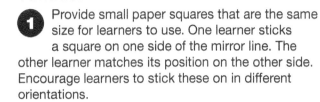

Learning objective

Code	Learning objective
3Gp.02	Sketch the reflection of a 2D shape in a horizontal or vertical mirror line, including where the mirror line is the edge of the shape.

Resources

plain paper (per pair); ruler (per learner); small paper squares (per learner) (for the variation)

What to do

- Ask learners to draw a line down the centre of a piece of plain paper.
- Tell them this is the mirror line.
- One learner draws a shape on one side of the mirror line.
- The other learner draws the same shape on the other side of the mirror line so that it is a reflection.
- They continue to do this until they have made a pattern.
- Do they notice that their pattern is symmetrical?

Variation

1 Provide small paper squares that are the same size for learners to use. One learner sticks a square on one side of the mirror line. The other learner matches its position on the other side. Encourage learners to stick these on in different orientations.

Geometry and Measure – Position and transformation

Statistics and Probability – Statistics

Revise

Diagrams

Learning objectives

Code	Learning objective
3Ss.02	Record, organise and represent categorical and discrete data. Choose and explain which representation to use in a given situation: - Venn and Carroll diagrams [- tally charts and frequency tables - pictograms and bar charts].
3Ss.03	Interpret data, identifying similarities and variations, within data sets, to answer non-statistical and statistical questions and discuss conclusions.

What to do

- Display the **Venn diagram tool** showing one criterion. Add appropriate numbers inside the loop and also outside it.
- Ask learners to guess what the heading for the Venn diagram might be, for example odd numbers or multiples of 10.
- Ask questions that require learners to interpret the data in the Venn diagram.

- Repeat for another criterion. Invite a volunteer to create a Venn diagram for the class to read and interpret.

Variations

▲2 Display the **Carroll diagram tool**. Repeat the above for Carroll diagrams with one criterion.

▲2 Display the **Venn diagram tool** or **Carroll diagram tool** with two criteria.

Tally charts

Learning objectives

Code	Learning objective
3Ss.01	Conduct an investigation to answer non-statistical and statistical questions (categorical and discrete data).
3Ss.02	Record, organise and represent categorical and discrete data. Choose and explain which representation to use in a given situation: - Venn and Carroll diagrams - tally charts and frequency tables [- pictograms and bar charts].
3Ss.03	Interpret data, identifying similarities and variations, within data sets, to answer non-statistical and statistical questions and discuss conclusions.

What to do

- On the board, draw a blank tally chart to show the numbers of learners who like different sports. Have five different sports to choose from and add 'other' as a sixth category.
- Ask learners to vote and make a tally for each vote.
- Ask different volunteers to quickly tell you the total of each tally.
- Ask questions that require learners to interpret the data in the tally chart.

Variation

▲2 Conduct another simple investigation related to a different suitable topic, for example, favourite pet, and record the data in a frequency table.

Revise

Charts and tables

Learning objectives

Code	Learning objective
3Ss.01	Conduct an investigation to answer non-statistical and statistical questions (categorical and discrete data).
3Ss.02	Record, organise and represent categorical and discrete data. Choose and explain which representation to use in a given situation: [– Venn and Carroll diagrams]　　　– tally charts and frequency tables　　　– pictograms [and bar charts].
3Ss.03	Interpret data, identifying similarities and variations, within data sets, to answer non-statistical and statistical questions and discuss conclusions.

What to do

- On the board, list five different fruits for learners to choose from, with 'other' as a sixth option. Learners choose their preferred fruit from the selection. Make a tally chart to record the numbers of learners who like the different fruits.
- As learners vote, make a tally beside the fruit they choose.

- Ask different volunteers to tell you, as quickly as they can, the total for each tally by counting in fives and adding extra marks, as appropriate.
- Ask questions that require learners to interpret the data in the tally chart.

Variation

2 Present the information in a frequency table, under the guidance of the learners.

Pictograms and bar charts

Learning objectives

Code	Learning objective
3Ss.01	Conduct an investigation to answer non-statistical and statistical questions (categorical and discrete data).
3Ss.02	Record, organise and represent categorical and discrete data. Choose and explain which representation to use in a given situation: [– Venn and Carroll diagrams]　　　[– tally charts and frequency tables]　　　– pictograms and bar charts.
3Ss.03	Interpret data, identifying similarities and variations, within data sets, to answer non-statistical and statistical questions and discuss conclusions.

What to do

- Use the information gathered about fruits in the previous revise activity: Charts and tables.
- Display the **Pictogram tool** and show the tally chart information as a pictogram, with each symbol representing one learner.
- Ask questions that require learners to interpret the data in the pictogram.

Variations

2 Use the **Pictogram tool** to display the data where each symbol represents two learners.

2 Use the **Bar charter tool** to represent the data.

Revise

Random or repeat

Learning objective

Code	Learning objective
2Sp.01	Use familiar language associated with patterns and randomness, including regular pattern and random pattern

Resources

squared paper (per pair); coloured pencils (per pair)

What to do

- Learners use coloured pencils to colour squares on squared paper. They make two rows of 12 coloured squares: one random pattern and one repeating pattern.
- They swap papers with another pair. They write down two facts about each pattern, including whether each is a repeating or a non-repeating (random) pattern.
- They continue the repeating pattern for six more squares.

Variation

Restrict the number of colours that learners can use to two or three and ask them to make two repeating patterns and two random patterns.

Will it or won't it?

Learning objective

Code	Learning objective
3Sp.01	Use familiar language associated with chance to describe events, including 'it will happen', 'it will not happen', 'it might happen'.

Resources

mini whiteboard and pen (per pair)

What to do

Prior to the activity, on the board write as a list:

it will happen

it might happen

it will not happen

- Working in pairs, learners think of six events that might or might not happen; for example, it will rain tomorrow, we shall have a Maths lesson this afternoon and write them on paper or their whiteboards.
- They swap lists with another pair and assign the statements: 'it will happen', 'it might happen', 'it will not happen'.

Variation

Prior to the activity, on the board write as a list:

certain to happen

might happen

cannot happen

Ask pairs of learners to list three things that are certain to happen, three that might happen and three that cannot happen.

Unit 1: Counting and sequences (A)

Collins International Primary Maths Recommended Teaching and Learning Sequence: Term 1, Week 2

Learning objectives

Code	Learning objective
3Nc.01	Estimate the number of objects or people (up to 1000).
3Nc.02	Count on and count back in steps of constant size: [1-digit numbers,] tens or hundreds, starting from any number (from 0 to 1000).
3Nc.03	Use knowledge of even and odd numbers up to 10 to recognise and sort numbers.

Unit overview

In this unit, learners will be counting in steps of 10 and 100 using practical apparatus, so they understand that, when counting in tens, the ones digit stays the same and when counting in hundreds, the tens and ones digits stay the same. They will continue to explore odd and even numbers and develop the understanding that an odd number can be described as an even number add 1.

Learners will consolidate their understanding of estimation and how an estimate is a sensible guess. They also begin to see the application of estimating, and appreciate why it is an important skill.

Prerequisites for learning

Learners need to:
- know that numbers come in a particular order
- understand that counting can be applied to anything
- know that even numbers have 2, 4, 6, 8, or 0 in the ones position
- know that odd numbers have 1, 3, 5, 7 or 9 in the ones position
- understand that an estimate is a sensible guess.

Vocabulary

zero, one, ten, hundred, thousand, even number, odd number, shared equally, divided by 2, estimate, approximately, range

Common difficulties and remediation

Often learners rote count without understanding what is happening. It is important to encourage them to look for patterns, which will lead to them noticing things and developing a depth of understanding.

Again, with odd and even numbers, many learners simply remember the rule applied to the ones digits. Patterns are important here, to help learners develop an understanding that an even number is a multiple of 2 (the number can be equally divided by two) and an odd number is a multiple of 2 add 1 (the number cannot be equally divided by two).

Most learners are averse to estimating; they want to be correct. It is important they develop the habit of estimating from the beginning, so they are able to estimate answers to calculations automatically. The way it is introduced in this unit is helpful in encouraging learners to estimate.

Supporting language awareness

When counting in ones, tens and hundreds, display the words that learners will say, for example: one, two, three, ...; ten, twenty, thirty, ... ; hundred. Invite learners to touch the words displayed when counting.

When teaching about odd and even numbers, it is helpful to have visual representations of them on display. Showing even numbers as 'pairs' and odd numbers as 'pairs and an extra one' is helpful. There are visual representations of this in the Student's Book.

Estimating can be a tricky skill to master. Provide examples that learners can use, such as a jar of marbles. Give them possible ranges to choose from, written on card, that they can discuss with a partner.

Promoting Thinking and Working Mathematically

During this unit learners have plenty of opportunities to specialise (TWM.01) and generalise (TWM.02). For example, when counting in hundreds, they look for patterns, for example, 134, 234, 334. They should be able to choose an example in this count that will fit. For example, they can see that the tens and ones numbers are the same so any number with a different hundreds digit, such as 834, will fit the pattern. Learners form mathematical ideas (TWM.03) when they apply what they learn in number to measurement. Learners discuss similarities and differences between numbers and justify their decisions (TWM.04, TWM.05, TWM.06), for example the values of particular digits in a number. Learners are given opportunities to evaluate (TWM.07) and improve (TWM.08) estimates during the lesson on estimating.

Success criteria

Learners can:
- count on and count back in tens and hundreds
- recognise odd and even numbers
- sort odd and even numbers
- estimate numbers of up to 1000 objects.

Lesson 1: **Counting**

Number – Counting and sequences

Learning objective

Code	Learning objective
3Nc.02	Count on and count back in steps of constant size: [1-digit numbers,] tens or hundreds, starting from any number (from 0 to 1000).

Resources

Base 10 equipment: ones, tens, hundreds and a thousand cube (per pair); Resource sheet 1: 100 square (per learner) (for Same day intervention)

Revise

Use the activity *Clap counting* from Unit 1: *Counting and sequences (A)* in the Revise activities.

Teach 📗 🖥 📊 **[TWM.01/02]**

- Count together in tens from zero to 100 and back. Count in hundreds from zero to 1000 and back.
- Display **Slide 1** and discuss the house numbers. Agree they are three-digit numbers.
- 🗣 Ask: **What are these numbers? What is the value of the 1 in 123? Which is the greatest/smallest number? Which numbers are odd/even?**
- Together, count on, then back in tens from each door number. Then count on in hundreds, stopping before 1000, and then count back.
- **[TWM.01]** Display **Slide 2**. Ask: **What number is represented? What does each part represent?**
- Agree that there are two tens, which represents 20, and seven ones, which represents 7. Together 20 and 7 are equivalent to 27.
- Ask pairs of learners to use the Base 10 equipment to make 27. Then ask them to add a tens and another and another, until they reach 97. Ask: **What is happening to your number? Which part is changing? Which part stays the same? Can you explain why?**
- **[T&T]** When they reach 97, ask learners to discuss what will happen if another tens is added.
- 🗣 Ask: **What are 10 lots of 10? What should we exchange our tens with?** Learners make the exchange. They then take tens away to get back to 27.
- Display the **Base 10 tool** and demonstrate adding tens to confirm what learners have done.
- Repeat for other two-digit numbers, with learners using their Base 10 equipment.
- Direct learners to the Student's Book, showing 265. Use the **Base 10 tool** to demonstrate adding hundreds to 265 (up to 965) with learners using their Base 10 equipment. Ask what happens to the tens and the ones digits.
- Then start at 965 and take away hundreds to get back to 265.
- Now repeat, starting again at 265 and adding tens (up to 295) and back again. Repeat, but at 295

demonstrate how to add another 10, then count on from 305 to 355 and back.
- **[TWM.02]** Learners work through the paired activity.
- Discuss the Guided practice example in the Student's Book.

Practise 📙 **[TWM.05]**

- Workbook

Title: Counting

Pages: 6–7

- Refer to Activity 1 from the Additional practice activities.

Apply 👥 🖥 **[TWM.07]**

- Display **Slide 3**.
- Learners consider whether counting in tens and hundreds applies to money.

Review

- 🗣 Ask: **What have we been learning about? When we count in tens, which digit changes? Which stay the same? Why is that?**
- 🗣 Ask: **When we count in hundreds, which digit changes? Which stay the same? Why is that?**
- 🗣 Ask: **Can we count in units of 10 centimetres? What about units of 100 grams?** Agree that counting is the same, no matter what units we use.

Assessment for learning

- What is 10 more/less than 356? What is 10 more/less than that?
- What is 100 more/less than 492? What is 100 more/less than that?

Same day intervention

Support

- Learners count on in tens from one-digit numbers, and back in tens from numbers in the nineties, using a 100 square. Encourage them to notice that the tens digit increases or decreases by 1 each time and the ones digit remains the same.

Enrichment

- Learners count on in multiples of 10 and 100, for example: steps of 20, 30, ...; 200, 300, ...

Lesson 2: **Even and odd numbers**

Learning objective

Code	Learning objective
3Nc.03	Use knowledge of even and odd numbers up to 10 to recognise and sort numbers.

Resources

interlocking cubes (per pair); ruler (per pair); pencil and paper (per learner); ruler (per learner); Resource sheet 1: 100 square (per learner) (for Same day intervention)

Revise

Use the activity *A counting wave* from Unit 1: *Counting and sequences (A)* in the Revise activities.

Teach 📖 🖥

- Direct learners to the picture in the Student's Book. Ask: **What do you notice?** Agree that the cubes represent numbers, the numbers increase by one each time and they are different colours.
- Ask pairs to place a cube in front of them, to represent 1. They then place two joined together to represent 2, as in the Student's Book. They make 3 in the same way, as in the Student's Book.
- 🗣 Ask: **What do you notice? Can you see 2 and one more makes 3? Is this an even number or an odd number?**
- Ask them to make 4. Ask: **What do you notice this time?** Agree that they have two groups of 2, which is 4, an even number. Repeat for all numbers to 10.
- For each number, discuss the pairs of cubes that they see, and whether there is an extra cube. Ask**: Is the number even or odd?**
- Ask them to sort the numbers they have made into even and odd numbers.
- **[TWM.02]** Referring to the first two paired activity questions in the Student's Book, ask: **What do you notice about the even numbers? What do you notice about the odd numbers?**
- **[TWM.01]** Agree that the cubes for even numbers are in pairs and odd numbers are 'an even number add 1'.
- **[TWM.02]** Learners discuss the second pair of question in the paired activity in the Student's Book. Establish that all numbers that end with the digit 2, 4, 6, 8 or 0 are even and all numbers that end with the digit 1, 3, 5, 7 or 9 are odd.
- **[T&T]** Display **Slide 1**. Ask learners to work in pairs to identify the even and odd numbers on the slide.
- Discuss the Guided practice example in the Student's Book. Explain that this Carroll diagram has been used to sort numbers according to whether they are even or not even.

Practise 📓 [TWM.01/04]

- Workbook

Title: Even and odd numbers

Pages: 8–9
- Refer to Activity 2 from the Additional practice activities.

Apply 👥 🖥

- Display **Slide 2**. Ask learners to look at the lines and discuss in pairs whether odd and even numbers apply to lengths.
- Each learner then draws three different lines that are an odd number of centimetres long and three different lines that are an even number of centimetres long.
- Learners then swap papers and measure and label each others' six lines.

Review

- Discuss the learning that has taken place during the lesson.
- 🗣 Ask: **What is an even number? What is an odd number?**
- 🗣 Ask: **Can you make a generalisation for each?**
- 🗣 Ask: **Could we say that an odd number is an even number add 1? Why?**
- 🗣 Ask: **Could we say that even numbers increase in twos? Why?**

Assessment for learning

- Give me an example of an even number. Give me another, and another.
- Tell me an example of an odd number. Tell me another, and another.
- How could we describe an odd number?
- How could we describe an even number?

Same day intervention

Support

- Give learners Resource sheet 1: 100 square. Ask them to identify the even and odd numbers on it. They colour them in, evens in one colour and odds in another. Do they notice that the columns on the 100 square alternate, odd, even?

Number – Counting and sequences

Lesson 3: **More about even and odd numbers**

Number – Counting and sequences

Learning objective

Code	Learning objective
3Nc.03	Use knowledge of even and odd numbers up to 10 to recognise and sort numbers.

Resources

interlocking cubes (per pair); squared paper (per learner)

Revise

Use the activity *Even and odd numbers* from Unit 1: *Counting and sequences (A)* in the Revise activities.

Teach [SB] 🖵 [TWM.02]

- Ask learners to use interlocking cubes to make the odd and even numbers from 1 to 10 as they did in Lesson 2.
- 💬 Ask: **What do you know about even numbers? What do you know about odd numbers? What do even numbers end with? What do odd numbers end with?**
- [T&T] Ask learners to think about what is special about even numbers.
- Direct learners to the Student's Book. Ask them what they notice about the way the cubes are arranged. Agree that with the exception of 1, they show the same sets of cubes as in Lesson 2. The even numbers have been divided vertically into two groups. Each group has the same number of cubes. The odd numbers have also been divided vertically. This time the two groups are not the same.
- Ask learners to do this with their cubes.
- Display **Slide 1**. Ask: **What numbers do you think are shown here? Are they even numbers or odd numbers? How do you know?**
- [TWM.02] Introduce the paired activity questions in the Student's Book. Ask learners to identify the generalisation that could be made about even numbers. Agree that even numbers can be divided by 2. Repeat for odd numbers. Agree that when odd numbers are divided by 2, there will always be one left.
- Discuss the Guided practice example in the Student's Book.

Practise [WB]

- Workbook

Title: More about even and odd numbers

Pages: 10–11

- Refer to Activity 2 from the Additional practice activities.

Apply 👥 🖵 [TWM.02]

- Display **Slide 2**.
- Display the calculation and explain that the box represents a missing one-digit number.
- Learners discuss in pairs what numbers could go in the box.
- They work out the answer each time and think about which calculations have answers that are even numbers and which have answers that are odd numbers. What generalisations can they make?

Review

- Discuss the learning that has taken place during the lesson.
- 💬 Ask: **Can you give me a three-digit even number? And another, and another?**
- 💬 Ask: **How do you know they are even?**
- 💬 Ask: **Can you give me a three-digit odd number? And another, and another?**
- 💬 Ask: **How do you know they are odd?**
- 💬 Ask: **What rules do we know about odd and even numbers now?**

Assessment for learning

- Why can dividing by 2 tell us if a number is even or odd?
- If a number has a remainder of 1 when divided by 2, is the number even or odd?
- If a number has no remainder when it is divided by 2, is it an even or odd number?
- How do we know that number is an odd number?

Same day intervention

Enrichment

- Ask learners to explore what happens if any number is doubled. Expect them to be able to tell you that the result is always even. Ask them to explore what they need to do to make it odd. Observe whether they can make the generalisation that when they double a number add 1, they will always get an odd number.

Lesson 4: **Estimating**

Learning objective

Code	Learning objective
3Nc.01	Estimate the number of objects or people (up to 1000).

Resources

at least 80 counters or similar countable objects (per pair); bowl for counters (per pair); exercise book (per learner) (for the Workbook)

Revise

Use the activity *Estimating* from Unit 1: *Counting and sequences (A)* in the Revise activities.

Teach [SB] 🖥 [TWM. 02/07/08]

• Display **Slide 1**. Ask learners to estimate how many red circles there are. Allow them a few seconds and then hide the image. Establish that they didn't have time to count them all, so they had to estimate.

• Write some of their estimates on the board. Discuss the range of estimates from lowest to highest. Also write estimates of 12 and 1000.

• Show the circles briefly again.

• **[TWM.07/08] [T&T]** Ask learners to talk to each other about the estimates. Ask: **Which estimates do you think are sensible?**

• Agree that 12 and 1000 are definitely not sensible estimates.

• Display **Slide 2**. Ask: **What is different about the way the circles are shown now?**

• 🖪 Ask: **How many circles are there? How do you know? Why is this arrangement easier to count?**

• **[T&T]** Ask learners, in pairs, to tip some counters onto their tables. They each make an estimate. Then they arrange them in tens, as on the slide. They compare their estimates with the actual number. They repeat several times.

• **[TWM.08]** Direct learners to the Student's Book. Ask: **How many jellybeans do you think there are? What would be a sensible range for an estimate? Why? How can you check?**

• **[TWM.02]** Introduce the paired activity in the Student's Book. Ask learners to discuss situations in everyday life when it is important and useful to make estimates. For example, estimating if you have enough money in your purse to pay for shopping, estimating the time it will take to do something, estimating lengths, heights, distances, masses and capacities. Ask pairs of learners to share their suggestions with the class.

• Discuss the Guided practice example in the Student's Book.

Practise [WB] [TWM.08]

• Workbook

Title: Estimating

Pages: 12–13

• Refer to Activity 2 from the Additional practice activities.

Apply 👥 🖥 [TWM.07]

• Display **Slide 3**.

• Learners first give a range for their estimate of how many apples are on the tree and then give a specific number for their estimate.

• Then they write an explanation of why it is impossible to know exactly how many apples there are.

Review 🖥

• Discuss what has been learned during the lesson. Ensure you talk about ranges as well as estimates.

• 🖪 Display **Slide 4**. Ask: **What animal is this?** Agree that it shows a leopard.

• 🖪 Ask: **How many spots can you see on this leopard? How can you give a sensible estimate?**

• Invite suggestions, such as counting the spots in one area and multiplying by the number of similar sized areas. Encourage learners to give their answer as a range.

Assessment for learning

• What is an estimate?

• Why do we estimate?

• What do we mean by range?

• What does it mean if the range is from 10 to 90?

Same day intervention

Support

• Give learners up to 30 counters to estimate and a tens frame to organise them in, so that they can easily count in tens and add any extra ones.

Number – Counting and sequences

Additional practice activities

Activity 1 👥 ⚠️2

Learning objective
- Count on and count back in steps of constant size: tens or hundreds, starting from any number (from 0 to 1000)

Resources
eight pieces of A6 paper (per pair)

What to do
- Each learner takes four pieces of paper.
- One learner writes a three-digit number on each of their four pieces.
- The other learner writes instructions + 10, + 100, −10, −100 on their four pieces.
- Learners turn the papers face down and put them in two separate piles.
- Learners turn over one number and one instruction.
- They take turns to count on or back in tens or hundreds to a maximum of 990.
- They do this four times to use all the numbers and instructions.

Variation
1 Learners write four two-digit numbers to count on in tens or hundreds for eight counts.

3 Learners write four three-digit numbers and count on or back in hundreds and tens, working on all four numbers at the same time for a set number of counts. For example: The numbers are 254, 367, 477 and 821. Counting back in tens three times:

254, 367, 477, 821

244, 357, 467, 811

234, 347, 457, 801

224, 337, 447, 791

Activity 2 👥 ⚠️2

Learning objective
- Use knowledge of even and odd numbers up to 10 to recognise and sort numbers

Resources
paper (per learner)

What to do [TWM.05/06]
- Each learner writes down ten two- or three-digit numbers.
- They swap papers, so they have their partner's numbers.
- They circle all the even numbers.
- Together, they make a simple Carroll diagram for one criterion, to sort all the numbers into odd and even.
- They write appropriate headings in the table.
- They fill in their Carroll diagrams with their numbers.

Variation
3 Repeat the activity but this time learners make a Carroll diagram for two criteria, for example:

	odd	not odd
greater than 500		
not greater than 500		

Number – Counting and sequences

Unit 2: Counting and sequences (B)

Number – Counting and sequences

Learning objectives

Code	Learning objective
3Nc.01	Estimate the number of objects or people (up to 1000).
3Nc.02	Count on and count back in steps of constant size: 1-digit numbers, [tens or hundreds,] starting from any number (from 0 to 1000).
3Nc.03	Use knowledge of even and odd numbers up to 10 to recognise and sort numbers.
3Nc.05	Recognise and extend linear sequences, and describe the term-to-term rule.
3Nc.06	Extend spatial patterns formed from adding and subtracting a constant.

Unit overview

In this unit, learners will continue to develop their skills in estimating. They will be counting in one-digit steps of 1, 2, 3, 4, 5, 6, 8 and 9. They use their understanding of odd and even numbers to explore patterns that occur when counting in these steps. It is important to allow time for learners to explore counting in the ways suggested, even if this takes more than one lesson.

Learners will also be looking at sequences and patterns in numbers and making appropriate links to counting in steps of different sizes.

Prerequisites for learning

Learners need to:
* know that an estimate is a sensible guess
* understand that an even number is a multiple of 2
* understand that an odd number is not a multiple of 2
* know their multiplication tables for 2, 5 and 10
* be able to recognise and continue simple sequences and patterns.

Vocabulary

estimate, range, counting in steps, even, odd, sequence, extend, step count, pattern, increase, decrease

Common difficulties and remediation

Most learners dislike the idea of estimating – they prefer correct answers. It is important that they get into the habit of estimating as early as possibly, so that they estimate answers to calculations automatically. This is the second lesson in which they focus on estimating. (If you are following the recommended order of teaching, estimation was also taught in Unit 1, Term 1 Week 2.) It is good practice to encourage learners to estimate whenever it is appropriate, for example, considering how many learners are in a group or the class, how many minutes until lunch time.

Learners often count in steps of different sizes without making a link to multiplication facts. It is important to encourage them to make connections to the tables

they already know and to look for the patterns that arise when counting in different steps. You could encourage learners to list the numbers when counting in, for example, steps of 5.

Supporting language awareness

To help learners develop their estimating skills, it would be useful to provide containers, such as pots of paper clips or jars of marbles, so that they can practise estimating 'outside' maths lessons. As mentioned in Unit 1, you could provide possible ranges to choose from, written on cards, that they can discuss with a partner.

When learners are counting in steps of one-digit numbers it is helpful to display the multiplication facts that match the step count they are practising. In this way, learners can make the connection between counting in steps and multiplication.

Promoting Thinking and Working Mathematically

During this unit learners have plenty of opportunities to conjecture (TWM.03) and convince (TWM.04). For example, when counting in steps of different sizes, they explore whether the numbers they count are odd or even and prove their thinking. They will also specialise (TWM.01) and make generalisations (TWM.02) that, for example, the multiples of 5 alternate between odd and even numbers. When sorting numbers according to different criteria, they characterise (TWM.05) and classify (TWM.06). They also have opportunities to critique (TWM.07) and improve (TWM.08) when making estimates.

Success criteria

Learners can:
* estimate up to 1000 objects
* count on and back in steps of constant size
* link counting in steps to known multiplication facts
* recognise odd and even numbers
* explain how number sequences can be made by adding or subtracting a number
* explain how different number patterns arise.

Unit **2** Counting and sequences (B)

Lesson 1: **More about estimating**

Learning objective

Code	Learning objective
3Nc.01	Estimate the number of objects or people (up to 1000).

Revise

Use the activity *Estimating* from Unit 2: *Counting and sequences (B)* in the Revise activities to review and consolidate work on estimating.

Teach SB ▢ [TWM.07/08]

- Display **Slide 1**. Ask learners to estimate how many jellybeans they can see by mentally grouping the jellybeans into tens or hundreds. Allow a few seconds and then hide the image.
- [TWM.08] Write some of their estimates on the board. Inform learners that when estimating, they should consider the lowest and highest amount they think there might be. This will give a range. Say: **The range is from 100 to 200.** Ask them to review their estimates and change them if they want to.
- [T&T] Ask learners to talk to a partner about the estimates you have written on the board. Ask: **Which of the estimates are within the range? Which estimate do you think is the most sensible?**
- Now say: **The range is from 160 to 180.** ▢ Ask: **Which estimates fit with this new range?**
- Display **Slide 2**. Ask: **What is different about the way the jellybeans are set out now? How many beans are there? Why is it easier and quicker to count the beans when they are arranged like this?**
- Agree that they are set out in two groups of 5 jellybeans, making larger groups of 10 jellybeans. Now learners can count in tens and then add on those that are left over.
- Whose estimate was the closest?
- [TWM.07/08] Display **Slide 3**. Ask: **Do you think the boy's range is a good one?** Give learners a few minutes to discuss this. Take feedback, expecting learners to give reasons for their thoughts.
- Direct learners to the Student's Book. Discuss, with examples, why estimating is an important skill.
- Ask: **How do we write a range?** Ask learners to look again in the Student's Book and explain that ranges are written with the two 'outside' numbers separated by a dash, for example: 150–200, 160–180.
- Learners work through the paired activity.
- Discuss the Guided practice example in the Student's Book.

Practise WB

- Workbook

Title: More about estimating

Pages: 14–15

- Refer to Activity 1 from the Additional practice activities.

Apply 👥 ▢ [TWM.07/08]

- Display **Slide 4**. Ensure that learners know that this animal is a camel. Encourage those who have seen one to talk about where they saw it, or where they might expect to see one.
- They are given the minimum and maximum recorded masses of camels. They choose the best range, discuss it with a partner and write an explanation for their choice.

Review

- Discuss what has been learned during the lesson.
- Ask learners why they think it is important to develop the skill of estimating.
- Call out some numbers as 'estimates' and invite learners to suggest a good range for each one.

Assessment for learning

- Why do we estimate?
- Why is it helpful to give a range for an estimate?
- What would be a good range for an estimate of 934?
- If I give you a range from 760 to 770, what could the estimate be? What else could it be? Are there any more possibilities?

Same day intervention

Support

- Learners work on numbers with ranges within 10, for example 250–260.
- Give learners greater ranges for their estimates, for example 200–300.

Enrichment

- Give learners opportunities to explore estimates and ranges for the numbers of pages in books from around the classroom or a collection of items that you have available, such as a container of drawing pins.

Lesson 2: **Counting on and back**

Learning objectives

Code	Learning objective
3Nc.02	Count on and count back in steps of constant size: 1-digit numbers, [tens or hundreds,] starting from any number (from 0 to 1000).
3Nc.03	Use knowledge of even and odd numbers up to 10 to recognise and sort numbers.

Resources

coloured pencils (per learner) (for the Workbook)

Revise

Use the activity *Counting in steps* from Unit 2: *Counting and sequences (B)* in the Revise activities.

Teach [SB] 🖵 [TWM.03/04/05]

- Count in steps of 1 from 113 to 123 and then back.
- 🐧 Ask: **What can you tell me about the ones digits? Are they odd or even?** Elicit that they alternate between odd and even.
- Display **Slide 1**. Ask: **What do you notice about the numbers?** Accept any correct answers, eliciting that they are all three-digit numbers, greater than 200. Ask: **Why are the colours different?** Agree that the colours differentiate odd and even numbers.
- Count in twos from 246 to 266 and back to 246. Agree that these are all even numbers.
- Count in fives from 320 to 420 and back. Agree that the numbers alternate even and odd.
- **[TWM.03/05]** Repeat for counting in tens from 320 to 420 and back. Ask: **What do you notice?** Agree all the numbers end with zero. Compare counting in fives and tens, i.e., every other number is the same and as they count in fives, the 10 count is double.
- Link the counts they have made to the times tables facts for 2, 5 and 10. Numbers in the 2 times table are even, in the 5 times table they end with 0 and 5 and in the 10 times table they end with 0.
- **[T&T] [TWM.03/05]** Show **Slides 2**, **3** and **4**. Use these to discuss the patterns that can be seen when counting in 3s and 6s, then 4s and 8s and then 3s, 6s and 9s. For each slide ask: **What do you notice?** Expect learners to discuss any 'even and odd' patterns. Ask them to continue each count for another ten numbers. Use this as an opportunity to discuss links between counting in 3s and doubling for steps of 6, and counting in 4s and doubling for steps of 8.
- **[T&T]** Ask: **Can you say why we will never say 21 when counting in steps of 2 from zero? Can you say why we will always say 21 when counting in steps of 3 from zero?**
- Direct learners to Let's learn in the Student's Book and ask them to describe the patterns made by the cubes. Ensure to discuss the doubling links between 2 and 4 and between 3 and 6.
- **[TWM.04]** Learners work through the paired activity.
- Together, discuss the Guided practice section.

Practise [WB]

- Workbook

Title: Counting on and back

Pages: 16–17

- Refer to Activity 2 from the Additional practice activities.

Apply 👥 🖵 [TWM.01]

- Display **Slide 5**. Learners work out how many hens laid two eggs and how many laid three eggs.
- Encourage them to make jottings and to find two or three options. If they start by finding an odd number less than 23 that is in the sequence 'counting in threes' and subtract it from 23, they will be left with an even number. The simplest answer is one hen lays three eggs and ten hens lay two eggs.

Review

- Discuss with the class the patterns they have noticed when counting in steps of a constant size, including odd and even numbers.
- Ask learners to describe why they think that counting in twos and fours from zero gives sequences of even numbers.

Assessment for learning

- What do you notice when counting in steps of...?
- What step counts contain only even numbers?
- What step counts contain a mixture of odd and even numbers?
- Tell me some numbers you say when you count on in steps of... from zero.

Same day intervention
Support

- Focus on counting in steps of 2, 5 and 10 from one-digit numbers.
- Provide learners with Resource sheet 1: 100 square to help them.

Enrichment

- Ask learners to consider how counting in steps of 4/3 can help them to count in steps of 8/6.

Lesson 3: **Making sequences with numbers**

Learning objective

Code	Learning objective
3Nc.05	Recognise and extend linear sequences, and describe the term-to-term rule.

Revise

Use the activity *Sequences in counting* from Unit 2: *Counting and sequences (B)* in the Revise activities to familiarise learners with the concept of sequences.

Teach 〔SB〕 ▯

- Count in steps of 1 from any number. Discuss the pattern that emerges: the numbers alternate between even and odd.
- Write 7, 8, 9, 10 on the board. **[T&T]** Ask: **How could we describe this pattern?** Allow learners to discuss this with a partner. Take feedback.
- ⏃ Ask: **Can you see that we are adding 1 to the previous number each time, for example: 7, 8 (7 + 1), 9 (8 + 1), 10 (9 + 1)?**
- Display **Slide 1**. Together, count along the top blue row, in steps of 2, from 17 to 31. Discuss why these are not even numbers. Agree that the starting number is odd, so all the numbers will be odd.
- **[T&T]** Ask: **Look at the second blue row – what do you notice?** After a few minutes, establish that the pattern is formed by adding 2 to the previous number each time.
- ⏃ Together, count along the top green row, in steps of 5, from 37 to 72. Ask: **What do you notice about this pattern?** Agree that the numbers alternate between even and odd.
- **[T&T]** Ask: **What does the second green row show?** Establish that this shows how the pattern builds – 5 is added to the previous number each time.
- Explain that each of the sequences they have been looking at follows a rule of adding on the same number each time.
- Repeat for the next two sequences on the slide, but this time elicit that they are subtracting the same number each time.
- Direct learners to Let's learn in the Student's Book. Invite learners to describe the patterns made up from the sweets and then the 5 cent coins. Each time, reinforce that the step count is added to the previous number.
- Give learners a few minutes to discuss the paired activity based on the sequence of 10 cent coins. Agree that this is another way to show adding 10 to the previous number.
- Discuss the Guided practice example in the Student's Book.

Practise 〔WB〕 [TWM.01]

- Workbook

Title: Making sequences with numbers

Pages: 18–19

- Refer to Activity 2 from the Additional practice activities. Encourage learners to focus on identifying and describing the rule.

Apply 👥 ▯ [TWM.01/02]

- Display **Slide 2**. Learners discuss with a partner how two sequences are made up. They write a description to explain what is happening. Then they work out what the next number or numbers are in the sequence.
- The first is subtract 2 from the previous number. The second is add 4 to the previous number.

Review

- ⏃ Ask: **Can you describe one of the sequences we looked at during the lesson? Can you think of another?**
- Invite one or two learners to write on the board how these can be made.
- Write 3, 6, 9, 12 on the board.
- ⏃ Ask: **How is this sequence made up? What are the next three numbers?**

Assessment for learning

- What is the next number in the sequence that begins 63, 65, 67? How can we describe this sequence? How else?
- What are the next five numbers in the sequence that begins 754, 744, 734? How do you know? How can you describe to a friend what is happening? Is there another way?

Same day intervention
Support

- Focus on counting in twos and linking to adding 2 to the previous number.

Enrichment

- Ask learners to explore how the sequence can be made when counting in threes and fours. They should be able to build on their understanding that the sequence for counting in twos is add 2 to the previous number, for counting in fives is add 5 to the previous number, and so on.

Lesson 4: **Making patterns with numbers**

Learning objective

Code	Learning objective
3Nc.06	Extend spatial patterns formed from adding and subtracting a constant.

Resources

mini whiteboard and pen (per pair)

Revise

Use the activity *Patterns* from Unit 2: *Counting and sequences (B)* in the Revise activities.

Teach 🆂🅱 🖥 [TWM.01/02]

- Display **Slide 1**. Ask: **How many buttons are in the first group? What patterns can you see in how the five buttons are arranged that help you count them?** Establish that there are groups of 2, groups of 1 and 2, groups of 3 and 2, a group of 5. It is important to point out the different patterns in the arrangement.

- Ask: **How can you work out how many buttons there are in the second column of the pattern?** Establish that there are two lots of 5, so learners could add 5 and 5 or double one group of 5. Stress that the total number of buttons has increased by 5. Ask: **Can you see a pattern yet?**

- Repeat for subsequent columns of groups of 5. Highlight the addition statements on the slide and elicit that each time, the number of buttons in the pattern has increased by 5.

- **[T&T]** Learners work together to draw, on their whiteboard, the next three columns in the pattern (as dots), writing the addition statements under them.

- [TWM.01/02] Ask: **How else can you describe what is happening?** Agree that they are counting in fives, using the 5 times table and also adding 5 onto the previous amount each time.

- Direct learners to Let's learn in the Student's Book.

- **[T&T]** [TWM.01/02] Say: **Look at the first part of the top pattern.** Elicit that there are three triangles. Ask: **How is the second part related to the first?** Agree that the first is made of three triangles and, in the second, three triangles have been added. With learners, establish that the numbers of triangles form a sequence with the rule: 'add 3'. Ask: **How does this link to counting in threes?**

- Repeat for the second pattern.

- Give learners a few minutes to discuss and complete the paired activity.

- Discuss the Guided practice example in the Student's Book.

Practise 🆆🅱 [TWM.01]

- Workbook

Title: Making patterns with numbers

Pages: 20–21

- Refer to Activity 2 from the Additional practice activities. Encourage learners to focus on identifying and describing the rule.

Apply 👥 🖥 [TWM.01/02/03/04]

- Display **Slide 2**. Learners are given a problem and one possible solution: adding groups of 2.

- Learners find the other five possible solutions (groups of 1, 3, 4, 6 and 12). They draw and write number statements for each possibility.

Review

- Ask learners to continue a pattern that you begin. For example, you draw three circles and then they draw those and add another three circles to give six circles, so producing a sequence of 3, 6, 9.

Assessment for learning

- How else can we say five add five add five? Is there another way?

- If I had four triangles and then added two more groups of four triangles, how many would I have?

- What would that look like as a number statement?

Same day intervention

Support

- Focus on building patterns to ten groups in twos. Change the shapes used.

Enrichment

- Encourage learners to look for links connecting groups of 3 and groups of 6. For example, if they have a pattern increasing in groups of 3, they double to get groups of 6.

- Include examples of groups of 8 and see if they can connect these to groups of 4. Groups of 8 would be double groups of 4.

Number – Counting and sequences

Additional practice activities

Activity 1 👥👥 🔺2

Learning objective
• Estimate the number of objects or people (up to 1000)

Resources
paper (per learner)

What to do
• Learners work in groups of about four. Without letting the rest of the group see, one learner writes down a number to represent an estimate for up to 1000 objects.
• They then give a range to the rest of the group and the group needs to guess what the first learner's 'estimate' was.

• Repeat this so that each learner has the opportunity to write an 'estimate' on paper and give a range for the others to guess.

Variation
As above, but this time the learner who makes the estimate draws a number line. The other learners plot their guesses onto the number line.

Activity 2 👥 🔺2

Learning objectives
• Count on and count back in steps of constant size: 1-digit numbers, [tens or hundreds,] starting from any number (from 0 to 1000)
• Use knowledge of even and odd numbers up to 10 to recognise and sort numbers
• Recognise and extend linear sequences, and describe the term-to-term rule
• Extend spatial patterns formed from adding and subtracting a constant

Resources
paper (per learner)

What to do [TWM.03/04]
• Ask learners to make a list of the first ten numbers they say when counting in steps of 2 and steps of 4.
• With their partner they discuss what they notice and write it down.
• Take feedback.
• Did they notice that all numbers they said are even? Did they also notice that counting in steps of 4 leads to doubling the numbers they say when counting in steps of 2 from the same starting number?

• Give them time to review their recording of what they noticed. Check that they explain what they noticed.

Variation [TWM.03/04]
1 Repeat the above for step counting in fives and tens.

3 Learners make a list of the first ten numbers they say when counting in steps of 3 and steps of 6 and discuss, and write down, what patterns they notice. They do the same for steps of 4 and steps of 8.

Number – Counting and sequences

Unit 3: Reading and writing numbers to 1000

Collins International Primary Maths Recommended Teaching and Learning Sequence: Term 1, Week 1

Learning objectives

Code	Learning objective
3Ni.01	Recite, read and write number names and whole numbers (from 0 to 1000).
3Np.01	Understand and explain that the value of each digit is determined by its position in that number (up to 3-digit numbers).
3Nc.02	Count on and count back in steps of constant size: 1-digit numbers, tens or hundreds, starting from any number (from 0 to 1000).
3Nc.03	Use knowledge of even and odd numbers up to 10 to recognise and sort numbers.
3Nc.05	Recognise and extend linear sequences, and describe the term-to-term rule.

Unit overview

In this unit, learners extend their understanding from Stage 2 to reading and writing numbers in words and numerals to 1000. Learners may not have used the term 'place value' yet, although they will have learned to associate the value of a digit with its position in the number. It is necessary to develop what they know from Stage 2 about tens and ones and extend this to hundreds in order for them to read three-digit numbers and understand how they are formed.

To ensure learners develop a firm foundation in their understanding of place value, this unit explores one of the key ideas that underpins place value, namely the concept of 'unitising' – treating a group of objects as one 'unit'. In other words, learners realise that, for example, 10 ones can also be thought of as a unit of ten, and 10 tens can also be thought of as a unit of a hundred. It is not necessary at this level to introduce the term 'unitisation', as long as learners understand the idea and can identify the units they use. This work will help learners when they study place value in more depth in Unit 14: Place value and ordering.

Prerequisites for learning

Learners need to:
- know the value of the digits in numbers to 100
- know the value of combinations of digits to 100
- be able to read and write in words and numerals any number to 100.

Vocabulary

numeral, number, digit, unit, place value zero, place holder, even number, odd number, sequence, increase

Common difficulties and remediation

It is important that learners learn to write numbers in words because this helps reinforce the concept of place value. For example, 'four hundred and fifty-six' indicates that there are 4 hundreds, 5 tens and 6 ones (400 and 50 and 6). The word 'and' separates the hundreds from the tens and ones. Sometimes the word 'and' coincides with the lack of a tens number, such as in five hundred and six, and therefore shows

that there will be a zero as a place holder in the tens position. Gaining an understanding of this will help to alleviate the problem that some learners have of writing numbers as they sound, for example, 400506 (456). Learners will also gain an understanding of unitising, where a unit represents a group of an equal amount. Four hundred is 4 hundreds, fifty is 5 tens and six is 6 ones.

Supporting language awareness

It is important to display the numbers as both numerals and words in the classroom for learners to refer to; for example, 376 is equal to three hundred and seventy-six. Vocabulary cards are provided on Resource sheet 2. Sets of these cards can be printed out and laminated for learners to use within groups. They can also be enlarged and displayed in the classroom.

Promoting Thinking and Working Mathematically

During this unit learners have opportunities to choose an example and check to see if it satisfies or does not satisfy specific mathematical criteria (TWM.01). They present evidence to justify or challenge a mathematical idea or solution (TWM.04), identify and describe the mathematical properties of numbers (TWM.05) and organise them into groups according to their properties (TWM.06). Learners compare and evaluate mathematical ideas, representations or solutions to identify advantages and disadvantages (TWM.07) and refine mathematical ideas or representations to develop a more effective approach or solution (TWM.08).

Success criteria

Learners can:
- read and write any number to 1000 in numerals
- read and write any number to 1000 in words
- make the link between numerals and numbers written as words
- know the purpose of the word 'and' when saying a three-digit number.

Number – Integers and powers

Unit 3 Reading and writing numbers to 1000

Lesson 1: **Numerals and words (A)**

Learning objectives

Code	Learning objective
3Ni.01	Recite, read and write number names and whole numbers (from 0 to 1000).
3Np.01	Understand and explain that the value of each digit is determined by its position in that number (up to 3-digit numbers).
3Nc.03	Use knowledge of even and odd numbers up to 10 to recognise and sort numbers.

Resources

Resource sheet 2: Numbers in numerals and words (per group)

Revise

Use the activity *Reading numbers* from Unit 3: *Reading and writing numbers to 1000* in the Revise activities.

Teach 🟦 🖥 [TWM. 03/04]

- Give each group a set of cards from Resource sheet 2. Say words and digits for learners to show, for example: **Show me the word 'eighty'. Show me the digit '9'.** It is important that learners become familiar with these cards before they begin to use them to build numbers.

- 🖐 Ask learners to sort the cards into two piles, one for numerals and the other for words. Hold up a word card and ask learners to match it, then say the number with the class. Repeat for the numeral cards.

- Display **Slide 1**. Ask learners to read the numbers. Ask: **How are these numbers the same? How are they different?**

- [TWM.03/04] [T&T] Ask: **Which one is the odd one out? Why? Is there another?**

- Agree that all these numbers are alike because they represent quantities. Encourage learners to discuss in pairs which are even numbers and which are odd, which are three-digit numbers and which are two-digit numbers. Encourage them to think of different criteria for which is the odd one out, for example 36 is the only number without a 7 in the tens position; 277 is the only one with two digits that are the same. Extend this idea to, for example, 875, which could be the odd one out because it is the only number greater than 500, and 36 because it is the only one less than 50.

- Ask groups to use the word cards to make 479. Invite a learner to write the number on the board in words. Do the class agree?

- Repeat for the other numbers on the slide. Ask: **Which number can't you make?** Agree 277 because there is only one 7 card.

- Display **Slide 2**. Ask learners to say the numbers they can see and explain the use of zero as a place holder.

- Remind learners that they were unable to make 277 with their digit cards. Suggest they make a new number by substituting one of the sevens with zero as a place holder. Ask: **What two numbers could you make?** Agree 207 and 270.

- Write some other numbers on the board for learners to make with their word cards.

- Direct learners to Let's learn in the Student's Book and discuss the examples with them. Discuss the use of 'and' to separate the hundreds and tens numbers.

- Learners investigate the paired activity. Discuss the fact that tens numbers are specific to a number of tens, for example thirty is always 3 tens. One, five and four are interchangeable between the hundreds and ones positions.

- Discuss the Guided practice example in the Student's Book.

Practise 🟥 [TWM.01]

- Workbook

Title: Numerals and words (A)

Pages: 22–23

- Refer to Activity 1 from the Additional practice activities.

Apply 👥 🖥 [TWM.04]

- Display **Slide 3**. Learners work out which number Felix cannot make using the 1 to 9 digit cards and explain why.

Review

- Invite volunteers to write six different numbers on the board, in words, such as two hundred and sixty-four, nine hundred and fifty-two.

- Ask the class to check the spelling to ensure the volunteers have written the numbers correctly.

- Check also that the word 'and' has been included in the correct position.

- Invite other learners to write the corresponding numerals.

Assessment for learning

- What digit should we use if there are no tens?
- How will you write six hundred and forty-nine in numerals?
- How is 135 written in words?

Same day intervention
Support

- Focus on writing two-digit numbers until learners are confident with them, then introduce hundreds. Focus on numbers with 1 hundred initially.

Number – Integers and powers

Lesson 2: **Numerals and words (B)**

Learning objectives

Code	Learning objective
3Ni.01	Recite, read and write number names and whole numbers (from 0 to 1000).
3Np.01	Understand and explain that the value of each digit is determined by its position in that number (up to 3-digit numbers).

Resources

Resource sheet 2: Numbers in numerals and words (per group/pair); Base 10 equipment: hundreds, tens, ones (per group); place value counters or Resource sheet 8: Place value counters or coloured counters in three colours (per pair); selection of newspapers and magazines (per pair); scissors and glue (per pair); poster paper (per pair)

Revise

Use the activity *Numerals and words* from Unit 3: *Reading and writing numbers to 1000* in the Revise activities.

Teach [SB] 🖥

- Display **Slide 1**. Ask: **What can you see on the slide?** Establish that the same number is represented in three different ways.
- Ask groups to make the number in the three ways shown, using Base 10 equipment and cards.
- 🗣 Ask: **What does the 8 represent?** Agree 8 tens or 80.
- 🗣 Ask: **What does the 9 represent?** Agree 9 ones.
- **[T&T]** Ask learners to discuss what they think the 3 represents. Agree 3 hundreds or 300. It is not necessary to mention place value at this stage. Learners should simply make links to what they already know.
- Ask learners to make a number from four hundred flats, two ten sticks and three one cubes. Discuss the idea that there are 4 hundreds, 2 tens and 3 ones which make 423.
- As a group they use the cards to make the written form of this number and then the numeral form.
- Display **Slide 2** and repeat. Stress that there are 6 hundreds, 5 tens and 4 ones.
- Direct learners to Let's learn in the Student's Book. Talk about the representation of the number shown. Agree that this representation is different from the Base 10 equipment but it shows a number by using place value counters. Stress that the number can be written in words or numerals, as before.
- Display **Slide 3**. Ask: **What do you notice about the first number?** Elicit that there are no ones. Remind learners about using zero as a place holder. Then repeat for the second number.
- Give learners the opportunity to complete the paired activity in the Student's Book. If learners are using coloured counters in three colours, rather than place value counters, tell them that they need to decide what value each colour will have.
- Discuss the Guided practice example in the Student's Book.

Practise [WB] [TWM.01/05]

- Workbook

Title: Numerals and words (B)

Pages: 24–25

- Refer to Activity 1 from the Additional practice activities.

Apply 👥 🖥 [TWM.04]

- Display **Slide 4**. Working in pairs, learners look through a selection of newspapers and magazines for examples of numbers to 1000 written as numerals or words.
- They cut out and glue the examples onto poster paper and write the equivalent number in words / numerals.
- Learners then discuss the possible reasons as to why there are more examples of numbers written as numerals than as words.

Review

- Write three different numbers on the board, such as 456, 789, 135. Invite learners to write them in words.
- Invite learners to write their own numbers, in numerals, on the board for others to say as words in units.

Assessment for learning

- In the number 467, how many tens in a hundred are there?
- How would we write 352 in numerals/as a word?

Same day intervention
Support

- Focus on tens and ones in two-digit numbers until learners are confident with these and then add hundreds. Focus on numbers with 1 unit of a hundred initially.

Number – Integers and powers

Unit 3 Reading and writing numbers to 1000

Lesson 3: **Reading and writing even numbers to 1000**

Number – Integers and powers

Learning objectives

Code	Learning objective
3Ni.01	Recite, read and write number names and whole numbers (from 0 to 1000).
3Np.01	Understand and explain that the value of each digit is determined by its position in that number (up to 3-digit numbers).
3Nc.02	Count on and count back in steps of constant size: 1-digit numbers, tens or hundreds, starting from any number (from 0 to 1000).
3Nc.05	Recognise and extend linear sequences, and describe the term-to-term rule.

Resources

mini whiteboard and pen (per pair); Resource sheet 2: Numbers in numerals and words (per pair)

Revise

Use the activity *Reciting, reading and writing even numbers* from Unit 3: *Reading and writing numbers to 1000* in the Revise activities to review units of equal amounts and even numbers.

Teach 🆂🅱 🖥 [TWM. 03/04/05]

- Ask: **Who can tell me what a sequence is?** Agree that a sequence is a set of shapes or numbers that follow a pattern. Explain to learners that they are going to look at sequences of numbers that increase or decrease when the same amount is added to or subtracted from the previous number.
- Display **Slide 1**. Ask learners to rearrange the eight numbers so that they make an increasing sequence starting from 252. Ask them to explain what is happening each time. Repeat for a decreasing sequence starting from 266.
- **[TWM.03/04/05]** 🗣 Ask: **What is the same about all of these numbers?** Agree that they are all three-digit numbers, they all have 2 hundreds, and they are all even numbers. Accept any other observations if correct.
- 🗣 Ask: **How do we know that these are even numbers?** Agree that most of them end with a digit that is a multiple of 2. Point to 260 and explain that, although zero can't be divided by 2, 60 can be divided into two equal groups of 30. Elicit that all of the numbers can be divided by 2.
- Display **Slide 2**. Ask pairs of learners to write down, as numerals, the four numbers shown by the Base 10 equipment on their whiteboard. Ask: **What is the same about 324 and 424? What is different about 324 and 424? What is the next number in the sequence?**
- Ask learners to write down the next three numbers in the sequence.
- Choose two or three numbers from the sequence and ask learners to write these numbers in words.
- **[T&T]** Ask: **How is this sequence different from the previous sequence? How is it the same?** Agree that it is different because the numbers are increasing in hundreds instead of 2 ones. The hundreds digit is increasing each time and

the tens and ones are staying the same. It is the same because these are three-digit numbers and all the numbers are even.

- Direct learners to Let's learn in the Student's Book.
- Give learners some time to work on the paired activity in the Student's Book.
- Discuss the Guided practice example in the Student's Book.

Practise 🆆🅱 [TWM.04/05]

- Workbook

Title: Reading and writing even numbers to 1000

Pages: 26–27

- Refer to Activity 2 from the Additional practice activities.

Apply 👥 🖥 [TWM.01]

- Display **Slide 3**. The problem requires learners to work out if there is another possible even number from the word cards presented.
- Encourage learners to discuss why this is not possible.

Review

- Ask pairs of learners to use the word cards from Resource sheet 2 to make some even numbers.
- Invite individuals to write examples, in numerals, on the board and get the class to make them with their word cards and read them back.

Assessment for learning

- How do we know if a number is even?
- What is the difference between writing numbers in numerals and writing numbers in words?
- What is the value of 6 hundreds?
- What is the value of 5 tens?
- What is the value of 8 ones?

Same day intervention

Support

- Focus on reading and writing two-digit even numbers.

Lesson 4: **Reading and writing odd numbers to 1000**

Learning objectives

Code	Learning objective
3Ni.01	Recite, read and write number names and whole numbers (from 0 to 1000).
3Np.01	Understand and explain that the value of each digit is determined by its position in that number (up to 3-digit numbers).
3Nc.02	Count on and count back in steps of constant size: 1-digit numbers, tens or hundreds, starting from any number (from 0 to 1000).
3Nc.05	Recognise and extend linear sequences, and describe the term-to-term rule.

Resources

mini whiteboard and pen (per pair); Resource sheet 2: Numbers in numerals and words (per pair)

Revise

Use the activity *Reciting, reading and writing odd numbers* from Unit 3: *Reading and writing numbers to 1000* in the Revise activities.

Teach [SB] [TWM. 04/05/06]

- Recap the sequences from Lesson 3. Agree that the sequences they looked at were sets of numbers that increased or decreased when the same amount was added or subtracted from the previous number.
- Display **Slide 1**. Ask learners to rearrange the eight numbers so that they make an increasing sequence starting from 461. Ask them to explain what is happening each time (increasing by 2). Repeat for a decreasing sequence starting from 475.
- [TWM.04/05/06] Ask: **What is the same about all of these numbers?** Agree that they are all three-digit numbers, they all have 4 hundreds, and that they are all odd numbers. Accept any other observations if correct.
- Ask: **How do we know that these are odd numbers?** Agree that they end with a digit that is not a multiple of 2 and therefore cannot be divided equally by 2 – there will always be 1 left over.
- Display **Slide 2**. Ask pairs of learners to write down, as numerals, the four numbers shown by the Base 10 equipment on their whiteboard. Ask: **How is 223 similar to 323? How is 223 different from 323? What is the next number in the sequence?** Write down the next three numbers in the sequence.
- Choose two or three numbers from the sequence and ask learners to write these numbers in words.
- [T&T] Ask: **How is this sequence different from the previous sequence? How is it the same?** Agree that it is increasing in hundreds instead of 2 ones. The hundreds digit is increasing each time and the tens and ones are staying the same. It is the same because these are three-digit numbers and all of the numbers are odd.
- Direct learners to Let's learn in the Student's Book. Discuss what is happening in the sequence and why the numbers remain odd (each number has 7 ones). Encourage them to write the numbers in numerals and words.
- Give learners some time to work on the paired activity in the Student's Book. Ask learners who

finish quickly to try to make a decreasing sequence. Then take feedback as a class.
- Discuss the Guided practice example in the Student's Book.

Practise [WB] [TWM.04/05]

- Workbook

Title: Reading and writing odd numbers to 1000

Pages: 28–29

- Refer to Activity 2 from the Additional practice activities.

Apply [👥] [TWM.05/06]

- Display **Slide 3**. The problem requires learners to work out if it is possible to make a different odd number from the word cards shown.
- Encourage learners to discuss how this is possible. Agree that they could simply swap the nine and seven to make seven hundred and sixty-nine.

Review

- Ask pairs of learners to use the word cards from Resource sheet 2 to make some odd numbers.
- Invite individuals to write examples, in numerals, on the board and get the class to make them with their word cards and read them back.
- Recap units of a hundred, ten and one.

Assessment for learning

- How do we know if a number is odd?
- What is meant by 7 hundreds? How is this different from 7 tens?
- What is the value of 7 tens?
- What is the value of 7 ones?
- What number is created with 7 hundreds, 1 ten and 1 one? Is this an even or odd number? How do you know?

Same day intervention

Support

- Focus on reading and writing two-digit odd numbers.

Number – Integers and powers

117

Number – Integers and powers

Additional practice activities

Activity 1 👥 ⚠2

Learning objective
• Recite, read and write number names and whole numbers (from 0 to 1000)

Resources
Resource sheet 2: Numbers in numerals and words (per pair); eight small pieces of paper (per pair); mini whiteboard and pen (per pair)

What to do
• Give each pair a complete set of cards from Resource sheet 2.
• They sort them into numeral cards and word cards.
• They write the words 'hundred and' on the eight pieces of paper.

• Their task is to use the cards to make as many three-digit numbers as they can, by placing the cards side by side to form the number. Their numbers should be either a numeral or in words. They can use each card only once.
• They write each number on paper or on whiteboards.

Variation
1 Learners complete the task, using tens and ones numbers only.

Activity 2 👥👥 ⚠2

Learning objective
• Recite, read and write number names and whole numbers (from 0 to 1000)

Resources
Resource sheet 2: Numbers in numerals and words (per group); mini whiteboard and pen (per learner)

What to do [TWM.04]
• Give each group of four learners a set of cards from Resource sheet 2.
• They sort the cards into three separate piles: all the digit cards together, all the multiples of 10 cards together, and all the 'teen' cards together.
• They then take turns to pick up two or three cards to make three-digit numbers, for example, if they pick '5' and 'thirteen' they lay down the 5 to the left of the 'thirteen' and say as a group: 'five **hundred and** thirteen'. If they pick up 'one', 'fifty' and '7' they say as a group: 'one **hundred and** fifty-seven'.

• Each learner records each number, as a numeral and in words, on paper or on whiteboards.

Variation
1 Learners pick two cards, a unit of ten card and a unit of one card.

They make a two-digit number.

They read the number as a group and record it as a numeral and in words.

Unit 4: Addition and subtraction (A)

Collins International Primary Maths Recommended Teaching and Learning Sequence: Term 1, Week 4

Learning objectives

Code	Learning objective
3Ni.02	Understand the commutative [and associative] properties of addition, and use these to simplify calculations.
3Ni.03	Recognise complements of 100 and complements of multiples of [10 or] 100 (up to 1000).
3Ni.04	Estimate, add and subtract whole numbers with up to three digits (regrouping of ones or tens).
3Nc.04	Recognise the use of an object to represent an unknown quantity in addition and subtraction calculations.

Unit overview

In this unit, learners will use their understanding of the relationship between addition and subtraction, learned in Stage 2, to develop the concept of commutativity. They will investigate how addition is commutative: the order in which we add numbers doesn't matter, the total will always be the same. They will develop practically an understanding that subtraction is **not** commutative. They will connect their knowledge of number pairs to 10 to complements of 100 and complements of multiples of 100 up to 1000.

Learners will also be estimating, adding and subtracting pairs of two-digit numbers. These concepts are covered in Lesson 3 of this unit but, although learners will have been taught to add and subtract pairs of two-digit numbers in Stage 2, this is the first time they will be regrouping. Some learners may find this concept difficult initially, so you may decide to teach this lesson over two or even more days, perhaps focusing on addition one day and subtraction the next, or teaching TO $\pm$ O one day (e.g. 34 + 7 = and 24 – 7 =) and TO $\pm$ TO the next day (e.g. 26 + 28 = and 62 – 25 =).

Learners will solve missing number problems in which the missing number is represented by an object.

Prerequisites for learning

Learners need to:
- know that numbers can be added together in any order
- know that subtraction is the inverse of addition so, for example, 2 + 3 = 5, therefore 5 – 3 = 2
- know the number pairs to 10
- be able to add and subtract pairs of numbers with up to two digits without regrouping
- be able to carry out addition and subtraction practically, using a variety of manipulatives.

Vocabulary

commutative, inverse, add, sum, total, subtract, difference, estimate, unit

Common difficulties and remediation

It is important to ensure that learners have mastered the requirements for Stage 2 before extending to those for Stage 3. If learners do not have a secure understanding of the addition and subtraction facts to 10, then they are likely to struggle in later learning. It is imperative to provide manipulatives and resources, such as Base 10 equipment, to allow learners to gain basic understanding of the processes behind addition and subtraction. These also ease the cognitive load, enabling learners to work through calculations and problems without being overwhelmed by the number of steps they need to remember.

Supporting language awareness

It is important to provide the number pairs for all numbers to 10, so that learners may refer to them and use the facts to apply to pairs of multiples of 100 to 1000. For example, if they can see the fact of 7 + 3 = 10, they can use that to show 70 + 30 = 100 and 700 + 300 = 1000. These could be displayed in the classroom or given to learners who would benefit. It is also important to have available the vocabulary associated with addition and subtraction.

Promoting Thinking and Working Mathematically

During this unit learners have the opportunity to specialise (TWM.01) and generalise (TWM.02) when they explore unknowns, and to form mathematical ideas and present evidence to justify their ideas (TWM.03, TWM.04). When working out their own mental calculation strategies for answering addition and subtraction calculations, they will have the opportunity to characterise (TWM.05) and classify (TWM.06). They will also have opportunities to compare and evaluate solutions and refine them to develop more effective strategies (TWM.07, TWM.08).

Success criteria

Learners can:
- use the commutative property of addition
- explain why subtraction is not commutative
- recall pairs of numbers that total 100
- add and subtract multiples of 100
- add and subtract pairs of two-digit numbers, with and without regrouping
- recognise pictures and objects as representations of unknown quantities.

Number – Integers and powers

Unit (4) Addition and subtraction (A)

Lesson 1: **Commutativity**

Number – Integers and powers

Learning objective

Code	Learning objective
3Ni.02	Understand the commutative [and associative] properties of addition, and use these to simplify calculations.

Resources

Base 10 equipment: tens and ones (per pair)

Revise

Use the activity *Relationships* from Unit 4: *Addition and subtraction (A)* in the Revise activities.

Teach [SB] 🖵 [TWM.05]

- ϸ Display **Slide 1**. Ask learners to identify the two numbers shown by the tens.
- Ask: **What calculation can you make?** Agree 60 + 40 = 100. Ask pairs to make both numbers with Base 10 equipment.
- **[TWM.05]** Display **Slide 2**. **[T&T]** Ask: **Look at the second calculation. What is the same as the first calculation? What is different?** Agree that again it shows an addition and the numbers are the same. The difference is the order of the numbers. Ask learners to use their Base 10 equipment again, changing the order, to show this calculation.
- Ask: **Why is the total the same?** Agree that the two numbers being added are the same so the total will be the same. Say: **Addition can be done in any order. It doesn't matter which way round the numbers are added, the total will always be the same.** Ask learners to show this by recording 60 + 40 = 40 + 60.
- Introduce the term 'commutative' and explain that this is the word we use to mean that the numbers can be added in any order.
- Display **Slide 3**. Remind learners that subtraction is the inverse (or opposite) of addition. Establish that these two subtraction calculations are the inverses of the two addition calculations. Ask: **Are the answers the same?** Agree that they are not, these are different calculations although they are made with the same numbers. Learners record the two calculations: 100 – 40 = 60 and 100 – 60 = 40.
- **[T&T]** Display **Slide 4** and discuss each calculation. Neither is correct. Use this slide to establish that subtraction is not commutative.
- Direct learners to the Student's Book. Discuss the examples showing addition as commutative and subtraction as inverse but not commutative.
- Learners work through the paired activity.
- Discuss the Guided practice example in the Student's Book.

Practise [WB] [TWM.01/04]

- Workbook

Title: Commutativity

Pages: 30–31

- Refer to Activity 1 from the Additional practice activities.

Apply 👥 🖵

- Display **Slide 5**. Learners are given three numbers. The first two can be added together to give the third.
- Learners work with a partner to write four different calculations: two additions and two subtractions.
- They repeat this for Pieter's numbers.

Review

- ϸ **What is meant by commutative? Is subtraction commutative? Why not?** They should say that addition is commutative and that subtraction is not.
- Write three two-digit multiples of 10 on the board and invite learners to write the different ways of adding them together.
- Ask them to decide with a partner which order is the most efficient way to add. There is no correct answer to this – it depends on the learner's opinion and what strategies for addition they use.
- Repeat for the inverse subtraction calculations. Highlight that subtraction is not commutative.

Assessment for learning

- What is meant by the word 'commutative'?
- Can you give an example to show that addition can be done in any order?
- Why is subtraction not commutative?
- Can you give an example to show that subtraction cannot be done in any order?
- What happens when we add three numbers together?

Same day intervention
Support

- Use one-digit numbers with practical apparatus that make 10 to show commutativity and inverses in addition and subtraction.

Lesson 2: **Complements of 100 and multiples of 100**

Learning objective

Code	Learning objective
3Ni.03	Recognise complements of 100 and complements of multiples of [10 or] 100 (up to 1000).

Resources

Base 10 equipment: ones, tens and 1 hundred (per pair); mini whiteboard and pen (per learner); coloured pencils (per learner) (for the Workbook)

Revise

Use the activity *Number pairs to 10 and 20* from Unit 4: *Addition and subtraction (A)* in the Revise activities to revise number pairs for 10, 20 and 100.

Teach 📖 🖥

- 🗒 Display **Slide 1**. Ask: **What addition fact does the Base 10 equipment show?** (6 + 4 = 10) **What is the addition fact related to 6 + 4 = 10?** (4 + 6 = 10) **What are the two corresponding subtraction facts?** (10 – 6 = 4 and 10 – 4 = 6)
- Display **Slide 2**. Ask: **What is the difference between the numbers represented on this slide and those on Slide 1?** Establish that these are multiples of 10. Ask: **How does knowing that 6 + 4 = 10, help you to work out the answer to 60 + 40?** Agree that if learners know the number facts for 10, they can apply this to multiples of 10, for example, 6 + 4 = 10 so, 60 + 40 = 100.
- Ask: **If you know that 60 add 40 equals 100, what other addition fact do you know?** (40 + 60 = 100) **What two subtraction facts do you also know?** (100 – 60 = 40 and 100 – 40 = 60)
- Display **Slide 3**. Repeat above for numbers with 5 in the ones position (55 + 45 =). Learners record the associated addition and subtraction facts. Give learners other examples, asking them to work in pairs to represent the addition using Base 10 equipment. Learners write the two addition and two subtraction facts. Ensure that all facts total 100.
- Display **Slide 4**. Repeat for numbers with any digit in the ones position (76 + 24 =). Again, give examples where the total is 100 and ask learners to represent the calculation using Base 10 equipment and to write the two addition and two subtraction facts.
- Display **Slide 5**. Ask: **What four addition and subtraction facts do you know from this representation?** Discuss the fact that these are hundreds. The same process applies. There are two addition and two subtraction facts that are known. Ensure learners write the four associated facts. Give other multiples of 100 for learners to add and subtract.
- Direct students to the Student's Book. Discuss the Let's learn section, working through the examples as a class.
- Give learners the opportunity to work through the paired activity and discuss the Guided practice.

Practise 📘 [TWM.04]

- Workbook

Title: Complements of 100 and multiples of 100

Pages: 32–33

- Refer to the first variation in Activity 1 from the Additional practice activities.

Apply 🖥 👥 [TWM.07/08]

- Display **Slide 6**. The problem requires learners to work systematically to find different solutions. They work with multiples of five from 5 to 95 to find all the possible combinations of pairs of two-digit numbers that total 100.

Review

- 🗒 Ask: **Why is knowing addition facts for 10 helpful for finding addition facts for multiples of 10 that total 100?**
- Invite individual learners to say a multiple of 10 (e.g. 20) and for the class to respond with the multiple of 10 that totals 100 (i.e. 80). Repeat for two-digit numbers with 5 ones (e.g. 45 + 65 =) and then any number of ones (e.g. 63 + 37 =).
- Finally ask learners how knowing addition facts for any number to 10 helps them know addition and subtraction facts involving multiples of 100, such as 300 + 700 =.

Assessment for learning

- If we know that 20 + 80 = 100, what other facts do we know?
- How does knowing 6 + 3 = 9 help us to find the sum of 600 and 300?
- If we know that 100 – 15 = 85, what addition and subtraction facts do we know?

Same day intervention
Support

- Focus on number pairs to 10 and apply these to multiples of 10 that total 100 and multiples of 100 that total 1000. For example, if 3 + 7 = 10, then 30 + 70 = 100 and 300 + 700 = 1000.

Number – Integers and powers

Lesson 3: **Addition and subtraction of 2-digit numbers**

Learning objective

Code	Learning objective
3Ni.04	Estimate, add and subtract whole numbers with up to three digits (regrouping of ones or tens).

Number – Integers and powers

Revise

Use the activity *Number pairs to 10 and 20* from Unit 4: *Addition and subtraction (A)* in the Revise activities. You may wish to use the interactive tool Ants on the Bus. Learners count how many ants are on the bus at any given time. Sometimes they get on the bus, sometimes they get off.

Teach [SB] 💻 📊 [TWM.07/08]

- 📖 Display **Slide 1**. Ask: **What is a sensible estimate for the sum of these two numbers? Why?** Agree 'about 50'.
- Display **Slide 2** and discuss how 45 has been partitioned into 4 tens and 5 ones (40 + 5). Display **Slide 3** and discuss how the six ones have been added, making a total of 11 ones. Display **Slide 4** and discuss how the 11 ones have been exchanged for 1 ten and 1 one. Display **Slide 5** and discuss how the 4 tens and 1 ten have been added and then tens and ones have been combined to make a total of 51.
- [T&T] [TWM.07/08] Ask: **Is there another way you could add these two numbers together?** Invite learners to demonstrate alternative strategies, discussing which they think is the most efficient strategy.
- Repeat above, adding one-digit numbers to other two-digit numbers, using the **Base 10 tool** if necessary.
- 📖 Display **Slide 6**. Ask: **What is a sensible estimate for the difference? Why?** Agree 'about 43'.
- Display **Slide 7** and discuss how 53 has been partitioned into 5 tens and 3 ones (50 + 3). Ask: **Is it possible to subtract 8 from the 3?** Agree that it is not. Display **Slide 8** and discuss how 53 can be partitioned into 4 tens and 13 ones. Display **Slide 9** and discuss how it is now possible to subtract 8 ones from 13 ones which leaves 5 ones, and that we now recombine the tens and ones to give us the answer of 45.
- [T&T] [TWM.07/08] Ask pairs to discuss other strategies for working out the answer to 53 – 8 =. Which methods do they prefer? Which are the most efficient? Repeat, subtracting one-digit numbers from other two-digit numbers, using the **Base 10 tool** if necessary.
- Direct learners to the Student's Book. Discuss the similarities and differences between the examples on this page and the slides. Establish they all involve

exchanging, but that the examples in the Student's Book involve adding and subtracting pairs of two-digit numbers.
- [TWM.07/08] Work through the process shown for addition. At each step, emphasise the way in which the calculation can be recorded – either using the expanded written method (shown in purple) or the compact formal method (shown in blue). Discuss other strategies that could be used.
- Repeat, adding other pairs of two-digit numbers using the **Base 10 tool** if necessary.
- Repeat the whole process for subtraction.
- Discuss the Guided practice example in the Student's Book.

Practise 📒

- Workbook

Title: Addition and subtraction of 2-digit numbers

Pages: 34–35

- Refer to Activity 2 from the Additional practice activities.

Apply 👥 💻 [TWM.07/08]

- Display **Slide 10**. Learners think of three different strategies for 68 + 57 =.
- What strategies did the learners think of? Invite individuals to demonstrate their methods.

Review

- Ask learners to add and subtract pairs of two-digit numbers. Include examples that involve regrouping as well as some that do not. Encourage learners to use the most efficient strategies.

Assessment for learning

- How would you solve 29 + 36? What other strategy could you use?
- What is 84 – 29? Why did you use this method?

Same day intervention
Support

- Focus on partitioning to add and subtract.
- Encourage learners to use manipulatives.

Lesson 4: **Unknowns!**

Learning objective

Code	Learning objective
3Nc.04	Recognise the use of an object to represent an unknown quantity in addition and subtraction calculations.

Resources

mini whiteboard and pen (per learner); collection of 3D shapes (per learner)

Revise

Use the activity *Missing numbers* from Unit 4: *Addition and subtraction (A)* in the Revise activities to review halving and simple division by 2.

Teach [SB] 🖥 [TWM.01]

- 🗣 Show **Slide 1**. Ask: **What shapes can you see? What are the similarities and the differences between them?** As a brief recap, agree that they are all 3D shapes: the first six are cubes, two are cuboids and two are cylinders.

- [T&T] Ask: **What is the value of each cube? How do you know?** Agree that they can halve $10 or divide $10 by 2 to find the value of one cube. Write $5 on each of the two cubes in the top calculation. Say: **So, one cube equals $5.** Write $5 on the cube in the second row on the slide.

- [T&T] Ask: **If you know the value of one cube, how can you work out the value of three cubes?** Write $5 on each of the three cubes in the second calculation and the total $15. Ask: **What about four/ five/six?**

- [T&T] Ask: **How can you work out the value of the cuboid?** Agree that, because they know the relationship between addition and subtraction, they can work out that the cuboids must be $3 each ($3 + $3 = $6, therefore $6 – $3 = $3).

- Write $3 on each of the cuboids.

- [T&T] [TWM.01] Ask: **How can you use what you already know to find the value of a cylinder? What is the value of the cube?** Agree that it is $5, and write $5 on the cube. Agree that the two cylinders are equivalent to $13 – $5, which is $8, so each cylinder must be $4. Write $4 on the cylinder before checking the calculation with the class.

- Ask learners to use their collection of 3D shapes to make up problems like this with unknown quantities and to ask their partner to work out what the missing values must be.

- Direct learners to the Student's Book and discuss the examples shown.

- Give learners some time to work on the paired activity in the Student's Book. Take feedback to find out how learners tackled the problem.

- Discuss the Guided practice example in the Student's Book.

Practise [WB]

- Workbook

Title: Unknowns!

Pages: 36–37

- Refer to the variation in Activity 2 from the Additional practice activities.

Apply 👥 🖥 [TWM.03/04]

- Display **Slide 2**. Learners are given part of the information to a problem and the answer.

- They work out the missing information so it gives the correct answer.

- They then make up their own similar problems before asking their partner to solve them.

- Encourage learners to explain to the class their strategies for making up the problems so that they work. Encourage the class to comment on what the learners did.

Review

- 🗣 Ask: **What have we been learning about today?**

- Draw out the fact that objects and shapes can be used to represent missing quantities and that learners can use addition and corresponding subtraction facts to find the missing quantities. Finish the lesson by playing the game Coconut calculations on the interactive tool. Select Level 3. Learners have to find the missing numbers as quickly as they can.

Assessment for learning

- 24 subtract something gives a difference of 12. How can we work out the missing number (subtrahend)?

- Something add 50 gives a total of 64. How can we work out the missing number (augend)?

- Something subtract 15 gives a difference of 20. How can we work out the missing number (minuend)?

Same day intervention

Support

- Ensure learners use equipment such as money or Base 10 equipment to set out the given amounts. With guidance, this will help them to find the unknown quantity.

Number – Integers and powers

Additional practice activities

Activity 1

Learning objectives
- Understand the commutative property of addition
- Recognise complements of 100

Resources
Resource sheet 3: Digit cards (per pair)

What to do
- Give pairs a set of digit cards from Resource sheet 3.
- They pick two cards and make a two-digit number and write this down on paper, for example, 5 and 3, 53.
- They then reverse the order of the cards and write the new number, for example, 35.
- They then use the two numbers to write two addition calculations that show addition is commutative, for example, 53 + 35 = 35 + 53.
- The focus is commutativity, so there is no need to expect learners to find the sum of the two numbers.
- They repeat this until all their digit cards have been used.

Variation

3 Learners pick three digit cards and make as many different two-digit numbers as they can.

2 Learners pick two cards and make a two-digit number and write this down on paper, for example, 5 and 3, 53. They then work out the complement of 100, i.e. 47 and write the corresponding addition and subtraction facts, i.e. 53 + 47 = 100, 47 + 53 = 100, 100 − 53 = 47, 100 − 47 = 53. They repeat this until all their digit cards have been used.

They choose three of their numbers and write down the six different addition calculations that can be made.

Activity 2

Learning objective
- Estimate, add and subtract pairs of two-digit numbers with regrouping

Resources
Base 10 equipment: tens and ones (per pair)

What to do
- Write four two-digit numbers on the board, for example: 78, 97, 69, 56.
- Learners work with a partner. They choose two of the numbers to add together.
- They estimate the total and then make both numbers with Base 10 equipment.
- They find the total by partitioning and record all their working.
- They consider alternative strategies and use one of these to check their answer.
- They repeat this with another pair of numbers and then another and another.
- Once they have spent some time adding, ask them to repeat this for subtraction.

- Ensure they make the minuend, estimate the answer and partition to find the difference.
- They check using an alternative strategy.

Variation

2 Repeat the activity above, but this time one number is the augend or minuend and the second number is the total or difference.

Learners create a missing number calculation and work out what the missing number must be.

Number – Integers and powers

Unit 5: Addition and subtraction (B)

Collins International Primary Maths Recommended Teaching and Learning Sequence: Term 1, Week 5

Learning objectives

Code	Learning objectives
3Ni.02	Understand the [commutative and] associative properties of addition, and use these to simplify calculations.
3Ni.03	Recognise [complements of 100 and] complements of multiples of 10 [or 100] (up to 1000).
3Ni.04	Estimate, add and subtract whole numbers with up to three digits (regrouping of ones or tens).
3Nc.04	Recognise the use of an object to represent an unknown quantity in addition and subtraction calculations.

Unit overview

In this unit, learners will use their understanding of the commutative property of addition to develop an understanding of its associative property. The associative property applies when three or more numbers are added together and a decision is made about which two to add first, to simplify the calculation. For example, given $3 + 8 + 7 =$, it is sensible to add 3 and 7 first because it is a number pair to 10, and then add the remaining 8 to give a sum of 18.

Learners will be extending their understanding of finding complements of 100 to 1000, by using numbers that contain units of 100 and also units of 10, for example: $170 + 300 =$, $170 + 330 =$. This draws on prior learning of number pairs to 10 and 100.

Learners will extend their understanding of estimating, adding and subtracting by adding and subtracting one-, two- and three-digit numbers to and from three-digit numbers. They will be given the opportunity to review and consolidate mental calculation strategies, considered in previous units, as and when appropriate. Learners will solve missing number problems where the unknown quantity is represented by an object.

Prerequisites for learning

Learners need to:
- understand that addition is commutative
- understand that subtraction is not commutative
- know number pairs to 10, multiples of 10 to 100 and multiples of 100 to 1000
- add and subtract pairs of two-digit numbers
- use mental strategies to add and subtract.

Vocabulary

commutative, associative, inverse, add, sum, total, subtract, difference, estimate

Common difficulties and remediation

It is imperative to provide manipulatives such as Base 10 equipment for learners to gain an understanding of regrouping. They need to understand that when there are, for example, 10 or more units of ten (or one), these can be exchanged for one unit of one hundred (or ten).

If learners cannot recall facts to 10 easily, provide printouts of the addition facts (Resource sheet 4: Number pairs to 10) to refer to. They can then use these to find number pairs for units of ten to make 100 and also units of a hundred to make 1000. For example, if they can see the fact $7 + 3 = 10$, they can use it to show $70 + 30 = 100$ and $700 + 300 = 1000$.

Supporting language awareness

It is useful to have the vocabulary associated with addition and subtraction displayed in the classroom. It is important to display the vocabulary, with examples of the appropriate number for each word, for all learners to see. Ensure the examples reinforce the inverse property of addition and subtraction.

Promoting Thinking and Working Mathematically

During this unit learners have opportunities to form mathematical ideas and present evidence to justify their ideas (TWM.03, TWM.04). When working out their own mental calculation strategies for answering addition and subtraction calculations they will have the opportunity to characterise (TWM.05) and classify (TWM.06). They will also have opportunities to compare and evaluate solutions and refine them to develop more effective strategies (TWM.07, TWM.08). Learners also specialise (TWM.01) and generalise (TWM.02) when they explore more about unknowns.

Success criteria

Learners can:
- use the associative property for addition
- add and subtract three-digit multiples of 10
- add and subtract a one-digit number to or from a three-digit number
- add and subtract a multiple of 10 to or from a three-digit number
- add and subtract a multiple of 100 to or from a three-digit number
- recognise pictures and objects as representations of unknown quantities.

Number – Integers and powers

Lesson 1: **Associative property**

Number – Integers and powers

Learning objective

Code	Learning objective
3Ni.02	Understand the [commutative and] associative properties of addition, and use these to simplify calculations.

Revise

Use the activity *Commutativity* from Unit 5: *Addition and subtraction (B)* in the Revise activities.

Teach [SB] 🖵 [TWM.08]

- **[T&T]** Display **Slide 1**. Ask: **In what order could these three numbers be added? Can you see two numbers that add to make 10?** Agree that 8 add 2 makes 10. Say: **If we add 8 and 2 first, we just need to add the 4 to 10.**

- 🖉 Ask: **If we add 4 and 2 together first and then add 8, will we get the same total?** Agree that it doesn't matter which way round we add the numbers, the total will be the same.

- Repeat for other numbers that involve making 10 and adding a third number onto the 10, for example, $6 + 7 + 4 =$, $3 + 6 + 7 =$. Ask learners to rewrite the calculations. Start with the numbers to be added first, then show the sum of just two numbers, for example: $6 + 7 + 4 = 6 + 4 + 7 = 10 + 7 = 17$.

- **[T&T] [TWM.08]** Now on the board write calculations that involve doubling, for example $5 + 7 + 7$. Ask: **What is an efficient way to add these numbers?** Accept any answers that are correct. If no one mentions doubling, suggest doubling 7 and then adding 5. Repeat with similar examples.

- **[T&T] [TWM.08]** Repeat with other calculations for example, $5 + 6 + 5 =$. Ask learners to decide which order they would add the numbers together and record this.

- Introduce the term 'associative' and explain that what they have been learning is the 'associative property of addition'. Explain that this is another way of saying that we can find different ways of grouping numbers when we add them and the total will still be the same.

- 🖉 Ask: **What is meant by the associative property of addition?** Expect learners to reply in a complete sentence.

- Direct learners to Let's learn in the Student's Book. Discuss the first example shown. Ask learners to use counters to make the calculation. They then rearrange them to make the second calculation and add them together as suggested.

- Learners work through the paired activity.

- Discuss the Guided practice example in the Student's Book.

Practise [WB] [TWM.04]

- Workbook

Title: Associative property

Pages: 38–39

- Refer to Activity 1 from the Additional practice activities.

Apply 👥 🖵 [TWM.08]

- Display **Slide 2**. Learners find the total amount using the strategy they have been learning ($80c + 20c = 100c$, $100c + 45c = 145c$).

- They write an explanation for Tommy on how to use this method.

Review

- 🖉 Ask: **Can you describe what is meant by the associative property of addition?**

- Expect learners to be able to give you examples of adding three numbers by making 10 and also by doubling one number.

- Write three one-digit numbers on the board and invite learners to explain the most efficient ways to add them together. It could be that they know other facts and want to use these, rather than making 10. Reinforce the idea that we are trying to look for ways to make addition efficient.

Assessment for learning

- What is meant by the associative property of addition?
- Can you give an example to show how we can add three numbers together efficiently?
- Can you give another?
- What would be the most efficient way to add 3, 6 and 7? Why?
- How can this help us to add 30, 60 and 70?

Same day intervention
Enrichment

- Include examples that involve two-digit numbers where learners can make 10, for example, $24 + 36 + 40$. 24 and 36 total a multiple of 10. 40 is than added to 60, using knowledge that 4 add 6 equals 10, so 40 add 60 equals 100.

Lesson 2: **Addition and subtraction of multiples of 10**

Learning objective

Code	Learning objective
3Ni.03	Recognise [complements of 100 and] complements of multiples of 10 [or 100] (up to 1000).

Resources

Base 10 equipment (per pair); coloured cubes (per pair)

Revise

Use the activity *Number pairs* from Unit 5: *Addition and subtraction (B)* in the Revise activities to review the recall of number pairs to 10 and then apply this to multiples of 10 to 100 and multiples of 100 to 1000.

Teach 📖 🖥 [TWM.01/02]

- ▦ [TWM.01/02] Display **Slide 1**. Ask: **What addition fact are these cubes showing? If we know this fact, what commutative and inverse facts do we know?** Agree 4 + 5 = 9, 9 – 5 = 4, 9 – 4 = 5.
- Establish that in this example there are 5 ones and 4 ones, which together make a total of 9 ones.
- ▦ Ask: **What calculations can we make if each cube represents ten?** (50 + 40 = 90) Ask: **What calculations can we make if each cube represents one hundred?** (500 + 400 = 900) For each question, expect learners to be able to tell you the addition and subtraction facts.
- [T&T] [TWM.01/02] Display **Slide 2**. Ask: **What calculation does this show? How can we use what we know about adding one-digit numbers to add these?** Give learners time to discuss this with their partner. To clarify their thinking, they can make the two numbers with Base 10 equipment.
- Agree that if they know that 2 and 1 make 3, then 200 and 100 make 300. If they know that 6 and 2 make 8, then 60 and 20 make 80. The total of 260 and 120 is 380.
- Discuss the fact that they can make each number 10 times smaller and add them as the two-digit numbers 26 and 12. They then make the result 10 times greater.
- Ask learners to tell you the commutative addition and the two subtraction calculations (120 + 260 = 380, 380 – 120 = 260, 380 – 260 = 120).
- Repeat with other numbers where the augend remains the same and the addend changes, for example, 260 and 130, 260 and 140, 260 and 150.
- Display **Slide 3**. Work through this and other examples of subtracting three-digit multiples of 10 from other three-digit multiples of 10 in the same two ways.
- Direct learners to Let's learn in the Student's Book. Discuss the models shown and recap the different methods for finding the total and difference.
- Learners work through the paired activity.

- Discuss the Guided practice example in the Student's Book.

Practise 📝 [TWM.04]

- Workbook

Title: Addition and subtraction of multiples of 10

Pages: 40–41

- Refer to Activity 1 from the Additional practice activities.

Apply 👥 🖥 [TWM.01]

- Display **Slide 4**. Learners use the given numbers to write and answer six calculations.
- They will need to work systematically to find all the different solutions.

Review

- ▦ Ask: **What have we been learning about today?**
- Ask learners how using addition number facts helps them to find the totals of any three-digit multiples of 10.
- Ensure that they explain both the methods used in the lesson.
- Repeat for subtraction.
- Say some different three-digit multiples of 10 for learners to add and subtract.
- Invite learners to make up their own numbers for the class to add and subtract. Finish the lesson by playing Busy Ant Bounce from the interactive tools. Learners need to select the correct number to complete the calculation.

Assessment for learning

- If we know that 23 add 12 equals 35, what three-digit addition do we know?
- If we know that 65 subtract 12 equals 53, what three-digit subtraction do we know?
- How does knowing 27 subtract 13 equals 14 help us to know the answer to 270 subtract 130?
- Can you explain in another way?

Same day intervention
Support

- Focus on adding and subtracting three-digit multiples of 10 that do not involve regrouping, for example: 240 + 110 = and 460 – 120 =.

Number – Integers and powers

Unit 5 Addition and subtraction (B)

Lesson 3: Addition and subtraction with 3-digit numbers

Number – Integers and powers

Learning objective

Code	Learning objective
3Ni.04	Estimate, add and subtract whole numbers with up to three digits (regrouping of ones or tens).

Resources

Base 10 equipment: hundreds, tens and ones (per learner) (for the Workbook)

Revise

Use the activity *Using mental calculation strategies* from Unit 5: *Addition and subtraction (B)* in the Revise activities.

Teach SB 🖥 [TWM.07/08]

- ⟐ Display **Slide 1**. Ask: **What is a sensible estimate for the total of these two numbers? Why?** Agree 'about 250'.
- Display **Slide 2** and discuss how 245 has been partitioned into 2 hundreds, 4 tens and 5 ones (200 + 40 + 5). Display **Slide 3** and discuss how the six ones have been added making a total of 11 ones. Display **Slide 4** and discuss how the 11 ones have been exchanged for 1 ten and 1 one. Display **Slide 5** and discuss how the 4 tens and 1 ten have been added and then hundreds, tens and ones have been combined to make a total of 251.
- **[T&T] [TWM.08]** Ask: **Is there another way you could add these two numbers together?** Invite learners to demonstrate alternative strategies, discussing which they think is the most efficient strategy.
- Repeat above, adding one-digit numbers to other three-digit numbers, using the **Base 10 tool** if necessary.
- ⟐ Display **Slide 6**. Ask: **What is a sensible estimate for the difference? Why?** Agree 'about 443'.
- Display **Slide 7** and discuss how 453 has been partitioned into 4 hundreds, 5 tens and 3 ones (400 + 50 + 3). Ask: **Is it possible to subtract 8 from the 3?** Agree that it is not. Display **Slide 8** and discuss how 453 can be partitioned into 4 hundreds, 4 tens and 13 ones. Display **Slide 9** and discuss how it is now possible to subtract 8 ones from 13 ones which leaves 5 ones, and that we now recombine the hundreds, tens and ones to give us the answer of 445.
- **[T&T] [TWM.07/08]** Ask pairs to discuss other strategies for working out the answer to 453 – 8. Which methods do they prefer? Which are the most efficient? Repeat, subtracting one-digit numbers from other three-digit numbers, using the **Base 10 tool** if necessary.
- Display **Slides 10 to 14** and then **Slides 15 to 18**, working through the examples of adding and

subtracting a tens number to and from a three-digit number. Display **Slides 19 to 21** and then **Slides 22 to 24** briefly working through the examples of adding and subtracting a hundreds number to and from a three-digit number.

- Direct learners to Let's learn in the Student's Book and discuss the examples.
- Learners work through the paired activity.
- Discuss the Guided practice example.

Practise WB [TWM.04]

- Workbook

Title: Addition and subtraction with 3-digit numbers

Pages: 42–43

- Refer to Activity 2 from the Additional practice activities.

Apply 👥 🖥 [TWM.07/08]

- Display **Slide 25**. Learners think of three different strategies for 460 + 90 =.
- What strategies did the learners think of? Invite individuals to demonstrate their methods.

Review

- Ask learners to add and subtract 1s, 10s and 100s to and from three-digit numbers. Include examples that involve regrouping as well as some that do not. Encourage learners to use the most efficient strategies.

Assessment for learning

- What is 385 + 40? Which is the most efficient method?
- What strategy would you use to add 9 to 385?
- What is 665 – 200? Why is this easy?

Same day intervention

Support

- Focus on adding and subtracting one-digit numbers to and from a three-digit number.
- Encourage learners to use Base 10 equipment.

Lesson 4: **More unknowns!**

Learning objective

Code	Learning objective
3Nc.04	Recognise the use of an object to represent an unknown quantity in addition and subtraction calculations.

Resources

mini whiteboard or paper and pen (per learner); collection of 3D shapes (per pair)

Revise

Choose an activity from Unit 5: *Addition and subtraction (B)* in the Revise activities.

Teach [SB] 🖥

- 📸 Display **Slide 1**. Ask: **What can you see?** Agree some cylinders, symbols and an amount of money.
- **[T&T]** Ask: **What is the value of each cylinder? How do you know?** Agree that to have one cylinder left after taking one away from $10 means that the two cylinders together must have the same value as $10. Therefore, one cylinder must be half of $10.
- Demonstrate the calculation $10 – $5 = $5 and write $5 on each cylinder.
- Point to the second row and ask: **What would these three cylinders be worth, if one is $5?** Agree the answer is $15 and write $5 on each of the three cylinders followed by the total: $15.
- Then point to the third row and ask: **What about the five cylinders? What do you know that would help you work this out?** Agree that they can count in steps of 5 or use multiplication facts. Write $5 on each of the five cylinders followed by the total: $25.
- Display **Slide 2** and change the initial price of $10. Ask: **What if $10 is changed to $8/$12/$20?**
- **[T&T]** Display **Slide 3**. Ask: **What can you see?** Agree it shows that the total cost when the bananas are added is $4. Ask: **How can we find the cost of two bananas? What about the cost of one?** Agree halving is a good method. Half of $4 is $2 and half of $2 is $1. The cost of one banana is $1.
- Ask learners to use their collection of 3D shapes to make up problems like this with unknown quantities and to ask their partner to work out what the missing values must be.
- Direct learners to Let's learn in the Student's Book and discuss the examples shown.
- Learners work through the paired activity.
- Discuss the Guided practice example in the Student's Book.

Practise [WB]

- Workbook

Title: More unknowns!

Pages: 44–45

- Refer to Activity 2 from the Additional practice activities.

Apply 👥 🖥 [TWM.03/04]

- Display **Slide 4**.
- Learners use the given information to work out the cost of the apple.
- Encourage them to work systematically and give a written explanation of how they found their answer. Learners need to find two numbers that add to 3. They should suggest 1 and 2. They should then see that 2 is double 1, so the price of the apple is $1.

Review [TWM.03/04]

- 📸 Ask: **What have we been learning about today?**
- Draw out the fact that objects and shapes can be used to represent missing quantities and that using addition and corresponding subtraction facts will enable the missing quantities to be found.
- Give some missing number calculations that involve adding and subtracting multiples of 10 or 100, using objects for the unknowns.

Assessment for learning

- Why is knowing number facts important? Can you think of another reason?
- 100 subtract something gives a difference of 40. How can we find the missing number?
- Something add 800 gives a total of 1000. How can we work out the missing number?
- Something subtract 15 gives a difference of 50. How can we work out the missing number?

Same day intervention

Support

- Ensure learners use equipment such as money or Base 10 equipment so that they can set out the given amounts. This will help them to find the missing quantities.
- Give examples that use number pairs to 10 and multiples of 10 to 100. In this way learners can focus on finding the missing numbers and not worry about calculating.

Number – Integers and powers

Additional practice activities

Activity 1

Learning objective
- Understand the associative property of addition, and use these to simplify calculations

Resources
Resource sheet 3: Digit cards (per pair); paper and pen (per learner)

What to do [TWM.07/08]
- Give pairs a set of digit cards from Resource sheet 3: Digit cards.
- Learners pick three cards, and use the three digits to write down an addition calculation, for example: 5, 8 and 3 would make 5 + 8 + 3 =.
- They then discuss the most efficient way to find the total, by choosing two numbers to add first. They rearrange the cards and record their new calculation, for example: 5 + 3 + 8 =.

- Learners record each part of their calculation to show the steps they take to find the total, for example 5 + 3 = 8 and 8 + 8 = 16.
- Ask them to consider another way that they could do this, choosing another two numbers to add first.
- Learners do this and then reflect on which was the most efficient for them and why. You could ask them to record their reasoning.
- They repeat this several times, using three different digit cards each time.

Variation
3 Learners repeat the above activity but this time the digit cards represent tens, for example 5, 8 and 3 represent 50, 80 and 30.

Activity 2

Learning objective
- Estimate, add and subtract 1s, 10s and 100s to and from three-digit numbers with regrouping

Resources
Base 10 equipment: hundreds, tens and ones (per pair)

What to do [TWM.07/08]
- Write a three-digit number on the board, for example 278.
- Learners add 9 to 278. They then add 90 to 278.
- They estimate the totals first and then make both numbers, using Base 10 equipment.
- They find the total by partitioning.
- They discuss alternative strategies with their partner, for example, adding and subtracting 10 and 100 and adjust, to check their answer.

- Learners then subtract 9 from 278. They then subtract 90 from 278.
- They estimate the difference first, and then make 278 and partition to find the difference.
- They check by using an alternative strategy.

Variation [TWM.07/08]
1 Repeat the activity above, but focus on adding and subtracting ones where regrouping is required, for example, 278 + 8 = and 278 − 9 =.

Number – Integers and powers

Unit 6: Addition

Collins International Primary Maths
Recommended Teaching and
Learning Sequence: Term 3, Week 1

Learning objective

Code	Learning objective
3Ni.04	Estimate, add [and subtract] whole numbers with up to three digits (regrouping of ones or tens).

Unit overview

For all addition calculations, learners should be encouraged to consider mental calculation strategies first. These include:
- using known number facts and knowledge of place value
- partitioning into hundreds, tens and ones and recombining
- compensation.

The formal written method of column addition should be used for calculations that cannot be answered efficiently by mental calculation strategies. This method will be developed, alongside the expanded written method, in order to aid progression. Learners are encouraged to discuss similarities and differences between these methods.

The focus in this unit is the addition of two-digit numbers and three-digit numbers and the addition of pairs of three-digit numbers. This will involve regrouping. Regrouping is the term that describes the process of changing, for example, 10 ones into 1 ten or 10 tens into 1 hundred.

Initially, learners will be introduced to the written method without regrouping. They will then be given calculations that involve regrouping ones and then tens.

Learners will be expected to estimate simple calculations so that they recognise, without a formal calculation, when an answer is likely to be incorrect.

Prerequisites for learning

Learners need to:
- understand the expanded method for the addition of pairs of two-digit numbers
- understand that addition is commutative
- be able to add pairs of two-digit numbers with regrouping
- use mental calculation strategies to add.

Vocabulary

add, equals, sum, total, estimate, regroup, expanded written method, formal written method

Common difficulties and remediation

Providing manipulatives such as Base 10 equipment is imperative for learners to gain an understanding of the written method for addition. These will also help their understanding of regrouping.

It is important to work with learners on the expanded method for addition and to link this closely with the formal written method. Learners should not move on from the expanded method until they are ready. For most learners, the transition from one to the other is smooth and will happen in Stage 3.

It is helpful to link addition and subtraction because these are inverse operations. As learners practise written methods for addition, they should subtract the answer from the augend. Look for opportunities to do this in lessons. Provide calculators for subtraction because the focus of this unit is addition.

Supporting language awareness

It is important to be specific with vocabulary and use the word 'add' and 'subtract' appropriately. This is the operator. Introduce other vocabulary such as 'plus', 'sum' and 'total' as and when it arises.

Promoting Thinking and Working Mathematically

During this unit, learners will have opportunities to compare and evaluate solutions and refine them to develop more effective strategies (TWM.07, TWM.08) as they consider which methods to use to add numbers together. They will have opportunities to form mathematical ideas and present evidence to justify their ideas (TWM.03, TWM.04). When working out their own mental calculation strategies for answering addition calculations, they will have the opportunity to characterise (TWM.05) and classify (TWM.06). They will have opportunities to explore what they also know if they know a particular fact, which involves specialising (TWM.01) and generalising (TWM.02).

Success criteria

Learners can:
- use mental calculation strategies to answer additions when appropriate
- add a three-digit number and a two-digit number, with and without grouping
- add pairs of three-digit numbers, with and without grouping
- use the expanded written method for addition
- use the formal written method for addition.

Number – Integers and powers

Unit 6 Addition

Number – Integers and powers

Lesson 1: Add 3-digit numbers and tens (A)

Learning objective

Code	Learning objective
3Ni.04	Estimate, add [and subtract] whole numbers with up to three digits (regrouping of ones or tens).

Resources

Base 10 equipment: hundreds, tens and ones (per pair); squared paper (per learner) (for the Workbook)

Revise

Use the activity *Mental addition: partitioning* from Unit 6: *Addition* in the Revise activities.

Teach 🆂🅱 🖥 [TWM.01/02/07/08]

- ⬚ Display **Slide 1**. Ask: **What is a sensible estimate for the sum of 246 and 53?** Agree 250 + 50 = 300.
- **[TWM.01/02/07/08]** Ask: **How can we add 246 and 53?** Explore different mental strategies, such as partitioning. Ask pairs of learners to use Base 10 equipment and to find the sum using one of the strategies discussed.
- **[T&T]** Point to the three calculations on the slide and ask: **What do you notice about these three calculations? How are they the same? How are they different? How do they link to the Base 10 equipment?** Remind learners of the expanded written method they looked at in Stage 2. Explain that here the augend is a three-digit number.
- Invite learners to explain what is happening in each calculation. Confirm that the first calculation shows the digits with the highest place value added first, and the second shows those with the lowest place value added first. Introduce the term 'expanded written method' to describe these two calculations.
- Talk to learners about how the third calculation shows the same calculation but in a shorter form, without the expanded part. Point out that the expanded model that adds the digit with the lowest place value first links with the shorter written method. Tell learners that this is called the 'formal written method'.
- Work through this method, adding the ones first, then the tens and finally the hundreds.
- Repeat for other 'three-digit add two-digit' number calculations that do not need regrouping. Learners use Base 10 equipment to make the numbers. They estimate answers first, use a mental strategy to find the sums and then practise using the written methods.
- Ensure you demonstrate the expanded model alongside the formal written method.
- Direct learners to Let's learn in the Student's Book. Discuss the example shown.
- Learners complete the paired activity, then as a class discuss the Guided practice example.

Practise 🆆🅱

- Workbook

Title: Add 3-digit numbers and tens (A)

Pages: 46–47

- Refer to Activity 1 from the Additional practice activities.

Apply 👥 🖥 [TWM.01/04]

- Display **Slide 2**. Learners are given five digits to use for a three-digit add two-digit calculation aiming to find the highest and lowest possible totals.
- Encourage learners to think about where to position the digits; for example, 5 in the hundreds position for the highest total, and 1 in the hundreds for the lowest total.

Review

- ⬚ Ask: **What must we be sure to do before we calculate?** Agree, make an estimate.
- On the board, write: 425 and 63. Invite learners to suggest the best ways to add the numbers. Agree that a good method would be to keep 425 whole and partition and add 63.
- ⬚ Ask: **Why have we learned a written method for addition?** Establish that we use this method when a mental calculation is too difficult.

Assessment for learning

- Why estimate the answer to an addition?
- Can you give me an example of an addition calculation that uses a mental strategy/written method?

Same day intervention

Support

- Focus on the addition of pairs of two-digit numbers until learners are confident. Then move on to the expanded method of three-digit numbers and tens.

Lesson 2: **Add 3-digit numbers and tens (B)**

Learning objective

Code	Learning objective
3Ni.04	Estimate, add [and subtract] whole numbers with up to three digits (regrouping of ones or tens).

Resources

Base 10 equipment: hundreds, tens and ones (per pair); squared paper (per learner) (for the Workbook)

Revise

Use the activity *Mental addition: compensation* from Unit 6: *Addition* in the Revise activities to rehearse this mental calculation strategy, which will be included in this lesson.

Teach 🔲 🖥 [TWM.01/02/08]

- ⌂ **[TWM.01/02]** Display **Slide 1**. Ask: **What would be a sensible estimate for the sum of 246 and 29?** Agree that 246 is close to 250 and 29 is close to 30, so a good estimate would be 280.

- **[TWM.08]** Ask: **How can we add 246 and 29?** Explore different mental calculation strategies. Focus on compensation. Discuss the fact that 29 is close to 30, so they can add 30 and subtract 1. Ask pairs of learners to use Base 10 equipment to make the two numbers and to find the sum by exchanging 29 for three tens, adding those and subtracting 1 from 246. Repeat for adding 39 and 49.

- **[T&T]** Point to the two calculations on the slide and ask: **What do you notice about these two calculations? How are they the same? How are they different? How do they link to the Base 10 equipment?** Agree that the numbers are the same. The Base 10 equipment shows the numbers that are written as calculations. The calculations show two written methods of finding the sum.

- Invite learners to explain what is happening in each calculation. Confirm that the first shows how the ones, tens and hundreds can be added separately (the expanded written method). The second calculation shows the same calculation but in a short or compact form (the formal written method).

- Work through the formal written method, adding the ones first, then the tens and finally the hundreds. Model explicitly that the 15 ones are regrouped into 1 ten and 5 ones. The 1 ten is added to the other tens. Learners do this at the same time, using their Base 10 equipment.

- Display **Slide 2**. Work through this example in the same way. Estimate first, then consider mental calculation strategies. Suggest that this calculation might be completed more efficiently using a written method. Model the written methods on the slide, with learners using their Base 10 equipment. Repeat for other similar calculations.

- Direct learners to Let's learn in the Student's Book. Discuss the example shown.

- Learners complete the paired activity, then as a class discuss the Guided practice example.

Practise 🔳

- Workbook

Title: Add 3-digit numbers and tens (B)

Pages: 48–49

- Refer to the variation in Activity 1 from the Additional practice activities.

Apply 👥 🖥 [TWM.01/04]

- Display **Slide 3**. Learners use the five digits to make one three-digit number and one two-digit number.

- They form calculations in which the ones need regrouping. A combination of four or five with 6 in the ones will bring this about.

Review

- ⌂ Ask: **What have we been learning about in our lesson today?**

- Invite learners to demonstrate both the expanded and the formal written methods for adding three-digit and two-digit numbers. Ensure examples need only the ones regrouping.

- ⌂ Ask: **What mental calculation strategies have we been thinking about when adding?** Expect them to talk about, for example, partitioning and compensation.

Assessment for learning

- Can you give an example of a calculation that can be answered using compensation?

- Can you give an example of a calculation that it would be sensible to work out the answer using a written method?

Same day intervention
Support

- Focus on adding with regrouping, using the expanded method.

Number – Integers and powers

Unit 6 Addition

Lesson 3: Add two 3-digit numbers (A)

Number – Integers and powers

Learning objective

Code	Learning objective
3Ni.04	Estimate, add [and subtract] whole numbers with up to three digits (regrouping of ones or tens).

Resources

Base 10 equipment: hundreds, tens and ones (per pair); squared paper (per learner) (for the Workbook)

Revise

Use the activity *Mental addition: making 10s* from Unit 6: *Addition* in the Revise activities to review this strategy, which learners will explore during this lesson.

Teach 〔SB〕 ☐ [TWM.01/02/04/05/08]

- ⌐ **[TWM.01/02]** Display **Slide 1**. Ask: **What would be a sensible estimate for the sum of 246 and 127?** Suggest that 246 is close to 250 and 127 is close to 130, so a good estimate might be 380.
- **[T&T] [TWM.08]** Ask: **How can we add 246 and 127?** Explore different mental calculation strategies, for example partitioning and making 10.
- **[TWM.04/05]** Focus on making 10. Ask pairs of learners to use Base 10 equipment to make these numbers and to find the sum by taking 4 from 127 and adding it to 246 to make the calculation 250 + 123. Ask: **What do you think of this strategy? Why is it a good one? Why/Why not?**
- **[T&T]** Point to the two calculations on the slide and ask: **What do you notice about these two calculations?** Agree that these are similar to those that learners saw when adding three-digit and two-digit numbers together. These are different because both numbers have three digits.
- Invite learners to explain what is happening in both calculations.
- Work through the formal written method, adding the ones first, regrouping into 1 ten and 3 ones, then the tens and finally the hundreds.
- Repeat with other pairs of three-digit number calculations where the ones need regrouping but the tens do not. Learners use Base 10 equipment to make the numbers. They estimate answers first, decide whether to use a mental strategy or a written method to find the sum, then use their chosen strategy/method to work out the answer.
- Ensure you demonstrate the expanded written method alongside the formal written method.
- Display **Slide 2**. Repeat the process for this calculation.
- Direct learners to Let's learn in the Student's Book. Work through the example shown.

- Learners complete the paired activity, then as a class discuss the Guided practice example.

Practise 〔WB〕

- Workbook

Title: Add two 3-digit numbers (A)

Pages: 50–51

- Refer to Activity 2 from the Additional practice activities.

Apply 👥 ☐ [TWM.01/04]

- Display **Slide 3**. Learners find the missing digits that make up three calculations.
- When they have worked out the missing digits, they explain their strategies to the class.

Review

- Write 465 + 218 = on the board. Ask learners to decide how they would work out the answer and to explain why they chose that strategy/method.
- Ask for volunteer learners to use their chosen strategy/method to work out the answer. Repeat if appropriate.
- Reinforce the importance of estimating first and trying mental calculation strategies to find totals.

Assessment for learning

- How can we add 147 and 128? Is there another way? Which is the better strategy/method?
- What do we mean by regrouping? When might we need to do this?

Same day intervention

Support

- Focus on adding pairs of three-digit numbers, using the expanded method. Ensure learners use Base 10 equipment.

Lesson 4: **Add two 3-digit numbers (B)**

Learning objective

Code	Learning objective
3Ni.04	Estimate, add [and subtract] whole numbers with up to three digits (regrouping of ones or tens).

Resources

Base 10 equipment: hundreds, tens and ones (per learner); squared paper (per learner) (for the Workbook)

Revise

Use any activity from Unit 6: *Addition* in the Revise activities to reinforce a mental calculation strategy that requires more practice.

Teach [SB] 🖵 [TWM.08]

- 🖵 Display **Slide 1**. Ask: **How does this show 246 add 171?** Expect learners to tell you, for example, that the first part shows 2 hundreds, 4 tens and 6 ones.
- **[T&T] [TWM.08]** Ask: **What is a sensible estimate of the sum? Can you explain why?** Agree, for example, 246 is close to 250 and 171 is close to 200, therefore a sensible estimate would be around 450. Ask: **How could we use a mental calculation strategy to find the sum?** Discuss any suggestions that learners might have, including partitioning, compensation and adapting making 10 to making 100 (306 + 111 or 216 + 201).
- Suggest that a written method might be more efficient. Demonstrate the written methods shown on the slide: first the expanded method and then the formal method. Expect learners to work with you, using Base 10 equipment. Ask: **How are these numbers different from the pairs of three-digit numbers we added in Lesson 3?** Agree that no regrouping is required in the ones, but regrouping is required in the tens. Explain that there are 11 tens and these are regrouped into 1 hundred and 1 ten.
- Work through **Slide 2** in the same way. Ensure you include estimating the sum and asking learners to think about what mental strategy they might use to answer the calculation and whether or not it would be as efficient as using a written method. If learners suggest a suitable mental strategy, ask them to demonstrate this to the class. Then discuss the two written methods on the slide.
- Give learners pairs of three-digit numbers to add, where regrouping is required with the tens.
- Direct learners to Let's learn in the Student's Book and discuss the example shown.
- Learners complete the paired activity, then as a class discuss the Guided practice example.

Practise [WB]

- Workbook

Title: Add two 3-digit numbers (B)

Pages: 52–53

- Refer to the variation in Activity 2 from the Additional practice activities.

Apply 👥 🖵 [TWM.01/04]

- Display **Slide 3**. This is another problem that requires learners to find the missing digits that make up three calculations. This time they consider regrouping the tens.
- When they have worked out the missing digits, they explain their strategies to the class.
- As a class, work through the problem. Encourage learners to explain their strategies for finding the answer.

Review

- 🖵 Ask: **What have we been learning today?**
- 🖵 Ask: **What must we do when we are given an addition calculation?** Draw out the fact that when adding two numbers, learners must:
  - first estimate the answer
  - look carefully at the numbers and consider if it is suitable to use a mental strategy
  - use a written method if this is more efficient
  - check their answers.

Assessment for learning

- What is a sensible estimate for 365 + 251?
- Can you explain why?
- How would you find the sum of 365 and 251? Can you think of any other strategies/methods?
- Would you say the formal written method the best choice for 365 + 251? Why/why not?

Same day intervention
Support

- Ensure learners use concrete resources such as Base 10 to develop their understanding of the expanded and formal written methods.
- Move them on to the formal written method only when learners are confident with the expanded method.

Number – Integers and powers

Additional practice activities

Activity 1

Learning objective
- Estimate and solve addition of three-digit numbers and tens

Resources
Base 10 equipment: hundreds, tens and ones (per pair); pencil and plain or squared paper (per learner)

What to do [TWM.04]
- Write these digits on the board: 1, 2, 3, 4, 5.
- Learners use these digits to make a three-digit number and a two-digit number, for example: 123 and 45.
- They use Base 10 equipment to make their numbers.
- They each estimate the answer and then use a mental calculation strategy of their choice to work out the sum.
- They explain their choice to their partner.

- Learners then practise using the expanded written method and the formal written method to confirm the sum.
- They then make another three-digit number and a two-digit number by changing the digits from their first calculation, for example: 124 + 35 =. They repeat the process of estimating, using a mental calculation strategy and then recording both written methods.
- Learners continue to do this until they have made up to six possible combinations of digits.

Variation
3 Learners repeat the above activity but this time write 3, 4, 5, 6 and 7. They make numbers that require regrouping of the ones only, for example 634 + 57 =.

Activity 2

Learning objective
- Estimate and solve addition of pairs of three-digit numbers

Resources
Base 10 equipment: hundreds, tens and ones (per pair); pencil and plain or squared paper (per learner)

What to do [TWM.04]
- Write these digits on the board: 1, 2, 3, 4, 5, 6.
- Learners use these digits to make pairs of three-digit numbers to add together.
- Their numbers must involve regrouping of ones for example: 124 + 356 =.
- They use Base 10 equipment to make their numbers.
- They each estimate the answer and then use a mental calculation strategy of their choice to work out the sum.
- They explain their choice to their partner.

- Learners then practise using the expanded written method and the formal written method to confirm the sum.
- They then make another pair of three-digit numbers by changing the digits from their first calculation, for example: 125 + 346 =. They repeat the process of estimating, using a mental calculation strategy, and then recording both written methods.
- How many different combinations can they make that involve regrouping of ones?

Variation
3 Learners repeat the above activity but this time they make numbers that require regrouping of the tens only, for example: 263 + 154 =.

Number – Integers and powers

Unit 7: Subtraction

Collins International Primary Maths Recommended Teaching and Learning Sequence: Term 3, Week 2

Learning objective

Code	Learning objective
3Ni.04	Estimate[, add] and subtract whole numbers with up to three digits (regrouping of ones or tens).

Unit overview

As with addition, for all subtraction calculations learners are encouraged to consider mental calculation strategies first. These include:

- using known number facts
- counting on and back
- compensation
- partitioning.

The formal written method of column subtraction should be used for calculations that cannot be answered efficiently by using mental calculation strategies. This method will be developed, alongside the expanded written method, in order to aid progression. Learners are encouraged to discuss similarities and differences between these methods.

In this unit, learners draw on previous knowledge of subtraction to understand how the formal written method works. They will be encouraged to use written methods as and when appropriate. Therefore, examples have been chosen carefully in order for these methods to be used appropriately.

The focus in this unit is the subtraction of two-digit numbers from three-digit numbers and the subtraction of pairs of three-digit numbers. This will involve regrouping, which is the process of changing, for example, 1 ten into 10 ones and 1 hundred into 10 tens.

Initially, learners will learn the process of the written method without regrouping. They will then be given calculations that involve regrouping a unit of ten and then a unit of a hundred.

Learners will be expected to estimate answers first, consider the most appropriate method to use and check that their answers are correct.

Prerequisites for learning

Learners need to:

- understand the expanded method for the subtraction of pairs of two-digit numbers
- be able to subtract pairs of two-digit numbers with regrouping of 1 ten into 10 ones
- use mental calculation strategies to subtract.

Vocabulary

subtract, minuend, subtrahend, difference, equals, estimate, regroup

Common difficulties and remediation

It is essential to provide manipulatives such as Base 10 equipment for learners to understand the written method for subtraction, particularly the process of regrouping.

It is important to work with learners on the expanded method for subtraction and to link this closely with the formal written method. Learners should not move on from the expanded method until they are ready. For most learners, this will happen in Stage 3.

It is helpful to link addition and subtraction because these are inverse operations. As learners practise written methods for subtraction, they should add the difference to the subtrahend and see if the result is the minuend. Provide calculators for the addition because the focus of this unit is subtraction.

Supporting language awareness

It is important to be specific with vocabulary and use the word 'subtract' appropriately. This is the operator. Introduce other vocabulary such as 'take away' and 'difference' as and when it arises.

Promoting Thinking and Working Mathematically

During this unit, learners will have opportunities to compare and evaluate solutions and refine them to develop more effective strategies (TWM.07, TWM.08). They will also have opportunities to convince, characterise and classify (TWM.04, TWM.05, TWM.06) when developing mental calculation strategies and form mathematical ideas and present evidence to justify their ideas (TWM.03, TWM.04). They will explore what they also know if they know a particular fact, which involves specialising (TWM.01) and generalising (TWM.02).

Success criteria

Learners can:

- use mental calculation strategies to answer subtractions when appropriate
- subtract a two-digit number from a three-digit number, with and without grouping
- subtract pairs of three-digit numbers, with and without grouping
- use the expanded method for subtraction
- use the formal written method for subtraction.

Number – Integers and powers

Number – Integers and powers

Lesson 1: **Subtract 3-digit numbers and tens (A)**

Learning objective

Code	Learning objective
3Ni.04	Estimate[, add] and subtract whole numbers with up to three digits (regrouping of ones or tens).

Resources

Base 10 equipment: hundreds, tens and ones (per pair); squared paper (per learner) (for the Workbook)

Revise

Use the activity *Mental subtraction: partitioning* from Unit 7: *Subtraction* in the Revise activities.

Teach ⬚ 🖥 [TWM.08]

- 🖰 Display **Slide 1**. Ask: **What would be a sensible estimate for 467 subtract 43?** Agree 470 – 40 = 430.
- [TWM.08] Ask: **How can we subtract 43 from 467?** Explore different mental strategies, such as partitioning and compensation. Ask pairs of learners to use Base 10 equipment to find the difference using one of the strategies discussed.
- [T&T] Point to the two calculations on the slide and ask: **What do you notice about these two calculations? How are they the same? How are they different? How do they link to the Base 10 equipment?** Remind learners of the expanded written method they looked at in Stage 2. Explain that here the minuend is a three-digit number.
- Invite learners to explain what is happening in each calculation. Confirm that the first calculation shows the numbers have been partitioned and then the parts have been subtracted. Introduce the term 'expanded written method' to describe this calculation.
- Talk to learners about how the second calculation shows the same calculation but in a shorter form, without the expanded part. Tell learners that this is called the 'formal written method'.
- Work through this method, subtracting the ones first, then the tens and finally the hundreds. Check by adding the difference to the subtrahend to give the minuend.
- Repeat for other 'three-digit subtract two-digit' number calculations that do not need regrouping. Learners use Base 10 equipment to make the minuend and subtract the subtrahend. They estimate answers first, use a mental strategy to find the differences and then practise using the written methods.
- Ensure you demonstrate the expanded model alongside the formal written method.

- Direct learners to Let's learn in the Student's Book. Discuss the example shown.
- Learners complete the paired activity, then as a class discuss the Guided practice example.

Practise 📓

- Workbook

Title: Subtract 3-digit numbers and tens (A)

Pages: 54–55

- Refer to Activity 1 from the Additional practice activities.

Apply 👥 🖥 [TWM.03/04]

- Display **Slide 2**. Learners are given a problem with an incorrect solution.
- They discuss if they agree with the solution and explain why and then work out the correct solution.

Review

- 🖰 Ask learners to explain why they need to estimate an answer first. Then ask: **Why is it better to use a mental strategy if you can?**
- On the board, write: 275 and 63. Invite learners to explain the best ways to subtract 63 from 275. Agree that a good method would be to keep 275 whole and partition and subtract 63 (275 – 60 – 3 =).
- 🖰 Ask: **Why have we learned how to subtract using a written method?** Establish that we use this method when a mental calculation is too difficult.

Assessment for learning

- Can you give an example of a calculation that we can subtract using a mental method?
- Explain the expanded written method to me.
- Explain how the formal written method works.

Same day intervention
Support

- Focus on subtracting multiples of 10 from three-digit numbers initially and, when learners are confident, include ones. Ensure learners use Base 10 equipment or place value counters.

Lesson 2: **Subtract 3-digit numbers and tens (B)**

Learning objective

Code	Learning objective
3Ni.04	Estimate[, add] and subtract whole numbers with up to three digits (regrouping of ones or tens).

Resources

Base 10 equipment: hundreds, tens and ones (per pair); squared paper (per learner) (for the Workbook)

Revise

Use the activity *Mental subtraction: compensation* from Unit 7: *Subtraction* in the Revise activities to review this mental calculation strategy, which will be included in this lesson.

Teach 〔SB〕 💻 [TWM.08]

- 🔲 Display **Slide 1**. Ask: **What would be a sensible estimate for 265 subtract 29?** Agree that 265 is close to 270 and 29 is close to 30, so a good estimate would be 240.

- [TWM.08] Ask: **How can we subtract 29 from 265?** Explore different mental calculation strategies. Focus on compensation. Discuss the fact that 29 is close to 30, so they can subtract 30 and then add 1. Ask pairs of learners to use Base 10 equipment to make 265 and to find the difference by subtracting three tens, and then adding 1 unit of one to 235. Repeat with the same minuend for subtracting 39, 49 and 59.

- [T&T] Point to the two calculations on the slide and ask: **What do you notice about these two calculations? How are they the same? How are they different? How do they link to the Base 10 equipment?** Agree that the numbers are the same. The Base 10 equipment shows the minuend, from which 29 must be subtracted. The calculations show two written methods of finding the difference.

- Invite learners to explain what is happening in each calculation. Confirm that the first is the expanded method and the second calculation shows the same calculation but in a short or compact form (the formal written method).

- Work through the formal written method. Model explicitly that the 9 ones cannot be subtracted from the 5 ones. Therefore, 1 ten needs to be regrouped into 10 ones. There are now 15 ones and 9 can be subtracted. Learners do this with you, using their Base 10 equipment. Check by adding the difference to the subtrahend to give the minuend.

- Display **Slide 2**. Work through this example in the same way. Estimate first, then consider mental calculation strategies. Learners may suggest subtracting 50 and adding 3. This calculation might be more efficiently answered by using the written method. Model the written methods on the slide with learners using their Base 10 equipment. Repeat for other similar calculations.

- Direct learners to Let's learn in the Student's Book. Discuss the example shown.

- Learners complete the paired activity, then as a class discuss the Guided practice example.

Practise 〔WB〕 [TWM.04]

- Workbook

Title: Subtract 3-digit numbers and tens (B)

Pages: 56–57

- Refer to the variation in Activity 1 from the Additional practice activities.

Apply 👥 💻 [TWM.03/04]

- Display **Slide 3**. Learners consider a problem and its solution.

- The solution is correct. Expect learners to work with a partner to discuss this.

Review

- 🔲 Ask: **What have we been learning about today?**

- Invite learners to demonstrate both the expanded and the formal written methods for subtracting two-digit numbers from three-digit numbers. Ensure examples need only the ones regrouping.

- Review the mental calculation strategies that they have also used to subtract, for example, partitioning and compensation.

Assessment for learning

- Can you give an example of a subtraction calculation that can be answered by using compensation? Can you give another?

- Can you give an example of a subtraction calculation that can be more efficiently answered by a written method? Can you give another?

Same day intervention

Support

- Focus on subtracting with regrouping, using the expanded method.

Number – Integers and powers

Unit **7** Subtraction

Lesson 3: **Subtract two 3-digit numbers (A)**

Number – Integers and powers

Learning objective

Code	Learning objective
3Ni.04	Estimate[, add] and subtract whole numbers with up to three digits (regrouping of ones or tens).

Resources

Base 10 equipment: hundreds, tens and ones (per pair); squared paper (per learner) (for the Workbook)

Revise

Use the activity *Mental subtraction: counting on and back* from Unit 7: *Subtraction* in the Revise activities.

Teach ⬛ 🖵 [TWM.04/08]

- 🖑 Display **Slide 1**. Ask: **What would be a sensible estimate for the difference between 268 and 259?** Suggest that 268 is close to 270 and 259 is close to 260, so a good estimate might be 10.
- [T&T] [TWM.08] Ask: **How can we subtract 259 from 268?** Explore different mental calculation strategies. Then ask: **What do you notice about the numbers?** Agree that they are close together. Therefore, counting on or back would be a sensible strategy. Ask learners to draw a number line and position 259 at the beginning and 268 at the end. Ask learners to count on from 259 to 268 and back from 268 to 259. The difference in both cases is 9.
- Ask: **Should we use a written method to subtract 259 from 268? Why not?** Discuss that whenever we need to perform a calculation, we must first look at the numbers and decide whether it is best to use a mental strategy or a written method.
- [T&T] Display **Slide 2**. Ask: **What do you estimate the answer to be?** Point to the two calculations and ask: **What do you notice about these calculations?** Agree that these are similar to those that learners looked at when subtracting three-digit and two-digit numbers previously. The difference is that both numbers have three digits.
- Invite learners to explain what is happening in both calculations. Ask pairs of learners to use Base 10 equipment to make 457. Together, work through the formal written method, 1 ten for 10 ones. Check by adding the difference to the subtrahend to give the minuend.
- Repeat with other pairs of three-digit number calculations in which the ones need regrouping. Learners use Base 10 equipment to make the minuends. They estimate answers first, decide whether to use a mental strategy or a written method to find the difference, then use their chosen strategy/method to work out the answer.

- Ensure you demonstrate the expanded method alongside the formal written method.
- [TWM.04] Display **Slide 3**. Repeat the above for this calculation. Ensure you include counting on or back as a mental calculation strategy.
- Direct learners to Let's learn in the Student's Book. Work through the example shown.
- Learners complete the paired activity, then as a class discuss the Guided practice example.

Practise 🟥

- Workbook

Title: Subtract two 3-digit numbers (A)

Pages: 58–59

- Refer to Activity 2 from the Additional practice activities.

Apply 👥 🖵 [TWM.04]

- Display **Slide 4**. Learners find the missing digits that make up three calculations (applying the inverse relationship is a useful strategy).
- When they have worked out the missing digits, they explain their strategies to the class.

Review

- Discuss the subtraction methods that learners considered during the lesson.
- Invite learners to give examples of differences that can be found by counting on or back for the class to answer. Repeat for the written methods.

Assessment for learning

- How can we subtract 147 from 162?
- Is there another way?
- Which is the most efficient method?

Same day intervention
Support

- Focus on subtracting pairs of three-digit numbers, using the expanded method. Ensure learners use Base 10 equipment.

Lesson 4: **Subtract two 3-digit numbers (B)**

Learning objective

Code	Learning objective
3Ni.04	Estimate, [add] and subtract whole numbers with up to three digits (regrouping of ones or tens).

Resources

Base 10 equipment: hundreds, tens and ones (per learner); squared paper (per learner) (for the Workbook)

Revise

Use any activity from Unit 7: *Subtraction* in the Revise activities to reinforce a mental calculation strategy that requires more practice. Alternatively, use the Interactive Tool Busy Ant School to rehearse quick mental addition and subtraction of two-digit numbers.

Teach 🔲 🖵 [TWM.08]

- 🄿 Show **Slide 1**. Ask: **In the calculation 346 – 171, which number is the minuend? Which is the subtrahend?** Agree the visual representation shows the minuend of 346. The number to be subtracted (171) is the subtrahend.

- [T&T] [TWM.08] Ask: **What is a sensible estimate of the difference? Can you explain why?** Agree, for example, 346 is close to 350 and 171 to 170, therefore a sensible estimate would be around 180. Ask: **How could we find the difference using a mental calculation strategy?** Discuss any suggestions that the learners might have, including compensation, partitioning and counting on or back.

- Suggest that a written method might be more efficient for these numbers. Demonstrate the written methods shown on the slide, first the expanded method and then the formal method. Expect learners to work with you, using Base 10 equipment.

- Ask: **How is this calculation different from the pairs of three-digit numbers we subtracted in Lesson 3?** Agree that no regrouping is required in the units of ten, but regrouping is required in the units of a hundred. Explain that one of the hundreds has been regrouped into 10 tens, so there are now 14 tens and 7 tens can be subtracted. Check by adding the difference to the subtrahend to give the minuend.

- Work through **Slide 2** in the same way. Ensure you include estimating the difference and asking learners to think about what mental strategy they might use to answer the calculation and whether or not it would be as efficient as using a written method. If learners suggest a suitable mental strategy ask them to demonstrate this to the class. Then discuss the two written methods on the slide.

- Give learners pairs of three-digit numbers to subtract where regrouping is required in the units of a hundred.

- Direct learners to Let's learn in the Student's Book and discuss the example shown.

- Learners complete the paired activity, then as a class discuss the Guided practice example.

Practise 🔲

- Workbook

Title: Subtract two 3-digit numbers (B)

Pages: 60–61

- Refer to the variation in Activity 2 from the Additional practice activities.

Apply 👥 🖵 [TWM.04]

- Display **Slide 3**. Learners discuss the best strategy/method for finding the difference between two numbers.

- Encourage learners to discuss the appropriateness of using each of the five strategies listed.

- Take feedback on which strategies/methods individual learners used and why they chose that strategy/method.

Review

- 🄿 Ask: **What have we been learning about in our lesson?**

- 🄿 Ask: **What must we always do when we calculate a subtraction?** Draw out the fact that when subtracting, learners must:
  - first estimate the answer
  - look carefully at the numbers and consider if it is suitable to use a mental strategy
  - use a formal written method if this is more efficient
  - check their answers.

Assessment for learning

- What is a sensible estimate for 456 subtract 392? Can you explain why?

- How would you find the difference between 456 and 392? Can you think of any other strategies/methods?

- Would you say the formal written method is more efficient for 456 – 392? Why/why not?

Same day intervention

Support

- Ensure learners use concrete resources such as Base 10 to make the minuends of the calculations to develop their understanding of the expanded and formal written methods.

- Move them on to the formal written method only when learners are confident with the expanded method.

Number – Integers and powers

Additional practice activities

Activity 1 ‍‍ ⚠2

Learning objective
- Estimate and solve subtraction of three-digit numbers and tens

Resources
Base 10 equipment: hundreds, tens and ones (per pair); pencil and plain or squared paper (per learner)

What to do [TWM.04]
- Write these digits on the board: 1, 2, 3, 4, 5.
- Learners use these digits to make a three-digit number subtract a two-digit number calculation, for example, 154 – 23 =. Tell them that the ones digit in the two-digit number must be less than ones digit in the three-digit number.
- They make their minuend using Base 10 equipment.
- They each estimate the answer and then use a mental calculation strategy of their choice to work out the difference. Learners' calculations should not involve regrouping at this stage.
- They explain their choice of calculation strategy to their partner.
- Learners then practise using the expanded written method and the formal written method to confirm the difference.

- They then make another three-digit number and a two-digit number by changing the digits from their first calculation (ensuring that the ones digit in the two-digit number is less than ones digit in the three-digit number), for example, 154 – 32 =. They repeat the process of estimating, using a mental calculation strategy and then recording both written methods.
- Learners continue to do this until they have made up to six possible combinations of digits.

Variation
3 Learners repeat the above activity but this time write 3, 4, 5, 6 and 7. They make numbers that require regrouping of 1 ten to 10 ones, for example, 674 – 35 =. They should ensure that the ones digit in the two-digit number is greater than the ones digit in the three-digit number.

Activity 2 ‍‍ ⚠2

Learning objective
- Estimate and solve subtraction of pairs of three-digit numbers

Resources
Base 10 equipment: hundreds, tens and ones (per pair); pencil and plain or squared paper (per learner)

What to do [TWM.04]
- Write these digits on the board: 1, 2, 3, 4, 5, 6.
- Learners use these digits to make pairs of three-digit numbers to subtract.
- Their numbers must involve regrouping of 1 unit of ten, for example 632 – 514 =.
- They use Base 10 equipment to make their minuends.
- They each estimate the answer and work out the difference, using a mental calculation strategy of their choice.
- They explain their choice to their partner.

- Learners then practise using the expanded written method and the formal written method to confirm their difference.
- They then make another pair of three-digit numbers by changing the digits from their first calculation, for example, 623 – 514 =. They repeat the process of estimating, using a mental calculation strategy and then recording both written methods.
- How many different combinations can they make that involve regrouping of ones?

Variation
3 Learners repeat the above activity but this time learners make numbers that require regrouping of 1 hundred to 10 tens, for example 623 – 541 =. They should ensure that they make calculations that require one regrouping only.

Number – Integers and powers

Unit 8: Multiplication and division

Collins International Primary Maths Recommended Teaching and Learning Sequence: Term 2, Week 3

Learning objectives

Code	Learning objective
3Ni.05	Understand and explain the relationship between multiplication and division.
3Ni.06	Understand and explain the commutative and distributive properties of multiplication, and use these to simplify calculations.

Unit overview

This unit focuses on the relationship between multiplication and division. It is important that learners understand that these two operations are inverse operations, for example $5 \times 3 = 15$ and $15 \div 5 = 3$. This will help them, for example, to derive division facts from their multiplication tables and check answers to calculations. There is also a focus on learners developing an understanding that multiplication is commutative. They will learn that numbers can be multiplied in any order and the product will always be the same. So, for example, $3 \times 5 = 5 \times 3$ and $2 \times 7 \times 5 = 2 \times 5 \times 7$. Learners will practise this, using the multiplication facts they know (initially, the 2, 5 and 10 times tables). They will learn that division is not commutative, so $15 \div 3$ is not equal to $3 \div 15$. Learners are introduced to the distributive law for multiplication, which corresponds well with the grid method for multiplication which learners will be introduced to in Unit 12: Multiplication. The distributive law links multiplication and addition; for example, to multiply 13 by 5, we can multiply 10 by 5 and 3 by 5 and add the two products together. In later years, brackets are used. At Stage 3, learners practise this informally, using numbers to 20 that are multiplied by 2, 3, 4 or 5.

Prerequisites for learning

Learners need to:
- use the symbols ×, ÷ and =.
- understand multiplication as an array and know that two calculations can be made from that array
- understand multiplication as adding equal groups
- understand division as subtracting equal groups
- know that number facts on either side of the equals symbol have the same value.

Vocabulary

multiply, multiplication, divide, division, array, groups, inverse, product, quotient, commutative, partition, distributive

Common difficulties and remediation

It is important to provide manipulatives, such as counters or cubes to enable learners to understand that multiplication is commutative. Learners could make arrays, for example, one that is 4 by 2. This would enable them to see that $4 \times 2 = 8$ and $2 \times 4 = 8$. Division is not commutative but it is important that we don't tell learners that $3 \div 15$ cannot be done (it can, giving a quotient of one fifth or 0.2).

Supporting language awareness

It is good practice to say key vocabulary out loud together as a class, for example:
- 2 multiplied by 5 equals 10 and 5 multiplied by 2 equals 10 because multiplication is commutative.

Promoting Thinking and Working Mathematically

During this unit learners will have opportunities to think and work mathematically. For example, there are many opportunities in this unit for learners to discuss what they notice about the images on the slides and also what is the same and what is different about what they see. This means that they will specialise (TWM.01), make generalisations (TWM.02), make conjectures (TWM.03), convince (TWM.04) characterise (TWM.05) and classify (WTM.06).

Success criteria

Learners can:
- demonstrate an understanding that multiplication is commutative
- demonstrate an understanding that division is not commutative
- demonstrate an understanding of the relationship between multiplication and division
- demonstrate an understanding of the distributive law for multiplication.

Unit **8** Multiplication and division

Lesson 1: **Multiplication and division**

Learning objective

Code	Learning objective
3Ni.05	Understand and explain the relationship between multiplication and division.

Resources

counters (per pair), mini whiteboard and pen (per pair)

Revise

Use the activity *Counting stick (1)* from Unit 8: *Multiplication and division* in the Revise activities to review multiplication and corresponding division facts for 1, 2, 5 and 10.

Teach [SB] 🖥 [TWM.01]

- 🄟 Display **Slide 1**. Ask: **What can you see?** Expect learners to tell you that there are two rows of five counters or five columns of two counters. Ask: **How many counters can you see altogether? How can you work that out?** Agree that they could count all the counters individually but that this is not efficient. Encourage them to consider that they could add five lots of 2 or two lots of 5. Elicit that they could use the multiplication facts 2 × 5 and 5 × 2. Ask learners, working in pairs, to make this array with counters and then draw a picture of it. They then record the multiplication calculations.

- **[T&T]** Ask: **There are 10 counters. How many rows or groups of 5 are there? Can we make up a division statement for this?** Agree that there are two rows or groups of 5. Ask learners to move groups of 5 physically away from their arrays, to practise division by subtraction. Write the corresponding division statement: 10 ÷ 5 = 2 on the board. Ask learners to make the same array again, then remove counters in twos. They write the division statement 10 ÷ 2 = 5.

- 🄟 **[TWM.01]** Display **Slide 2**. Ask: **What do you notice this time?** Draw out the fact that there are two groups of ten counters. Invite learners to explain the multiplication and division calculations and how they relate to the counters. Ask learners to make this array. They draw it and write the two multiplication and two division calculations: 2 × 10 = 20 and 10 × 2 = 20; 20 ÷ 2 = 10 and 20 ÷ 10 = 2.

- Explain that division is the inverse of multiplication, and that 'inverse' means 'opposite'. Introduce the terms 'product' and 'quotient' by saying: **The answer to a multiplication is the product. The answer to a division is the quotient.** Learners are not expected to use these terms.

- Ask learners to make their own arrays, draw pictures of them and to identify and record the multiplication and division facts.

- Direct learners to the Student's Book. Discuss the example shown, encouraging learners to use counters to make the arrays for themselves.

- Learners work through the paired activity.

- Discuss the Guided practice example in the Student's Book.

Practise 🔲

- Workbook

Title: Multiplication and division

Pages: 62–63

- Refer to Activity 1 from the Additional practice activities.

Apply 👥 🖥 [TWM.01/02]

- Display **Slide 3**.

- Learners are given two facts and they generate other related facts from each of them. Encourage them to discuss this with a partner.

- As a class, work through the task. Ask learners to share the other facts they find.

Review

- Write a multiplication or division fact on the board and ask learners to find other related facts. Discuss commutativity (although this concept will be explored in detail in Lesson 3) and inverse facts, and repeated addition and subtraction.

- 🄟 Ask: **Can you explain how multiplication and division are linked?** Encourage learners to use the words 'inverse' or 'opposite'.

Assessment for learning

- What does it mean when we say division is the opposite of multiplication?

- Can you give two other operations that are opposites or inverses?

- If you know 4 × 5 = 20, what division fact do you know?

- If you know 20 ÷ 2 = 10, what multiplication fact do you know?

Same day intervention

Support

- It is important that learners understand the inverse relationship between multiplication and division. Work with any that do not understand, using multiplication facts for 1, 2, 5 and 10. Ensure that they make arrays.

Number – Integers and powers

Lesson 2: **Checking multiplication and division**

Learning objective

Code	Learning objective
3Ni.05	Understand and explain the relationship between multiplication and division.

Resources

counters (per pair)

Revise

Use the activity *Counting stick (1)* from Unit 8: *Multiplication and division* in the Revise activities to rehearse multiplication and corresponding division facts for 1, 2, 5 and 10.

Teach 🖥️ 💻

- ▶ Display **Slide 1**. Ask: **What can you see?** Expect learners to tell you that there are five rows of ten squares. Ask: **How many squares can you see altogether? How can you work that out?** Agree that they could add ten lots of 5 or five lots of 10, or that they could use multiplication facts (5 × 10 = or 10 × 5 =). Ask learners, working in pairs, to make this array with counters by making a row of 10, five times. Ask them to record the two multiplication calculations.

- [T&T] Ask: **There are 50 squares. How many groups of 10 are there? How many groups of 5? Can we make up division statements for this?** Agree, five groups of 10 and ten groups of 5. Ask learners to move these groups away, physically, from their arrays. Invite individuals to write the division statements of 50 ÷ 10 = 5 and 50 ÷ 5 = 10 on the board.

- Display **Slide 2**. Focus on the first image and ask: **What can you see?** Expect learners to tell you that there are 6 rows of 5 squares, and 30 squares altogether.

- Discuss the second image. Ask: **How does this show the division calculation?** Agree that six groups of 5 are taken away from the 30 (30 ÷ 5 = 6). Emphasise the fact that they can use this knowledge to check that their multiplication is correct.

- Direct learners to Let's learn in the Student's Book. Discuss the example shown.

- Work through the paired activity and the Guided practice together. It is important learners understand that they can check one operation by using the inverse operation.

Practise 📘 [TWM.04]

- Workbook

Title: Checking multiplication and division

Pages: 64–65

- Refer to Activity 1 from the Additional practice activities.

Apply 👥 💻 [TWM.04]

- Display **Slide 3**. Learners decide whether or not Kwame is able to work out a division if he knows his multiplication facts for 5.

- Expect learners to work with a partner to discuss this.

- Agree that if Kwame knows that 5 × 10 = 50, which he does, he can work out the division.

Review

- Invite learners to demonstrate how they can use division to check a multiplication.

- Invite learners to demonstrate how they can use multiplication to check a division.

- Recap why this is and ask learners to explain what an inverse operation is.

Assessment for learning

- How could you check that 5 × 6 = 30? Why could you do that?

- How could you check that 50 ÷ 5 = 10? Why could you do that?

Same day intervention

Support

- Focus on one multiplication table, for example 5.

- Learners make arrays and check multiplication by using division and check division by using multiplication.

Number – Integers and powers

Unit **8** Multiplication and division

Lesson 3: **Commutativity**

Number – Integers and powers

Learning objective

Code	Learning objective
3Ni.06	Understand and explain the commutative [and distributive] properties of multiplication, and use these to simplify calculations.

Resources

counters (per pair); Resource sheet 5: Spots grid (per learner); scissors (per learner); 2 cm squared paper (per learner)

Revise

Use the activity *Counting stick (1)* from Unit 8: *Multiplication and division* in the Revise activities to rehearse multiplication and corresponding division facts for 1, 2, 5 and 10.

Teach 🔲 🖥 [TWM.05/06]

- 🔲 [TWM.05/06] Display **Slide 1**. Ask: **What is the same about these representations? What is different?** Agree that both are arrays and are made from 15 counters. Agree that one has three rows of five counters and the other has five rows of three counters. Establish that they show the repeated addition calculation statements of $5 + 5 + 5 = 15$ and $3 + 3 + 3 + 3 + 3 = 15$

- [T&T] Ask: **What do these two representations tell us?** Agree that one shows that 5 multiplied by 3 equals 15 and the other that 3 multiplied by 5 equals 15. Emphasise that it doesn't matter which way around numbers are multiplied, the answer (product) is always the same. Remind them this is the same with addition. Say: **Addition is commutative and multiplication is also commutative.** Reinforce by asking learners to say what this means in their own words.

- Ask learners, working in pairs, to make the first array, with counters, on a piece of paper. Now ask them to turn the paper to make the second. They then draw a picture of both and write the two multiplication facts.

- Focus on the repeated subtraction and division statements that can be made: $15 \div 5 = 3$ and $15 \div 3 = 5$. Discuss the fact that division is not commutative, so $15 \div 5$ is not equal to $5 \div 15$. Both have different answers (quotients). $15 \div 5 =$ has a quotient of 3; $5 \div 15 =$ has a quotient of one third.

- Give each learner a copy of Resource sheet 5: Spots grid and a pair of scissors. Ask them to cut out an array of three rows of 10. They place them to show the multiplication fact of $10 \times 3 = 30$, and then turn them to show the commutative fact of $3 \times 10 = 30$.

- They then draw the representation and write the facts. Ask: **What else does your array show?** Agree it shows the two division facts of $30 \div 10 = 3$ and $30 \div 3 = 10$. They also write these facts.

- Repeat this for six rows of 5 and seven rows of 5.

- Direct learners to Let's learn in the Student's Book. Work through the example shown.

- Work through the paired activity and the Guided practice together. Some of the examples lead into new multiplication tables. Encourage any learners who comment on this.

Practise 🔲 [TWM.04/05]

- Workbook

Title: Commutativity

Pages: 66–67

- Refer to Activity 2 from the Additional practice activities.

Apply 👥 🖥 [TWM.04/05]

- Display **Slide 2**. Learners spot the mistakes.

- Give time for them to talk to their partner about why these are mistakes and what needs to be done to correct them.

Review

- 🔲 As a class, explore question 7 from the Workbook. Ask: **Why is it only possible to make one multiplication fact from these arrays?**

- Agree that the two numbers being multiplied are the same.

- Invite learners to the board to draw other arrays that give only one fact. These will all be square arrays.

- Discuss what it means if we say multiplication is commutative.

Assessment for learning

- If we know $5 \times 4 = 20$, what other multiplication fact do we know?

- If we have an array that has three rows of 4, what two multiplication facts can we make?

- Multiplication is commutative. What does that mean?

- Why is there only one multiplication fact for $6 \times 6 = 36$?

Same day intervention
Support

- Prepare arrays for learners from Resource sheet 5: Spots grid. Start with arrays in the 2 times table.

Lesson 4: **Using place value to multiply**

Learning objective

Code	Learning objective
3Ni.06	Understand and explain the [commutative and] distributive properties of multiplication, and use these to simplify calculations.

Resources

Base 10 equipment: ones and tens (per learner)

Revise

Use the activity *Distribution* from Unit 8: *Multiplication and division* in the Revise activities.

Teach 🆂🅱 🖥 [TWM.03/04]

- 🄿 Display **Slide 1**. Ask: **What can you see?** Agree that the image shows two groups of 13. Ask: **How can we find out how many there are altogether?** Establish that they could add 13 to 13. Remind them that multiplication is related to addition and elicit that they could multiply 13 by 2.
- [T&T] [TWM.03/04] Ask: **How could we multiply 13 by 2?** Give learners a few minutes to discuss with their partner. Encourage them to think of different ways: for example, they could double, or they could work out 10 × 2 = 20 and count on another two threes. During feedback, encourage learners to show that they are correct.
- Ask learners to use Base 10 equipment to make 13 twice, as shown on the slide. Then to rearrange the equipment, putting the tens together and the ones together. Agree that they could multiply each group by 2 and then add the two products together. Explain that they have partitioned 13 into 10 and 3, multiplied and added.
- Repeat for 14, 15, 16 and 17, multiplying each by 2.
- Display **Slide 2**. Point to the first image of 15 and show how it has been partitioned into one 10 and 5 ones. Then point to the other groups of 15. Say: **There are five lots of 15 here altogether.** Ask: **How can we work out how many that makes altogether?** Elicit that they can use partitioning to work out the answer to 15 multiplied by 5. Show how the tens can be multiplied by 5 and the ones can be multiplied by 5. Say: **To find the product we add the two answers, 50 + 25 = 75.** Establish that partitioning numbers in this way can make it easier to multiply two-digit by one-digit numbers.
- Ask: **How can we multiply 26 by 5?** Let learners use Base 10 equipment to make 26. They multiply the 2 tens by 5 to get 100. They multiply the 6 ones by 5 to get 30. They add the two products, 100 and 30, together. As they do this, encourage them to explain to a partner what they are doing and show that they are correct. Repeat for 27, 28 and 29.

- Say: **When we partition numbers, to multiply them, we are using the distributive law for multiplication.**
- Direct learners to Let's learn in the Student's Book and discuss the example shown.
- Work through the paired activity and the Guided practice example together.

Practise 🆆🅱 [TWM.03]

- Workbook

Title: Using place value to multiply

Pages: 68–69

- Refer to Activity 2 from the Additional practice activities.

Apply 👥 🖥 [TWM.04/05/06]

- Display **Slide 3**. Learners are given three calculations. They discuss what is the same and what is different. They then discuss why the answer is the same each time.

Review

- Write 5 × 3 = 3 × 5 on the board and discuss commutativity and the fact that with multiplication, it doesn't matter in which order numbers are multiplied.
- Discuss how partitioning to use the distributive property of multiplication can simplify a calculation.
- Write a few calculations on the board. As a class, find the product each time by partitioning the numbers into tens and ones.

Assessment for learning

- What is 34 partitioned into tens and one?
- How can we work out 30 multiplied by 5?
- What is 14 × 5? How can you calculate this?

Same day intervention

Support

- Give learners numbers with one unit of ten and any number of ones to multiply by 2 and 3. Ensure they partition the numbers first.

Enrichment

Learners solve calculations involving numbers less than 20 multiplied by 3 or 4.

Number – Integers and powers

Additional practice activities

Activity 1 👤 ⚠2

Learning objective
• Understand the relationship between multiplication and division

Resources
counters (per learner)

What to do
• Ask learners to take 12 counters and to make as many different arrays as they can.
• Their arrays should show $1 \times 12 = 12$ / $12 \times 1 = 12$, $2 \times 6 = 12$ / $6 \times 2 = 12$ and $3 \times 4 = 12$ / $4 \times 3 = 12$.
• They record one multiplication calculation and the inverse division calculation for each array they make. For example, two of the calculations they should record are $3 \times 4 = 12$ and $12 \div 4 = 3$.

• Once they have done this, they make all the arrays they can for 20 counters.

Variation
1 Begin by making arrays with six counters, then eight, nine and ten.

Activity 2 👥 ⚠2

Learning objective
• Understand and explain the commutative properties of multiplication, and use these to simplify calculations

Resources
Resource sheet 5: Spots grid (per learner); scissors (per learner); glue (per learner)

What to do [TWM.04]
• Give learners a copy of Resource sheet 5: Spots grid and a pair of scissors.
• Ask them to cut out four different arrays.
• They stick these in their books and write the two commutative number calculations.

• They then compare their arrays with a partner and discuss what is the same about them and what is different.

Variation
1 Give learners different arrays to stick into their books.
They then write the two facts that the arrays show.

Number – Integers and powers

Unit 9: Times tables (A)

Collins International Primary Maths Recommended Teaching and Learning Sequence: Term 2, Week 4

Learning objectives

Code	Learning objective
3Nc.02	Count on and count back in steps of constant size: 1-digit numbers, [tens or hundreds,] starting [from any number] (from 0 [to 1000]).
3Nc.05	Recognise and extend linear sequences, and describe the term-to-term rule.
3Ni.07	Know [1,] 2, [3,] 4, 5, [6,] 8[, 9] and 10 times tables.
3Ni.10	Recognise multiples of 2, 5 and 10 (up to 1000).

Unit overview

In this unit, learners focus on developing the mental calculation strategies of using known multiplication and corresponding division facts and doubling. It is very important that learners learn their tables. Being able to recall them quickly enables learners to focus on new mathematical ideas and at later stages, be able to multiply combinations of one-, two-, three- and four-digit whole numbers as well as decimals. It also enables them to confidently use and apply mathematics to solve real-life problems.

Understanding of these multiplication facts develops from initially counting in steps of different sizes, which was a focus of Unit 1: Counting and sequences (A) and Unit 2: Counting and sequences (B).

Learners will develop an understanding and be able to recognise the relationship between multiples of 5 and 10 as well as the relationship between the 2, 4 and 8 times tables.

Learners use the language of doubling to make connections between times tables; for example, they learn the 4 times table by doubling the multiples in the 2 times table. They learn the 8 times table by doubling the 4 times table.

At the same time as learners work on the multiplication facts, they learn the corresponding division facts.

They will also learn about multiples, focusing on those for 2, 5 and 10. A multiple is the product of two numbers that are multiplied together.

Prerequisites for learning

Learners need to:
- know the 1, 2, 5 and 10 times tables
- know the corresponding division facts for the 1, 2, 5 and 10 times tables
- understand that multiplication is repeated addition
- understand that multiplication and division are inverse operations.

Vocabulary

multiply, product, divide, times table, equivalent, multiple, groups, equal, step count

Common difficulties and remediation

Some learners have difficulty recalling tables. They can often feel overwhelmed by the quantity of facts they need to learn. It is important, therefore, to stress the fact that multiplication can be done in any order – the commutative property of multiplication (which learners were introduced to in Unit 8: Multiplication and division). If they know one fact, they automatically know a corresponding second fact. So, in reality, they only need to learn half the tables facts.

Supporting language awareness

It is good practice to count in steps (from zero) of one particular size and then recite the times table for this number. It is also helpful to remind learners that they already know some of the tables facts for the new tables; for example, if they know the 2, 4, 5 and 10 times tables, they only need to learn these new facts when they learn the 8 times table: $8 \times 3 = 24$, $8 \times 6 = 48$, $8 \times 7 = 56$, $8 \times 8 = 64$ and $8 \times 9 = 72$.

Promoting Thinking and Working Mathematically

During this unit, learners will have opportunities to think and work mathematically; for example, there are many opportunities in this unit for learners to discuss what they notice about the images on the slides. They discuss what is the same and what is different about patterns made by the multiples explored. They also use what they know to generate other facts. This means that they will specialise (TWM.01), make generalisations (TWM.02), make conjectures (TWM.03), convince (TWM.04), characterise (TWM.05) and classify (TWM.06).

Success criteria

Learners can:
- recall multiplication facts for the 5 and 10 times tables and the corresponding division facts
- recall multiplication facts for the 2, 4 and 8 times tables and the corresponding division facts
- understand what a multiple is
- recognise multiples of 2, 5 and 10.

Number – Counting and sequences, Integers and powers

Lesson 1: **5 and 10 times tables**

Number – Counting and sequences, Integers and powers

Learning objectives

Code	Learning objective
3Nc.02	Count on and count back in steps of constant size: 1-digit numbers, [tens or hundreds,] starting [from any number] (from 0 [to 1000]).
3Ni.07	Know [1, 2, 3, 4,] 5[, 6, 8, 9] and 10 times tables

Resources

mini whiteboard and pen (per learner); interlocking cubes (per learner)

Revise

Use the activity *Counting stick (2)* from Unit 9: *Times tables (A)* in the Revise activities to review multiplication and corresponding division facts for 1, 2, 5 and 10.

Teach 🔲 💻 [TWM.01/02]

- 🔳 Display **Slide 1**.
- **[TWM.01/02]** Ask: **What do you notice? What else? Is there anything else?** Expect learners to tell you that there are two number lines. One goes up in jumps of 5 (shows multiples of 5) and the other goes up in jumps of 10 (shows multiples of 10).
- Point out that there are ten jumps on the top line and five jumps along the tens line, so each jump along the tens line is double a jump on the fives line. Relate this to the multiplication tables for 5 and 10, and establish that the answers to the 10 times table (multiples of 10) are double the answers to the 5 times table (multiples of 5).
- As a class, recite the 5 and 10 times tables. Write on the board different multiplications from these tables, for example 5 × 6 =, and learners write the product on their whiteboard. Then ask them to write the commutative multiplication fact and the associated division facts, for example, 6 × 5 = 30, 30 ÷ 5 = 6 and 30 ÷ 6 = 5.
- Next, write a 5 and 10 times table on the board with the same multiplier, for example 5 × 6, 10 × 6, and ask learners to write the products. Then ask them to write the commutative multiplication fact and the associated division facts, for example, 6 × 5 = 30, 30 ÷ 5 = 6 and 30 ÷ 6 = 5; 6 × 10 = 60, 60 ÷ 10 = 6 and 60 ÷ 6 = 10.
- Ask: **How can doubling and halving help when we need to multiply by 5 or 10?** Establish that to multiply by 5, we can multiply by 10 and halve the product. Call out some numbers for learners to multiply by 5 and 10, using this strategy, for example: 20 × 5 = 20 x 10 ÷ 2.
- Direct learners to Let's learn in the Student's Book. Discuss the example shown.
- Learners work through the paired activity.
- Discuss the Guided practice example in the Student's Book.

Practise 🔳

- Workbook

Title: 5 and 10 times tables

Pages: 70–71

- Refer to Activity 1 from the Additional practice activities.

Apply 👥 💻 [TWM.01]

- Display **Slide 2**. Learners list some 5 and 10 times table facts that have the same product.
- Encourage them to challenge themselves and work with different numbers from the ones in the lesson, for example 18 × 5 = 9 × 10.
- Although there are only four pairs that are within the 5 and 10 times tables up to the 10th multiple, there are an infinite number of other examples, so give learners a maximum number to find.
- Discuss the products that learners have found. Invite them to share their facts.

Review

- 🔳 Ask: **What is the relationship between 5 and 10?**
- Ask learners to suggest some two-digit numbers to multiply by 5 by multiplying by 10 and halving.
- To reinforce the inverse nature of multiplication and division, write some 5 and 10 times table facts on the board. Ask learners to state the corresponding divisions.

Assessment for learning

- If you know 5 × 7 = 35, what else do you know?
- Can you explain the relationship between 5 and 10? How does knowing this help us to multiply any number by 5?
- What is 18 multiplied by 5? Can you explain what you did to work it out?

Same day intervention
Support

- Learners make towers of interlocking cubes which they break to show physically that, for example, 5 × 2 is equal to 10 × 1 or 5 × 4 is equal to 10 × 2.

Lesson 2: **2, 4 and 8 times tables**

Learning objectives

Code	Learning objective
3Nc.02	Count on and count back in steps of constant size: 1-digit numbers, [tens or hundreds,] starting [from any number] (from 0 [to 1000]).
3Ni.07	Know [1,] 2, [3,] 4, [5, 6,] 8[, 9 and 10] times tables.

Revise

Use the activity *Counting stick (2)* from Unit 9: *Times tables (A)* in the Revise activities to review multiplication and corresponding division facts.

Teach 🔲 🖥 [TWM.01/02]

- ⚲ **[TWM.01/02]** Display **Slide 1**. Ask: **What do you notice? What else? Is there anything else?** Expect learners to tell you that there are three number lines. Elicit that the first one shows counting in twos, the second counting in fours and the third counting in eights. Link this to times tables facts, for example: $8 \times 2 = 16$, $3 \times 4 = 12$, $3 \times 8 = 24$.

- **[T&T]** Ask learners to work in pairs and draw number lines to match those on the slide. They then extend the number lines for 4 and 8 to the 10th multiples. Once pairs have done this, discuss the multiples for 4 and 8 that extend beyond those shown on the slide.

- **[T&T]** Referring to the numbers on **Slide 1** ask: **What can you tell me about the numbers 8, 16 and 24?** Elicit that they all appear in the 2, 4 and 8 times tables. Explore the fact that the 4 times table facts are double the 2 times table facts and the 8 times table facts are double the 4 times table facts, when they are multiplied by the same number.

- **[T&T]** Ask: **How can knowing this help us learn our 4 and 8 times tables?** After a few minutes, establish that they could double the numbers in the 2 times table for the 4 times table and double again for the 8 times table, or that for the 8 times table they could first multiply by 4 and then double.

- Direct learners to Let's learn in the Student's Book. Discuss the examples shown. Make sure to emphasise the link between the multiplication and related division facts for the 4 and 8 times tables.

- Learners work through the paired activity.

- Once learners are secure with the methods outlined in the Student's Book, say different numbers to 10 and ask learners to multiply by 4 by doubling the answers to the 2 times table. Also ask questions related to the division facts for the 4 times table.

- Repeat so that learners double and double again or multiply by 4 and double for multiplying by 8. Also ask questions related to the division facts for the 8 times table.

- Discuss the Guided practice example in the Student's Book.

Practise 📝

- Workbook

Title: 2, 4 and 8 times tables

Pages: 72–73

- Refer to the variation in Activity 1 from the Additional practice activities.

Apply 👥 🖥 [TWM.01]

- Display **Slide 2**. Learners list some 2, 4 and 8 times table facts that have the same product.

- Encourage them to challenge themselves and work with different numbers from those in the lesson, for example $2 \times 20 = 4 \times 10 = 8 \times 5$.

- Although there are only two sets that are within the 2, 4 and 8 times tables up to the 10th multiple, there are an infinite number of other examples, so give learners a maximum number to find.

- Discuss the products that learners have found. Invite them to share their facts.

Review

- ⚲ Ask: **What is the relationship between 2, 4 and 8?**

- Ask learners to think of some teen numbers to multiply by 4 using doubling.

- To reinforce the inverse relationship between multiplication and division, write some multiplication facts for the 2, 4 and 8 times table facts on the board and ask learners to tell you the corresponding division fact. Finish the lesson by playing Splat-a-fact on the Interactive tool. Select Level 2.

Assessment for learning

- If you know $2 \times 4 = 8$, what else do you know?

- Can you explain the relationship between 2, 4 and 8? How does knowing this help us to multiply any number by 12?

- What is 8 multiplied by 12? Can you explain what you did to work that out?

Same day intervention
Support

- Focus on the relationship between 2 and 4.

- Ensure you give learners cubes to build towers of 2 and 4 to aid their understanding of their relationship.

Number – Counting and sequences, Integers and powers

Lesson 3: **Multiples**

Learning objective

Code	Learning objective
3Ni.10	Recognise multiples of 2, 5 and 10 (up to 1000).

Revise

Use the activity *Bingo* from Unit 9: *Times tables (A)* in the Revise activities to review multiplication and corresponding division facts.

Teach 🔲 🖳 [TWM.01/02]

- ⏳ [TWM.01/02] Display **Slide 1**. Ask: **What do you notice? What else do you notice?** Encourage learners to talk about the number patterns.

- Say: **The numbers in each column are multiples of the number at the top of the column.** Establish that a multiple is the product of two numbers. Tell learners that all the numbers in the '2' column are multiples of 2, those in the '5' column are multiples of 5 and those in the '10' column are multiples of 10. Ask learners to say, for example, '25 is a multiple of 5', so that they can practise saying the word 'multiple'.

- [T&T] Ask: **How many numbers can you see that appear more than once?** Give learners a few minutes to discuss with a partner. During feedback, ask learners to tell you what the numbers they found are multiples of; for example, 10 is a multiple of 2, 5 and 10.

- Ask: **What do you notice about all the multiples of 2 and 10?** Agree that they are even. Ask: **What do you notice about all the multiples of 5?** Agree that they alternate between odd and even.

- Draw out the fact that multiples of 2 include all even numbers, multiples of 5 have 5 or 0 in the ones position and multiples of 10 all end with 0. Ask learners to use this knowledge to tell you whether numbers such as 455, 660 and 888 are multiples of 2, 5 and/or 10.

- Ask learners to write down multiples of 2 that are between 100 and 1000. Repeat for multiples of 5 and then multiples of 10.

- Direct learners to Let's learn in the Student's Book.

- Learners work through the paired activity.

- Discuss the Guided practice example in the Student's Book.

Practise 🔲 [TWM.01/04]

- Workbook

Title: Multiples

Pages: 74–75

- Refer to Activity 2 from the Additional practice activities.

Apply 👥 🖳 [TWM.01/02]

- Display **Slide 2**. Learners find the numbers that are multiples of 2, 5 and 10.

- They then say what they notice about these multiples.

- Discuss the fact that all the numbers that are multiples of 2, 5 **and** 10 end with zero. Ask: **Can you tell me any more numbers that must be multiples of 2, 5 and 10?**

Review

- ⏳ Ask: **What are the first five multiples of 2? What about 5? What about 10?**

- Ask learners to say a number and to tell you what it is a multiple of. Encourage learners to say numbers up to 1000.

Assessment for learning

- Can you tell me a multiple of 2?
- Can you tell me a multiple of 5?
- Can you tell me a multiple of 10?
- Can you give me a multiple of 2, 5 and 10?
- Is it possible to have a multiple of 2 and 5 that is not a multiple of 10? Why?
- Can you explain the meaning of multiple?

Same day intervention

Support

- Focus on recognising multiples up to the tenth multiple.

Enrichment

Focus on learners recognising multiples beyond the tenth multiples, i.e. two- and three-digit multiples of 2, 5 and 10.

Lesson 4: **Counting in steps**

Learning objectives

Code	Learning objective
3Nc.02	Count on and count back in steps of constant size: 1-digit numbers, [tens or hundreds,] starting [from any number] (from 0 [to 1000]).
3Nc.05	Recognise and extend linear sequences, and describe the term-to-term rule.

Resources

cubes (per pair)

Revise

Use the activity *Cross it out* from Unit 9: *Times tables (A)* in the Revise activities to review multiplication and corresponding division facts for 1, 2, 5 and 10.

Teach SB 🖵

- 🖰 Display **Slide 1**. Ask: **What do you notice? What else?** Expect learners to tell you that there are three sequences that show step counts. One going up in twos, one going up in fours and the third one going up in fives. As a class count in twos to 20 and back, then in fours to 40 and back and finally in fives to 50 and back.

- Encourage them to tell you what they notice about the numbers of cubes. Ask: **What is the same and what is different about them?** Draw out the idea that the numbers of cubes that increase in steps of two and four are all even. The cubes increasing in fives increase in steps alternating in odd and even numbers. Ask: **Why do you think this is?** Agree that 2 and 4 are even numbers so all steps are even; 5 is odd, so the steps alternate.

- **[T&T]** Ask: **How can we describe the columns of cubes?** Take feedback. Establish that they show the first few facts in the 2, 4 and 5 times tables. Point to different columns and ask learners to tell you the fact.

- Display **Slide 2**. Ask: **Can you match each step count with the correct multiplication statement?** Invite individuals to draw a line to match a multiplication statement with its step count.

- Ask pairs of learners to choose their own numbers to step count and make their own towers of cubes to show their choice. When they have built their towers, they write the step count and the multiplication fact.

- Direct learners to Let's learn in the Student's Book.

- Learners work through the paired activity.

- Discuss the Guided practice example in the Student's Book.

Practise WB [TWM.04/05]

- Workbook

Title: Counting in steps

Pages: 76–77

- Refer to the variation in Activity 2 from the Additional practice activities.

Apply 👥 🖵 [TWM.05/06]

- Display **Slide 3**.

- Learners are given numbers to sort into two groups. They choose their own criteria.

- You could encourage them to write an explanation for their decision, for example: This group are in the 4× table, this group are in the 5× table; this group are even numbers, this group are odd numbers.

- Encourage them to discuss how to sort with a partner.

- They should think of at least two ways to sort the numbers.

Review

- As a class, look at the Challenge 3 task from the Workbook (question 7). Ask learners who completed this correctly to explain how they did it.

- 🖰 Ask: **If I didn't know the answer to 5 × 8, how could I use step counting to help me?** Establish that if 5 × 5 is known, then counting on another three steps will give the product of 5 × 8.

- Recap the idea that counting in steps links to times tables facts.

Assessment for learning

- If we are counting in twos from zero, what comes after 16?

- If we are counting in tens from zero, what comes after 60?

- If we are counting in fours from zero, what comes after 24?

- What times table fact would we have if we counted in eights to 40?

Same day intervention

Support

- Give learners number lines to help them count in steps. When they write the tables fact, make sure they know to write the number of jumps first and then the step number.

Number – Counting and sequences, Integers and powers

Additional practice activities

Activity 1 👥 ⚠2

> **Learning objective**
> • Know 2, 4, 5, 8 and 10 times tables

Resources
Resource sheet 6: Numbers 1 to 10 (one set per learner)

What to do
• Give each learner a set of 1–10 cards from Resource sheet 6.
• One learner takes the 5 and 10 cards out of their pack and puts them face up on the table.

• The other learner puts their complete set face down in a pile on the table.
• Learners take turns to pick a card from their pile and multiply the number first by 5 and then by 10.
• They write in their books each times table fact they make. Ask them to also write the corresponding division.

⚠2 Variation
Learners repeat the activity, this time using the 2, 4 and 8 cards instead of the 5 and 10 cards.

Activity 2 👥 ⚠2

> **Learning objective**
> • Recognise multiples of 2, 5 and 10 (up to 1000)

Resources
Resource sheet 3: Digit cards (per pair); Resource sheet 6: Numbers 1 to 10 (per pair) (for variation)

What to do [TWM.01]
• Give pairs of learners a set of digit cards. They use these to create two- and three-digit numbers that are multiples of 2, 5 and 10.
• They read each number they make to their partner and tell them if it is a multiple of 2, 5 or 10. For example, they make 352, read it to their partner and say that it is a multiple of 2. Encourage them to explain how they know.

• Once they have done this, they repeat the task but this time make numbers that are multiples of 2, 5 and 10. For example, they make 750, read it to their partner and say that it is a multiple of 2, 5 and 10. Encourage them to explain how they know.

Variation
3 Ask learners to use the number cards 1, 2, 4, 5, 8 and 10. They pick one and count in that number of steps from zero. Their partner writes the steps on paper and then together they highlight the numbers that are multiples of 2, 5 and 10.

Number – Counting and sequences, integers and powers

Unit 10: Times tables (B)

Collins International Primary Maths
Recommended Teaching and
Learning Sequence: Term 2, Week 5

Learning objectives

Code	Learning objective
3Nc.02	Count on and count back in steps of constant size: 1-digit numbers, [tens or hundreds,] starting [from any number] (from 0 [to 1000]).
3Ni.07	Know 1, 2, 3, 4, 5, 6, 8, 9, and 10 times tables.

Unit overview

In this unit, learners continue to focus on developing the mental calculation strategies of using known multiplication and corresponding division facts, and doubling. It is very important that learners learn their tables. Being able to recall them quickly enables learners to focus on new mathematical ideas and at later stages, be able to multiply combinations of one-, two-, three- and four-digit whole numbers as well as decimals. It also enables them to confidently use and apply mathematics to solve real-life problems.

Understanding of these multiplication facts develops from initially counting in steps of different sizes, which was a focus of Unit 1: Counting and sequences (A) and Unit 2: Counting and sequences (B).

Learners will develop an understanding and be able to recognise the relationship between multiples of 3, 6 and 9. They will also look at the patterns of the products of these numbers.

Learners use the language of doubling to make connections between times tables; for example, they learn the 6 times table by doubling the multiples in the 3 times table. They make the link that to multiply by 9, they can triple the numbers in the 3 times table.

In addition, as they work on these facts, they learn the corresponding division facts. At the end of this unit there is an opportunity to focus on all the times table facts that learners need to know by the end of Stage 3.

Prerequisites for learning

Learners need to:
- know the 1, 2, 4, 5, 8 and 10 times tables
- know the corresponding division facts for the 1, 2, 4, 5, 8 and 10 times tables
- understand that multiplication is repeated addition
- understand that multiplication and division are inverse operations.

Vocabulary

multiply, product, multiple, times table, divide, quotient, equivalent, commutative, inverse

Common difficulties and remediation

Some learners have difficulty recalling tables. They can often feel overwhelmed by the number of facts they need to learn. It is important, therefore, to stress the fact that multiplication can be done in any order – the commutative property of multiplication (which learners were introduced to in Unit 8: Multiplication and division). If they know one fact, they automatically know a corresponding second fact. So, in reality, they only need to learn half the tables facts.

Supporting language awareness

It is good practice to count in steps (from zero) of one particular size and then recite the times table for this number. It is also helpful to remind learners that they already know some of the tables facts for the new tables; for example, if they know the 2, 3, 4, 5, 8 and 10 times tables, they only need to learn these new facts when they learn the 6 times table: $6 \times 6 = 36$, $6 \times 7 = 42$ and $6 \times 9 = 54$.

Promoting Thinking and Working Mathematically

During this unit, learners will have opportunities to think and work mathematically; for example, there are many opportunities in this unit for learners to discuss what they notice about the images on the slides. They discuss what is the same and what is different about patterns made by the multiples explored. They also use what they know to generate other facts. This means that they will specialise (TWM.01), make generalisations (TWM.02), make conjectures (TWM.03), convince (TWM.04) characterise (TWM.05) and classify (TWM.06).

Success criteria

Learners can:
- recall multiplication facts for the 3, 6 and 9 times tables and the corresponding division facts
- recall multiplication facts for the 1, 2, 4, 5, 8 and 10 times tables and corresponding division facts
- use doubling, halving and tripling.

Number – Counting and sequences, Integers and powers

Unit 10 Times tables (B)

Lesson 1: **3 and 6 times tables**

Number – Counting and sequences, Integers and powers

Learning objectives

Code	Learning objective
3Nc.02	Count on and count back in steps of constant size: 1-digit numbers, [tens or hundreds,] starting [from any number] (from 0 [to 1000]).
3Ni.07	Know [1, 2,] 3, [4, 5,] 6, [8, 9 and 10] times tables.

Resources

mini whiteboard and pen (per learner)

Revise

Use the activity *Counting in steps of 1, 5 and 10* from Unit 10: *Times tables (B)* in the Revise activities to review and consolidate the multiplication facts from Unit 9.

Teach 🔲 💻 [TWM.01/02]

- 🔲 [TWM.01/02] Display **Slide 1**. Ask: **What do you notice? What else do you notice?** Expect learners to tell you that the first number line shows the sequence formed by counting in steps of 3 and the second number line shows the sequence formed by counting in steps of 6. Ask learners to count in steps of 3 to 30 and link this with the multiplication facts for the 3 times table.

- Give learners the first part of a multiplication fact and ask them to write the product on their whiteboards. Repeat until you have asked all the facts up to 3 × 10 = 30. As they do this, expect them to also write the commutative and corresponding division facts. Mix up the facts so you don't ask them in order.

- [T&T] [TWM.01/02] Ask: **What do you notice about the number line that shows counting in steps of 6?** Give learners a few minutes to discuss this. Take feedback. Establish that the steps of 6 are double the steps of 3. Together, count in steps of 6 to 60 and link with the multiplication facts for the 6 times table.

- Give learners the first part of a multiplication fact and ask them to write the product on their whiteboards. Repeat until you have asked all the facts up to 6 × 10 = 60. Again, expect them to also write the commutative and corresponding division calculations, and mix up the facts so you don't ask them in order.

- Ask: **How can knowing the 3 times table help us to work out the answers for the 6 times table?** Agree that they can double a fact for 3 to give the fact for multiplying 6 by the same number. For example, if they know that 3 × 4 = 12, they double 12 to give 6 × 4 = 24.

- Call out some multiplication facts for 3 and ask learners to double them to find the corresponding multiplication fact for 6. For example, 3 × 2 = 6, double to get 6 × 2 = 12. They refer again to the number lines if they need to.

- [T&T] [TWM.01/02] Display **Slide 2**. Ask: **What relationship can you see between the 2 times**

table and the 6 times table? Give learners the opportunity to discuss this with their partner. After a few minutes, take feedback. Agree that the number lines show the sequences for counting in twos and counting in sixes. Establish that the 6 times table facts are three times the corresponding 2 times table facts. Establish that adding 'three lots' of a number is the same as tripling. Work through some examples, such as: **If we know that 2 × 4 = 8, we can triple 8 to get 6 × 4 = 24.**

- Direct learners to Let's learn in the Student's Book. Discuss the examples shown.

- Learners work through the paired activity.

- Discuss the Guided practice example in the Student's Book.

Practise 🔲 [TWM.04]

- Workbook

Title: 3 and 6 times tables

Pages: 78–79

- Refer to Activity 1 from the Additional practice activities.

Apply 👥 💻 [TWM.05/06]

- Display **Slide 3**. Learners identify the number that is the odd one out, and explain why (all the numbers are in the 3 and 6 times tables, except for 15).

Review

- 🔲 Ask: **What is the relationship between 3 and 6?**

- Ask learners to suggest some numbers to multiply by 3 and then double to multiply by 6.

- To reinforce inverse, write some 3 and 6 times table facts on the board. Learners tell you the corresponding division.

Assessment for learning

- If you know 3 × 4 = 12, what else do you know?
- What is the relationship between 3 and 6?

Same day intervention

Support

- Give learners cubes to build towers of 3 and 6 to aid their understanding of their relationship.

Lesson 2: **9 times table**

Learning objective

Code	Learning objective
3Ni.07	Know [1, 2, 3, 4, 5, 6, 8,] 9 [and 10] times table[s]

Resources

mini whiteboard and pen (per learner)

Revise

Use the activity *Counting in steps of 2, 4 and 8* from Unit 10: *Times tables (B)* in the Revise activities to review and consolidate the multiplication facts from Unit 9. Focus on steps of 8.

Teach 🆂🅱 🖵 [TWM.01/02]

- **[T&T]** 🔁 **[TWM.01/02]** Display **Slide 1**. Ask: **What can you see? What else? Can you see any patterns in the numbers?** Expect learners to tell you that the number line shows the sequence formed by counting in steps of 9. Give learners the opportunity to explore any patterns they can see in the numbers on the number line. After a few minutes, take feedback. Establish that the tens number increases by 1 for each step and the ones number decreases by 1 for each step. Say: **Try adding the digits in each number. What can you tell me?** Elicit that the digit sum for each multiple is 9. Say: **Look at the digits in each number.** Do they notice that, starting at both ends and comparing the numbers, the digits are reversed, for example: 18 and 81, 27 and 72.
- Ask learners to count in steps of 9 to 90 and link this with the multiplication facts for the 9 times table.
- Give learners the first part of a multiplication fact and ask them to write the product on their mini whiteboards. Repeat this until you have asked all the facts up to 9 × 10 = 90. As they do this, expect them also to write the commutative and corresponding division calculations. Mix the facts up so you don't ask them in order.
- **[T&T] [TWM.01/02]** Display **Slide 2**. Ask: **What do these number lines show you?** Elicit that they are the sequences for counting in threes and counting in nines, and that they correspond to the 3 times table and the 9 times table. Ask: **What relationship can you see between the 3 times table and the 9 times table?** Let learners discuss this with a partner. After a few minutes, take feedback. Agree that the 9 times table facts are three times the corresponding 3 times table facts. Tell learners that adding 'three lots' of a number is the same as tripling, or multiplying by 3. Work through some examples. Say: **If we know that 3 × 4 = 12, we can triple 12 to get 9 × 4 = 36.** Let them check this by using repeated addition if they need to.

- Direct learners to Let's learn in the Student's Book. Discuss the example shown.
- Learners work through the paired activity.
- Discuss the Guided practice example in the Student's Book.

Practise 🆆🅱 [TWM.04]

- Workbook

Title: 9 times table

Pages: 80–81

- Refer to the variation in Activity 1 from the Additional practice activities.

Apply 👥 🖵 [TWM.03/04]

- Display **Slide 3**. Learners need to decide whether they agree or disagree with two statements.
- Encourage them to explain why they agree or disagree, giving a written explanation that includes examples.

Review

- 🔁 Ask: **What is 9 multiplied by 6?**
- To reinforce commutativity and the relationship between multiplication and division, also ask learners to tell you what else they know, if they know 9 × 6 = 54.
- Repeat for other multiplication facts for the 9 times table.

Assessment for learning

- If you know 9 × 3 = 27, what two division facts do you know?
- If you know 36 ÷ 9 = 4 what two multiplication facts do you know?
- How could you multiply a number by 9? How else?

Same day intervention

Support

- Ensure learners have the 3 times tables facts available to refer to and that they are secure with multiplying by 3 (tripling). Also ensure that they look at the commutative fact for each one.

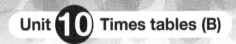

Unit **10** Times tables (B)

Lesson 3: **Multiplication and division facts (1)**

Learning objective

Code	Learning objective
3Ni.07	Know 1, 2, 3, 4, 5, 6, 8, 9 and 10 times tables.

Resources

Resource sheet 7: Times tables grid (per learner); red, blue and green coloured pencils (per learner); mini whiteboard and pen (per learner)

Revise

Use the activity *Counting in steps of 2, 4 and 8* from Unit 10: *Times tables (B)* in the Revise activities, applying this to any facts you wish to review from Lessons 1 and 2.

Teach 📖 🖥️ [TWM.01/02]

- 📝 [TWM.01/02] Display **Slide 1** and direct learners to the multiplication grid in the Student's Book. Ask: **What can you see?** Expect learners to tell you that this is a multiplication grid for all the facts up to and including 10 × 10 = 100. Ask: **Why are there three columns of red numbers, three of green and two of blue?** Establish that these reinforce the links between the different times tables facts; for example, 2s, 4s and 8s are red. Agree that the facts for 1 are in a different colour because these occur in every table. Inform learners that the facts for 7 are in a different colour because they are not related to any of the other numbers. Explain that they will learn the multiplication facts for 7 later.

- [T&T] Ask: **How can this table help us to find a multiplication or division fact? Can you give some examples?** Agree that if they pick a number from the white row and a number from the white column, then find where the row and column cross, they will find the multiplication fact for the numbers they chose. They can then use their knowledge of commutativity, and division being the inverse of multiplication, to find a second multiplication statement and then the two corresponding division facts.

- Give each learner a copy of Resource sheet 7: Times tables grid and ask them to shade the columns for 2, 4 and 8 red, 5 and 10 blue and 3, 6 and 9 green. Ask them to choose some facts to write on paper or their whiteboard. Ensure they write the commutative multiplication fact and the two inverse division facts as well.

- [TWM.01/02] Display **Slide 2** and direct learners to the trios in the Student's Book. Ask: **What do you think these images show?** Explain that these are called trios and they show multiplication and division facts, for example: 6 × 8 = 48, 8 × 6 = 48, 48 ÷ 8 = 6 and 48 ÷ 6 = 8.

- Learners work through the paired activity.

- Discuss the Guided practice example in the Student's Book.

Practise 📒

- Workbook

Title: Multiplication and division facts (1)

Pages: 82–83

- Refer to Activity 2 from the Additional practice activities.

Apply 👥 🖥️ [TWM.01]

- Display **Slide 3**. Learners find possible missing numbers to make the multiplication facts correct.

- There are various options for these. Encourage learners to find as many possibilities as they can within a specific time limit.

- Once they have, ask them to make up some examples of their own.

- Discuss the activity. Invite learners to share the facts they made up.

Review

- Referring to lessons 1 and 2, remind learners how using a number line, as shown in the Student's Book, can help them to work out different multiplication facts. Invite learners to demonstrate, for example, 6 × 7 =, by multiplying 3 by 7 and then multiplying the product by 2. Ask them to tell you why this works. Establish that multiplying by 6 is the same as multiplying by 3 and then 2 as 3 × 2 = 6.

- To reinforce inverse, write some times table multiplication facts on the board and ask learners to tell you the corresponding division facts.

Assessment for learning

- How can multiplying by 3 help us to multiply by 6?
- If you know that 8 × 5 = 40, what else do you know?

Same day intervention

Support

- Focus on the multiplication facts for 3, 6 and 9.

Lesson 4: **Multiplication and division facts (2)**

Learning objective

Code	Learning objective
3Ni.07	Know 1, 2, 3, 4, 5, 6, 8, 9 and 10 times tables.

Resources

interlocking cubes (per pair); mini whiteboard and pen (per pair)

Revise

Use the activity *Counting in steps of 3, 6 and 9* from Unit 10: *Times tables (B)* in the Revise activities to review and reinforce these multiplication and division facts. You may wish to focus on the 9 times table.

Teach 〔SB〕 ⬜ [TWM.01/02]

- [T&T] [TWM.01/02] Display **Slide 1**. Ask: **What do you notice? What else do you notice? Is there anything else you notice?** Establish that each tower represents the first multiple in the times tables that they need to learn. Ask learners to tell you the facts represented on the slide, for example, 1 × 1 = 1, 2 × 1 = 2.
- As they do this, encourage them to give you the commutative facts where possible, for example, 1 × 2 = 2 and the two division facts, for example, 2 ÷ 1 = 2 and 2 ÷ 2 = 1.
- [T&T] Working in pairs, learners choose a times table to focus on, and using the interlocking cubes, make towers of cubes to show the multiples of that number. They then write down the facts, including the commutative multiplication and corresponding division facts. Ask them to draw cubes to show their facts. If there are enough cubes available, they could build the full set up to the tenth multiple.
- Take feedback from the activity, inviting pairs to demonstrate what they have done. For example, if they have chosen facts for the 6 times table, they show their drawings or towers. They then show one fact for the class to give the four facts. So, if they show 5 towers of 6, the class would say, 6 × 5 = 30, 5 × 6 = 30, 30 ÷ 6 = 5 and 30 ÷ 5 = 6.
- Direct learners to Let's learn in the Student's Book.
- Learners work through the paired activity.
- Discuss the Guided practice example in the Student's Book.

Practise 〔WB〕 [TWM.01/04/05]

- Workbook

Title: Multiplication and division facts (2)

Pages: 84–85

- Refer to Activity 2 from the Additional practice activities.

Apply 👥 ⬜

- Display **Slide 2**. Learners work out the missing numbers for the trio. The product is 24.
- There are several correct answers and learners are asked to find all the possibilities.
- They write the two multiplication and two corresponding division statements for each possibility.

Review

- ▣ Write three numbers on the board, for example: 72, 9, 8. Ask: **What two multiplication statements can you make? What two divisions?**
- Repeat with different numbers.
- ▣ Ask: **What is 6 multiplied by 4?** To reinforce commutativity, and the relationship between multiplication and division also ask learners to tell you what else they know, if they know 6 × 4 = 24.
- Repeat for other multiplication facts for the times table up to 10 × 10. Finish the lesson by playing Splat-a-fact on the Interactive tool. Select Level 2.

Assessment for learning

- If you know that 36 divided by 4 is 9, what else do you know? What else?
- What can you tell me about multiples of 4? What else?
- What can you tell me about multiples of 6? What else?

Same day intervention
Support

- Focus on the multiplication and division facts that learners have difficulty recalling. Ensure to emphasise the link between related facts, for example, focus on fives and tens, or twos and fours.

Number – Counting and sequences, Integers and powers

Additional practice activities

Activity 1

Learning objective
- Know the 3, 6 and 9 times tables

Resources
Resource sheet 6: Numbers 1 to 10 (two sets per pair)

What to do
- Give pairs of learners two sets of number cards.
- They shuffle them together and place them in a pile, face down on the table.
- They take turns to pick a card and multiply the number shown by 3 and 6.

- They write in their books each times table fact that they make. Ask learners to also write the corresponding division.
- Encourage them to multiply by 6 by doubling the fact for 3.

Variation
2 Focus on the facts for the 9 times table and their corresponding division facts.

Activity 2

Learning objective
- Recall multiplication and division facts for the 1, 2, 3, 4, 5, 6, 8, 9 and 10 times tables

Resources
Resource sheet 3: Digit cards (per learner)

What to do
- Give each learner a set of digit cards. They remove the zero, shuffle them and place them in a pile face down on the table.
- With their partner they each take a card from their pile and place it face up on the table.
- They then make two multiplication and two division facts, which they write on paper.
- For example, if they pick 5 and 9, the multiplication facts would be $5 \times 9 = 45$ and $9 \times 5 = 45$. The two division facts would be $45 \div 5 = 9$ and $45 \div 9 = 5$.

- They keep doing this until they have used up all their cards.
- They repeat this but for other facts. If they pick two that have been used together before, they swap one for another card so that they can make two new multiplication and corresponding division facts.

Variation
1 Ask learners to use one pack of cards. They pick one card at a time and multiply the number by 2, then by 4 and, if they can, by 8, by doubling.

Number – Counting and sequences, Integers and powers

Unit 11: Multiplication

Collins International Primary Maths
Recommended Teaching and
Learning Sequence: Term 2, Week 6

Learning objectives

Code	Learning objective
3Ni.06	Understand and explain the [commutative and] distributive properties of multiplication, and use these to simplify calculations.
3Ni.08	Estimate and multiply whole numbers up to 100 by 2, 3, 4 and 5.

Unit overview

This unit focuses on multiplying whole numbers up to 100 by 2, 3, 4 and 5. Learners should be encouraged to estimate the answer first, before they work out the product. They use this estimate to judge whether the product of their multiplication is likely to be correct. There is also an expectation for learners to consider using a mental strategy for each calculation.

The mental strategies learners should consider include:

- counting on in steps of constant size
- using known multiplication facts and related facts involving multiples of 10
- doubling
- knowledge of place value and partitioning
- using the distributive law.

Once learners have found the product of their multiplication, they must check their answer by multiplying the two numbers the other way around, or by dividing the product by the multiplier to see if they arrive at the multiplicand. Learners check their answers by using division. They can use calculators to do this. Checking with division can be carried out using calculators, because the focus of this unit is multiplication.

During this unit they will explore:

- repeated addition on number lines
- using arrays with place value counters or similar manipulatives
- the grid method
- partitioning.

Prerequisites for learning

Learners need to:

- know the 1, 2, 3, 4, 5, 6, 8, 9 and 10 times tables
- know the corresponding division facts for the 1, 2, 3, 4, 5, 6, 8, 9 and 10 times tables
- understand that multiplication is repeated addition
- understand that multiplication and division are inverse operations
- be able to use the basic functions of a calculator in order to check answers.

Vocabulary

multiplication, multiplicand, multiplied by, multiplier, product

Common difficulties and remediation

When learning to multiply two-digit numbers by one-digit numbers, some learners may have difficulty in recalling tables facts quickly. Therefore, provide a list of the facts so they can focus on the process of multiplying; Resource sheet 7: Times tables grid is a useful resource for this purpose. It is also very important to make the link that multiplication is repeated addition. This is not the most efficient way to multiply but, for any learners having problems with multiplying two-digit numbers by 2, 3, 4 and 5, it is the method to focus on until they are ready to move on.

Supporting language awareness

It is good practice to model the accurate vocabulary for multiplication and have it displayed in the classroom. 'Multiplicand' is the name of the number we start with, 'multiplier' is the number we multiply by and 'product' is the answer. When working on written methods, we should use the term 'multiplied by' because it shows the operation.

Promoting Thinking and Working Mathematically

During this unit, learners will have opportunities to work on thinking and working mathematically; for example, when noticing differences and similarities between images on the slides, they will be making conjectures (TWM.03), convincing (TWM.04), characterising (TWM.05) and classifying (TWM.06). Particularly, they will have opportunities to critique (TWM.07) and improve (TWM.08) when they estimate, and decide on the most appropriate strategy to use when they check. They will also have the opportunity to specialise (TWM.01) and generalise (TWM.02) when they use what they know to generate other facts.

Success criteria

Learners can:

- multiply two-digit numbers by 2, 3, 4 and 5
- decide on appropriate strategies to multiply
- estimate and check answers.

Number – Integers and powers

Lesson 1: **Multiplying by repeated addition**

Number – Integers and powers

Learning objective

Code	Learning objective
3Ni.08	Estimate and multiply whole numbers up to 100 by 2, 3, 4 and 5.

Resources

mini whiteboard and pen (per learner); calculator for checking purposes only (per learner)

Revise

Use the activity *Multiplying by 2, 3, 4 and 5 (A)* from Unit 11: *Multiplication* in the Revise activities to review the use of a number line.

Teach 🔲 💻 [TWM.01]

- 📖 [TWM.01] Display **Slide 1**. Ask: **What do you notice on the slide? What is a good estimate of 24 × 2? Why?** Agree that it will be between 40 and 50. Introduce the new vocabulary: 'multiplicand', which in this calculation is 24; this is multiplied by the 'multiplier', which is 2; which equals the 'product'. Ask learners to repeat after you: **Multiplicand multiplied by multiplier equals product.**

- [T&T] Ask: **What mental calculation strategy could we use to work out 24 × 2?** Give learners a few minutes to discuss this. Take feedback. Referring to the number line, agree that they could use repeated addition and add 24 + 24, or double 24. Establish that when multiplying by 2, a good strategy is to double. Establish that 24 + 24 =, or double 24, is 48.

- Discuss the idea of checking: they could subtract 24 from 48, to see if they get 24, or they could halve 48.

- [TWM.01] Display **Slide 2**. Ask: **How is this the same as the first slide? How is it different?** Agree that the number being multiplied (the multiplicand) is the same but the multiplier is different. Refer to the number line and elicit that this has three jumps instead of two. It shows repeated addition. Discuss what would be a good estimate for the product (between 60 and 90).

- [T&T] Ask: **How could we use a mental calculation strategy to find the product?** Agree that they could double 24 and then add another 24. Learners can make jottings to keep track of their thinking.

- Repeat several times with different numbers for multiplicands; for example, 32 × 3 = by doubling 32 and adding one more 32. Learners make jottings to help keep track of their thinking. Encourage them to check their multiplications by dividing the product by the multiplier. They can use a calculator for this.

- Display **Slide 3**. Repeat the previous procedure to work out 24 × 4 =. Encourage the mental calculation strategy of doubling and doubling again. Learners should check by halving and halving again or using division on a calculator. Doing both would be a good idea.

- [T&T] Display **Slide 4**. Repeat the previous procedure to work out 24 × 5 =. Ask: **What mental calculation strategy can we use to multiply a number by 5?** After a few minutes, establish that as 5 is half of 10, they could halve the multiplicand and multiply by 10 or multiply by 10 and halve the product. Practise this with several examples. Ensure to discuss checking using the inverse, dividing by 10 and doubling or doing this on a calculator.

- Direct learners to Let's learn in the Student's Book.
- Learners work through the paired activity.
- Discuss the Guided practice example in the Student's Book.

Practise 📙 [TWM.07]

- Workbook

Title: Multiplying by repeated addition

Pages: 86–87

- Refer to Activity 1 from the Additional practice activities.

Apply 👥 💻 [TWM.08]

- Display **Slide 5**. Learners use the given digits to create a product as close to 100 as they can. They need to show they have the correct answer by finding all the possible products. Encourage them to work systematically.

- Invite learners to share their findings. Did they all manage to work out that the calculation they needed was 23 × 4 = 92?

Review

- Recap the different mental calculation strategies for multiplying by 2, 3, 4 and 5.
- As doubling and halving are important for mental calculation, practise doubling and halving different numbers.

Assessment for learning

- How can you work out 32 × 2? Is there another way?
- Can you tell me a quick way to multiply a number by 4/5?

Same day intervention

Support

- Focus on multiplying by 2 by doubling and on multiplying by 4 by doubling and doubling again.

Lesson 2: **Multiplying with arrays**

Learning objective

Code	Learning objective
3Ni.08	Estimate and multiply whole numbers up to 100 by 2, 3, 4 and 5.

Resources

place value counters or Resource sheet 8: Place value counters (per learner); mini whiteboard and pen (per learner); calculator for checking purposes only (per learner)

Revise

Use the activity *Multiplying by 2, 3, 4 and 5 (A)* from Unit 11: *Multiplication* in the Revise activities. Focus on the variation activity.

Teach 🔲 💻 [TWM.01/02]

- 🗣 Recap the strategies used for multiplying a two-digit number by 2, 3, 4 and 5 from Lesson 1, including the use of the number line. Tell learners that they will be multiplying by 2, 3, 4 and 5 again but this time they will be using another method.
- **[TWM.01/02]** Before showing **Slide 1** ask: **What is a good estimate of the product of 43 and 2? Why?** Agree an estimate of between 80 and 90. Display **Slide 1**. Ask: **What can you see? What calculation does the number line represent?** Discuss the place value counters and how these make an array of two rows of 43. Ask: **How are the place value counters the same as the number line? How are they different?** Ask learners to set out their place value counters or those from Resource sheet 8 to show what is on the slide. Ask: **How can we use this array to work out the product?** Agree that they could count up the tens, which gives 80, and then the ones, to give 6, and combine to give a product of 86. Establish that they could simply double 43 by partitioning into 40 and 3, doubling each number and then recombining. Agree that this is a good way to check if their product is correct.
- **[T&T]** Before showing **Slide 2** ask: **What is a good estimate of the product of 43 and 3? Why?** Agree an estimate of between 120 and 130. Display **Slide 2**. Ask: **What is the same about this slide and the previous one? What is different? What's the calculation?** Agree that the representations are the same, the multiplicand is the same, but the multiplier is different (i.e. 3).
- Ask learners to use their place value counters to set out this calculation. Ask: **How many tens are there?** Agree 12. Ask: **What do we need to do to the 12 tens?** Agree that they need to exchange 10 of them for a 100 counter. They do this and then add the tens and ones to give a product of 129.
- **[T&T] [TWM.01/02]** Before showing **Slide 3**, ask learners to give you an estimate for the product of 43 and 4. Display **Slide 3** and ask the questions

on the slide. Establish that they could double and double again as a mental strategy.

- Ask learners to use place value counters to make the calculation. They find the product for the ones first. Ask: **What do we need to do with the 12 ones?** Agree they need to exchange 10 ones for 1 ten. They find the product for the tens (160) and exchange 10 tens for 1 hundred, then add the product for the ones (12) to make 172. They check this, using division, on a calculator.
- Repeat for **Slide 4** to multiply 43 by 5. Remind learners that they could multiply by 10 and halve the product. This is a good way to check their answer.
- Direct learners to Let's learn in the Student's Book.
- Learners work through the paired activity.
- Discuss the Guided practice example in the Student's Book.

Practise 🔲 [TWM.02]

- Workbook

Title: Multiplying with arrays

Pages: 88–89

- Refer to the variation in Activity 1 from the Additional practice activities.

Apply 👥 💻 [TWM.04]

- Display **Slide 5**. Learners decide whether the statement made is always, sometimes or never true. Encourage them to give a written explanation that includes examples.

Review 📊

- Invite individual learners to use the **Place value counters tool** to show a multiplication calculation. The rest of the class work out what the calculation is and use their whiteboard to work out the answer.

Assessment for learning

- How do place value counters help us to work out the answer to a multiplication calculation?

Same day intervention
Support

- Focus on using counters to multiply by 2 and 3.

Number – Integers and powers

Unit **11** **Multiplication**

Lesson 3: **Multiplying by the grid method**

Number – Integers and powers

Learning objectives

Code	Learning objective
3Ni.06	Understand and explain the [commutative and] distributive properties of multiplication, and use these to simplify calculations.
3Ni.08	Estimate and multiply whole numbers up to 100 by 2, 3, 4 and 5.

Resources

mini whiteboard and pen (per learner); calculator for checking purposes only (per learner)

Revise

Use the activity *Multiplying by 2, 3, 4 and 5 (B)* from Unit 11: *Multiplication* in the Revise activities to reinforce work carried out in this unit.

Teach [SB] 🖵 [TWM.01/02]

- 📖 **[TWM.01/02]** Display **Slide 1**. Ask questions about the two representations: **What do you notice? What else? What is the same? What is different?** Accept any correct responses. Draw out the fact that these are different representations of the same calculation: 34 × 2 =. The first shows place value counters and in the second, the numbers are in a grid.

- **[T&T]** Ask: **Why are 30 and 4 written at the top of the grid?** Establish that this is the number being multiplied (the multiplicand). It has been partitioned into tens and ones. Elicit that the multiplier (2) has been written at the start of the second row. Ask: **Look at the 60 in the grid. Where can you see 60 in the place value array?** Expect learners to explain that the six tens counters represent 60. Ask: **What about the 8?** Agree that this corresponds to the eight ones counters.

- **[T&T]** Ask: **Can you explain what is happening in the grid?** Establish that the number 34 has been partitioned into 30 and 4 and each part has been multiplied by 2, or doubled. Ask: **Do you agree with the product? Can you explain why?**

- **[TWM.01/02]** 📖 Display **Slide 2**. Again, ask questions about the two representations: **What do you notice? What else? What is the same as the first slide? What is different?** Expect learners to tell you that the multiplicand is the same but the multiplier is different. This time they are multiplying by 3. Invite learners to describe how the grid and the counters show the same calculation. Invite learners to describe how the product has been found on the grid.

- **[T&T]** Ask learners to draw, on mini whiteboards, what the grid would look like for 35 × 3 = and for 36 × 3 =. Ask them to explain to a partner how they would use the grids to work out the products. Discuss how they could check, using division, on a calculator.

- Display **Slide 3** and repeat the previous steps to multiply 34 by 4.

- Display **Slide 4** and repeat the previous steps to multiply 34 by 5.

- Direct learners to Let's learn in the Student's Book. Discuss the examples shown and which method learners like.

- Learners work through the paired activity.

- Discuss the Guided practice example in the Student's Book.

Practise [WB] [TWM.04]

- Workbook

Title: Multiplying by the grid method

Pages: 90–91

- Refer to Activity 2 from the Additional practice activities.

Apply 👥 🖵 [TWM.04]

- Display **Slide 5**. Learners work out the product of 58 and 5 in three different ways. Encourage them to consider the methods they have been learning about and also the mental calculation strategy for multiplying by 5.

- They consider which they prefer and why.

- Discuss the activity. Invite learners to share the different methods they used to find the product. Include in this multiplying by 5 by halving and multiplying by 10 and vice versa.

Review

- Discuss with learners which of the methods they have learned they prefer. They are likely to have different preferences. Invite learners to share why they like one more than the others.

- Invite learners to demonstrate the grid method for calculations that you give them.

Assessment for learning

- How is the grid method the same as using place value counters? How is it different?

- How can using place value counters help us to use the grid method?

- Can you think of a way to multiply 26 by 4? Can you think of another way?

Same day intervention

Support

- Focus on using place value counters for multiplying by 3 and then translating this into the grid method.

Lesson 4: **Multiplying by partitioning**

Learning objectives

Code	Learning objective
3Ni.06	Understand and explain the [commutative and] distributive properties of multiplication, and use these to simplify calculations.
3Ni.08	Estimate and multiply whole numbers up to 100 by 2, 3, 4 and 5.

Resources

place value counters or Resource sheet 8: Place value counters (per learner); calculator for checking purposes only (per learner)

Revise

Use the activity *Multiplying by ten and halving* from Unit 11: *Multiplication* in the Revise activities to reinforce work carried out in this unit. Focus on the variation activity.

Teach 📖 🖥 [TWM.08]

- 📖 Before showing **Slide 1** ask: **What is a good estimate of the product of 16 and 3? Why?** Agree between 30 (10 × 3) and 60 (20 × 3). Display **Slide 1**. Briefly go through all the three methods that learners have been taught in this unit to multiply 16 by 3, then move on to the new method, partitioning, which is the focus of this lesson. Ask: **What do you notice about all these methods? What is the same about them? What is different?**

- Ask: **What do the last two pictures show?** The idea may be new to learners, so spend a little time exploring how the multiplicand has been partitioned and how each part is multiplied by the multiplier. The two products are then recombined to give the whole product. Indicate the second of these images and ask: **Why do you think the multiplicand has been partitioned into two eights?** Establish that partitioning in this way enables learners to use different multiplication facts.

- [T&T] [TWM.08] Ask: **Which of these methods do you prefer, to find the product of 16 and 3? Why?** Give learners a few minutes to discuss this with a partner. There is no correct answer to this but insist that learners explain their reasons.

- Before showing **Slide 5** ask: **What is a good estimate of the product of 26 and 3? Why?** Agree between 60 (20 × 3) and 90 (30 × 3). Display **Slide 5**. Ask learners to make 26 with their place value counters. Then they partition these into tens and ones. They multiply each part by 3, to give a total of three groups of 3 tens and 6 ones. They work out how many of each there are and make the necessary exchange of 10 ones to 1 ten. They then recombine to give the product of 72. Repeat for the other two calculations on the slide.

- [T&T] Ask: **How else could you partition the multiplicands? Would this make the multiplication easier?** Establish that they can partition in many ways, but tens and ones as shown is probably the most effective and efficient.

- Display **Slide 6**. Work through multiplying 26 by 2, 4 and 5 using the previous method. Ask learners to check, using calculators to divide. Can they

remember which number is divided and what the divisor could be? Each time ask learners to estimate the products first.

- Direct learners to the Student's Book and discuss the Let's learn section. Emphasise where it says learners should estimate, use a mental calculation strategy and check.

- Learners work through the paired activity.

- Discuss the Guided practice example in the Student's Book.

Practise 📒 [TWM.04]

- Workbook

Title: Multiplying by partitioning

Pages: 92–93

- Refer to the variation in Activity 2 from the Additional practice activities.

Apply 👥 🖥 [TWM.03/04]

- Display **Slide 7**. Learners are given the answer to a multiplication calculation. They work out what they need to multiply by 2, 3, 4 and 5 to give this product.

- They need to explain their thinking.

- Discuss the activity as a class. Invite learners to share their solutions and explanations.

Review

- Ask learners to recap all the methods for multiplication that they have explored in this unit.

- Write a two-digit number multiplied by 2, 3, 4 or 5 on the board and ask a volunteer learner to work out the answer using their preferred method. Encourage the learner to explain why they chose that method. Repeat several times asking for different volunteer learners.

Assessment for learning

- What mental calculation strategy can you use to multiply 32 by 2 32 by 3, 32 by 4, 32 by 5?

- Why did you use that method? Could you have used a better method? Why? Why not?

Same day intervention
Support

- Focus on using place value counters and the grid method for multiplying by three.

Number – Integers and powers

Additional practice activities

Activity 1

Learning objective
- Multiply two-digit numbers by 2, 3, 4 and 5

Resources
Resource sheet 3: Digit cards (per pair); mini whiteboard and pen for jottings (per learner) place value counters or Resource sheet 8: Place value counters (per pair) (for the variation)

What to do
- Give each pair of learners a set of digit cards.
- They shuffle them and place them in a pile, face down on the table.
- They take two and make a two-digit number.

- They use doubling to multiply the number by 2. They then multiply the number by 4 by doubling again.
- Next, they multiply the number by 5 by multiplying by 10 and halving.
- Finally, they multiply the number by 3. They use whichever method they prefer.
- Encourage learners to use paper or mini whiteboards to make jottings.

⚠ Variation
Focus on multiplying by 2, 3, 4 and 5, using place value counters.

Activity 2

Learning objective
- Multiply two-digit numbers by 2, 3, 4 and 5

Resources
Resource sheet 8: Place value counters (per pair); scissors (per learner); glue (per learner)

What to do
- Give each pair a set of place value counters from Resource sheet 8.
- Write some two-digit numbers on the board for learners to multiply by 2, 3, 4 and 5.
- Learners use the place value counters to represent the calculations and stick these in their books. They work out the products.

- They then write the grid method beside each representation to show how to use this method to find the product.
- They do this for four calculations, one multiplied by 2, the next by 3, the third by 4 and the fourth by 5.
- Then ask them to make up their own calculations and to represent them in the same ways.

Variation
⚠ Focus on multiplying by 2, 3, 4 and 5, using the partitioning method explored during the Lesson 4.

Unit 12: Division

Collins International Primary Maths Recommended Teaching and Learning Sequence: Term 2, Week 7

Learning objectives

Code	Learning objective
3Ni.09	Estimate and divide whole numbers up to 100 by 2, 3, 4 and 5.
3Np.03	Compose, decompose and regroup 2-digit [3-digit] numbers, using [hundreds,] tens and ones.

Unit overview

This unit focuses on dividing whole numbers up to 100 by 2, 3, 4 and 5. Initially, the answers (quotients) will be whole numbers, then progressing to those with remainders expressed as whole numbers.

As with all work on calculation, there is an expectation that learners will estimate the answer first before they work out the answer. They use this estimate to judge whether the quotient they have found is likely to be correct. There is also an expectation for learners to consider using a mental strategy for each calculation.

The mental strategies learners should consider include:

- counting back in steps of constant size
- using known multiplication and division facts and related facts involving multiples of 10
- halving
- knowledge of place value and partitioning
- knowing and applying the inverse relationship between multiplication and division.

Once learners have found the quotient to a calculation, they must check their answer by either multiplying the quotient by the divisor to see if they arrive at the dividend, or by dividing the dividend by the quotient to see if they arrive at the divisor. Checking with multiplication can be carried out using calculators, because the focus of this unit is division.

During this unit learners will explore using multiplication and division facts, arrays with place value counters or similar manipulatives, and partitioning, and they will link these to the written method.

Prerequisites for learning

Learners need to:

- know the 1, 2, 3, 4 and 5 times tables
- know the corresponding division facts for the 1, 2, 3, 4 and 5 times tables
- understand that division is repeated subtraction
- understand that multiplication and division are inverse operations
- be able to use the basic functions of a calculator in order to check answers.

Vocabulary

multiplication, division, dividend, divisor, quotient, exchange, division bracket, remainder

Common difficulties and remediation

If any learners have difficulty recalling tables facts quickly, provide a list of the multiplication facts so they can focus on the process of dividing; Resource sheet 7: Times tables grid is a useful resource for this purpose. It is also very important to make the link that division is repeated subtraction. This is not the most efficient way to divide but, for any learners having problems with dividing two-digit numbers by 2, 3, 4 and 5, it is a method to focus on until they are ready to move on.

Supporting language awareness

It is good practice to model the accurate vocabulary for division and have it displayed in the classroom. 'Dividend' is the name of the number we start with; 'divisor' is the number we divide by and 'quotient' is the answer. When working on written methods, we should use the term 'divided by' because it shows the operation.

Promoting Thinking and Working Mathematically

During this unit the learners will have opportunities to work on thinking and working mathematically; for example, when noticing differences and similarities between images on the slides, they will be making conjectures (TWM.03), convincing (TWM.04), characterising (TWM.05) and classifying (TWM.06). They will have opportunities to critique (TWM.07) and improve (TWM.08) when they estimate, and decide on the most appropriate strategy to use when they check. They will also have the opportunity to specialise (TWM.01) and generalise (TWM.02) when they use what they know to generate other facts.

Success criteria

Learners can:

- divide two-digit numbers by 2, 3, 4 and 5
- decide on appropriate strategies to divide
- estimate and check answers.

Number – Integers and powers

Unit  Division

Number – Integers and powers

Lesson 1: **Dividing using known facts**

Learning objectives

Code	Learning objective
3Ni.09	Estimate and divide whole numbers up to 100 by 2, 3, 4 and 5.
3Np.03	Compose, decompose and regroup 2-digit [3-digit] numbers, using [hundreds,] tens and ones.

Revise

Use the activity *Multiplication and division facts* from Unit 12: *Division* in the Revise activities to review and consolidate work covered in Unit 11: Multiplication.

Teach ⬛ 🖥

- 🖰 Display **Slide 1**. Ask: **What does it mean to divide?** Establish that it means finding out how many groups of the 'divisor' there are in the 'dividend'. Explain the new words, and say that the answer to a division is called the 'quotient'.
- Ask two questions, based on those on the slide: **What is a good estimate for 30 divided by 2?** Agree about 10, because learners know that 20 divided by 2 is 10. **What mental calculation strategy could we use to find the quotient?** Discuss the fact that to multiply by 2, they double so to divide by 2, they could halve. Invite a learner to consider how to partition 30 into numbers that can be halved easily (20 and 10), halve each part and recombine.
- **[T&T]** Ask: **What do you think the cubes show?** Give learners a few minutes to discuss this. Take feedback. Agree that they show counting on in twos to 30 and that this has been done in 15 steps. Alternatively, 2 has been taken from 30 fifteen times. Both examples give a quotient of 15.
- Discuss the idea that to check, they could double 15 to give the original number (the dividend) of 30. Practise with other dividends that are multiples of 2.
- Display **Slide 2**. Ask: **How is this the same as the first slide? How is it different?** Repeat the process as for the previous slide, this time for 36 ÷ 3 =. First ask learners to estimate the answer (30 ÷ 3 = 10), then partition 36 into numbers that divide evenly by 3 (30 and 6), then divide each part by 3 and recombine (10 + 2 = 12). Finally, refer to the cubes. Practise with other dividends.
- Display **Slide 3** and discuss what it shows. **[T&T]** Ask: **How could we use a mental calculation strategy to find the quotient?** Agree that because they could double and double again to multiply by 4, they could halve and halve again to find the quotient. Discuss checking by doubling. Practise with other dividends.
- Display **Slide 4**. Repeat the process as for the previous slides, this time for 60 ÷ 5 =. Each time, ask learners to estimate the quotient before dividing. The mental calculation strategy to encourage is

doubling and dividing by 10, which is the inverse of the strategy for multiplying by 5. This may be a new idea, so take time to explain it. Practise this with different dividends that are multiples of 5.

- Direct learners to Let's learn in the Student's Book. Discuss the strategies shown.
- Learners work through the paired activity.
- Discuss the Guided practice example in the Student's Book.

Practise 📖 [TWM. 04/07]

- Workbook

Title: Dividing using known facts

Pages: 94–95

- Refer to Activity 1 from the Additional practice activities.

Apply 👥 🖥 [TWM.01]

- Display **Slide 5**. Learners are given 12 as the answer to a division. They come up with a question with this as the answer; for example, 24 ÷ 2 =.
- Encourage them to use their multiplication facts and partitioning, as worked on in the lesson.
- Learners find as many possibilities as they can in the time allowed. Encourage them to work systematically.
- Invite learners to share their findings. How many possibilities did they find?

Review

- Recap the different mental calculation strategies for dividing by 2, 3, 4 and 5.
- 🖰 Ask: **What is the dividend? What is the quotient? What is the divisor?**

Assessment for learning

- How can you work out 60 ÷ 4? Is there another way?
- Can you explain how to divide a number by 2/5?

Same day intervention
Support

- Focus on dividing numbers by 2 by halving and then dividing numbers by 4 by halving the quotient after the number is divided by 2.

Lesson 2: Dividing by partitioning

Learning objectives

Code	Learning objective
3Ni.09	Estimate and divide whole numbers up to 100 by 2, 3, 4 and 5.
3Np.03	Compose, decompose and regroup 2-digit [3-digit] numbers, using [hundreds,] tens and ones.

Resources

mini whiteboard and pen (per learner); place value counters or Resource sheet 8: Place value counters (per pair); calculator for checking purposes only (per learner)

Revise

Use the activity *Using multiplication facts to divide* from Unit 12: *Division* in the Revise activities to rehearse using multiplication facts to divide.

Teach 🆂🅱 🖥

- 👭 Recap the strategies from Lesson 1 for dividing a two-digit number by 2, 3, 4 and 5. Tell learners that in this lesson they are going to use partitioning to divide by 2, 3, 4 and 5. Ask: **What do we mean by partitioning?** Agree that it means splitting a whole number into smaller numbers.

- Display **Slide 1**. Ask: **What can you see? What do you think the place value counters show?** Discuss the place value counters and how they show 54. Then discuss some of the different ways they have been partitioned to make a number of tens and another number. Ask learners to make 54 with their place value counters and then partition in the ways shown. Ask them to show the other possibilities of partitioning 54 into multiples of 10 and another number.

- **[T&T]** Before showing **Slide 2**, ask: **What is a good estimate for the quotient of 54 and 3? Why?** Agree between 10 and 20. Display **Slide 2**. Ask: **How does what we learned in Lesson 1 help us when we use place value counters to divide?** Agree that if we can partition 54 into two numbers, then we can use multiplication facts to divide. Work through the example together. Expect learners to use place value counters to follow this. Ask them to check their result using a calculator.

- Give learners other two-digit multiples of 3 to partition into numbers that can be divided exactly by 3; for example, 45 partitions into 30 and 15, 57 partitions into 30 and 27. Aim for learners to partition into multiples of 10 of the divisors where possible and another multiple of 3.

- Display **Slide 3**. Repeat the process above for partitioning to divide numbers by 4, using tables facts or halving and halving again. Make sure that all the numbers you give for learners to practise are multiples of 4, for example 56 (40 and 16), 64 (40 and 24), 68 (40 and 28). Learners can check with a calculator.

- Repeat the process for dividing by 2 and 5, with learners making and partitioning appropriate numbers that can be divided exactly.

- Direct learners to Let's learn in the Student's Book. Discuss the example shown.

- Learners work through the paired activity.

- Discuss the Guided practice example in the Student's Book.

Practise 🆆🅱

- Workbook

Title: Dividing by partitioning

Pages: 96–97

- Refer to the variation in Activity 1 from the Additional practice activities.

Apply 👥 🖥 [TWM.01]

- Display **Slide 4**. Learners show how Sally could partition 92 in three different ways. Encourage them to find at least two ways.

Review

- Call out some multiples of 5 that are greater than 60. Invite learners to explain how they could partition these numbers to make the division by 5 simpler.

- Encourage them to show how they would partition using the model shown on the slides.

Assessment for learning

- How can partitioning help us to work out the answer to a division calculation?

- How does knowing the multiplication and division facts help us to divide?

Same day intervention

Support

- Focus on estimating and dividing numbers less than 50 by 2 and 3, using the partitioning model. Encourage learners to physically make the numbers with place value counters.

Number – Integers and powers

Unit 12 Division

Lesson 3: **Dividing with arrays**

Learning objectives

Code	Learning objective
3Ni.09	Estimate and divide whole numbers up to 100 by 2, 3, 4 and 5.
3Np.03	Compose, decompose and regroup 2-digit [3-digit] numbers, using [hundreds,] tens and ones.

Resources

mini whiteboard and pen (per learner); place value counters or Resource sheet 8: Place value counters (per learner); calculator for checking purposes only (per learner)

Revise

Use the activity *Partitioning in different ways* from Unit 12: *Division* in the Revise activities to reinforce partitioning, which will be needed in this lesson.

Teach 🔲 💻 [TWM.01/02]

• ⏸ Display **Slide 1**. **What number is represented here?** Ask them to draw a grid like the one on the slide on their mini whiteboards or a piece of paper. They put their place value counters on the grid to show 36. Ask them to represent different numbers, such as 46, 53, 72, in their grids. Ask: **How are these similar to what we explored in Lesson 2?** [show partitioning into tens and ones].

• [T&T] Before showing **Slide 2**, ask: **What is a good estimate for the quotient of 36 and 3? Why?** Agree around 10. Display **Slide 2** and then **Slide 3**. Ask: **Why do you think that laying counters out like this can help us with division?** Establish that, to divide, we need to find how many groups of the divisor we can get out of the dividend. Ask learners to make 36 in their grids and work through the slides practically, establishing that there is one group of 3 tens, then look at the ones and establish that there are two groups of 3 ones, giving 36 ÷ 3 = 12.

• Before showing **Slides 4** and **5**, Ask: **What is a good estimate for the quotient of 69 and 3? Why?**

• Display **Slide 4** and then **Slide 5** and work through them as for Slides 2 and **3**. Establish that there are two groups of 3 tens, then look at the ones and establish that there are three groups of 3 ones, giving 69 ÷ 3 = 23.

• [T&T] [TWM.01/02] Before showing **Slides 6** and **7**, ask: **What is a good estimate for the quotient of 42 and 3? Why?** Agree around 10. Display **Slide 6** and then **Slide 7**. Ask: **How is this the same as the previous slides? How is it different?** Establish that the layout is similar but that this time, when learners make a group of 3 tens, there will be one left over. Ask: **What do you think we need to do with the 1 ten that is left over?** Agree that they need to exchange it for 10 ones, to make 12 ones. Learners do this practically with their resources. Check that they arrive at a quotient of 14.

• Give learners other numbers that are multiples of 3, where an exchange of 1 ten is needed to work

through practically with a partner, for example 45, 48, 72, 75. Learners can check with a calculator. Encourage learners to estimate quotients first.

• Ask: **Do you think that this method would work for dividing by 2, 4 and 5?** Display **Slide 8** and then **Slide 9** and talk through them, asking learners to divide the example given. There is a further example on **Slides 10** and **11**. Work through this and other examples with learners, to consolidate, as appropriate. Learners can check with a calculator.

• Direct learners to Let's learn in the Student's Book. Discuss the example shown.

• Learners work through the paired activity.

• Discuss the Guided practice example in the Student's Book.

Practise 📒 [TWM.02/04]

• Workbook

Title: Dividing with arrays

Pages: 98–99

• Refer to Activity 2 from the Additional practice activities.

Apply 👥 💻 [TWM.04]

• Display **Slide 12**. Learners explain to their partner how to divide, using the method learned during the lesson. They draw pictures to help them.

• Their partner then decides if their explanation is a good one.

Review

• Recap the model for division covered in this lesson.

• ⏸ Ask: **Do you think this is a helpful way to divide? Why?**

Assessment for learning

• How many groups of 4 tens can you make out of 8 tens?

• How can we check the answer to a division?

• Why are place value counters helpful for dividing?

Same day intervention

Support

• Give learners pre-prepared grids.

Number – Integers and powers

Lesson 4: **Division with remainders**

Learning objectives

Code	Learning objective
3Ni.08	Estimate and divide whole numbers up to 100 by 2, 3, 4 and 5.
3Np.03	Compose, decompose and regroup 2-digit [3-digit] numbers, using [hundreds,] tens and ones.

Resources

place value counters or Resource sheet 8: Place value counters (per learner)

Revise

Use the activity *Grouping* from Unit 12: *Division* in the Revise activities.

Teach 〔SB〕 🖵 📊 **[TWM.01/02]**

- 🔁 **[TWM.01/02]** Before showing **Slide 1**, ask: **What is a good estimate for the quotient of 42 and 3? Why?** Agree around 10. Display **Slide 1**. Ask: **What do you notice?** The example shown is a progression from Lesson 3. It shows the calculation 42 ÷ 3 = and how the groups of 3 are made from the dividend, as before. The formal written method at the bottom of the slide shows the same process. Explain that when we use the written method to divide, we always use the division bracket. Tell learners that the dividend is written inside the division bracket and the divisor is written outside, on the left.

- **[T&T]** Ask: **How do the place value counters link to the written model?** Establish that one group of 3 tens can be made from 4 tens. Say: **To show this, we write the numeral 1 on the division bracket above the 4 tens, so there is 1 ten in the quotient. We exchange the extra 1 ten for 10 ones. So we write 1 inside the division bracket, beside the 2 ones.** Demonstrate how in Step 4, the place value counters show that the 12 ones can be grouped into four groups of 3 ones. Say: **To show this, we write 4 on the division bracket above the 2 ones.**

- Using the **Place value counters tool**, draw a Tens and Ones grid on the board and place 7 tens and 2 ones on the grid. On the board write: 72 ÷ 3 = and underneath this write the formal written method: 3)‾72. Together work through the division, completing each step of the formal written method as you go.

- Repeat for the calculation: 94 ÷ 4 =. When it comes to dividing the 14 ones by 4, what do learners notice? Elicit that 4 does not divide exactly into 14, so the answer must be 3 with 2 left over. Tell them that what is left over is called the remainder. Ask learners to give an estimate for the quotient first. Agree that 94 ÷ 4 gives a quotient of 23 with a remainder of 2.

- Work through other examples that require regrouping of 1 ten and that produce remainders, such as 67 ÷ 5 =, 75 ÷ 3 =, 59 ÷ 4 =.

- Direct learners to the Student's Book and discuss the Let's learn section. Emphasise the vocabulary

used for division and the remainder. Expect learners to estimate quotients first.

- Learners work through the paired activity.

- Discuss the Guided practice example in the Student's Book.

Practise 〔WB〕 **[TWM.01]**

- Workbook

Title: Division with remainders

Pages: 100–101

- Refer to Activity 2 from the Additional practice activities. Focus on the variation where learners use numbers that will give remainders.

Apply 👥 🖵 **[TWM.01]**

- Display **Slide 2**. Learners are asked to find a solution to a problem. There are three possible solutions. To find them, learners use their knowledge of multiples of 4 and 5 to work out how many groups can be made.

- This highlights the inverse properties of multiplication and division.

- Invite learners to give their solutions and to share how they worked these out.

Review

- Invite learners to demonstrate, using the written method on the board, how to divide numbers less than 100 by 2, 3, 4 and 5. These should include numbers that divide to give whole numbers as well as remainders.

- Invite other learners to explain how they could check using multiplication.

Assessment for learning

- What is a remainder?
- What is a division bracket?
- How could we check that our answer is correct?

Support

- Focus on dividing by 2 and 3. Only move on to the written method when learners are confident in using place value grids with place value counters.

Number – Integers and powers

Additional practice activities

Activity 1

Learning objective
• Divide two-digit numbers by 2, 3, 4 and 5

Resources
mini whiteboard or paper for jottings (per learner)
place value counters or Resource sheet 8: Place value counters (per pair) (for the variation); scissors (per pair)

What to do
• Working in pairs, learners write down a selection of two-digit even numbers.
• They work together dividing each number by 2, using halving.

• If the quotient is an even number, they halve again, to divide the original number by 4.
• Learners then write down a selection of multiples of 10 and divide these by 5, by dividing by 10 and doubling or doubling and dividing by 10.

Variation

2 Repeat the activity above but provide learners with place value counters, so that they can demonstrate partitioning practically.

Activity 2

Learning objective
• Divide two-digit numbers by 2, 3, 4 and 5

Resources
Resource sheet 8: Place value counters (per learner); scissors (per learner); glue (per learner)

What to do
• Give each pair learner place value counters from Resource sheet 8.
• Write some two-digit calculations on the board for learners to divide by 2, 3, 4 and 5. Ensure that there will not be any remainders.
• Working in pairs, learners use their place value counters to complete the calculations. Then they stick these in their books. They work out the quotient by circling or looping groups of the divisor.

• Next, ask them to make up their own calculations to represent in the same way.

Variation

2 Repeat the above activity but give learners numbers to divide by 2, 3, 4 and 5 that will leave remainders.

Unit 13: Money

Collins International Primary Maths
Recommended Teaching and
Learning Sequence: Term 3, Week 3

Learning objectives

Code	Learning objective
3Nm.01	Interpret money notation for currencies that use a decimal point.
3Nm.02	Add and subtract amounts of money to give change.

Unit overview

In this unit, learners will continue to develop their understanding of money. In Stage 2 they began to recognise money notation, either in dollars and cents, or their local currency, and compare values of different combinations of coins and notes. In this unit they will be interpreting dollars and cents, or local currency, as money notation where, for example, the dollars appear to the left of the decimal point and cents to the right. Learners have not been introduced to decimals as yet, so cents as decimal fractions of a dollar are not mentioned. They will be using practical equipment such as money to find totals and change. They develop their understanding of finding totals and change by applying this to problem solving.

Prerequisites for learning

Learners need to:
- know the value of the coins they use in everyday life, for example ten 1 cent coins together are worth 10 cents
- know the value of combinations of notes or coins
- be able to combine amounts to 20c and multiples of 10c to one dollar.

Vocabulary

coins, notes, decimal point, place holder, total, spend, change, subtract difference

Common difficulties and remediation

Some learners find it difficult to remember the values of coins. It is often confusing to them when a coin is smaller than another yet has a greater value. It is important to have a supply of real coins available for learners to examine and sort into like types. It is helpful to encourage them to line these up according to value, smallest value to greatest. It is also useful to display visual clues, such as the example here, as a ready reference to their values compared with other coins.

 is the same as

Matching 1c, 10c and $1 with ones, tens and hundreds in Base 10 equipment can help learners to understand the relative values of these amounts

of money. Learners have used Base 10 equipment in previous units in Stage 3, when considering representing numbers in different ways and starting to move towards understanding place value. Ensure that you have equipment available for this.

Supporting language awareness

It is important to display photographs of the coins and notes, labelled with their values, and to refer to these as learners work through this unit. This includes dollars, which Cambridge International adopts as an internationally recognised currency, as well as local currency.

When referring to value in the context of money, the words 'increasing' and 'decreasing' are important. It would be worth displaying these words, with simple symbols to indicate what they mean, for example:

Increasing in value ↑ Decreasing in value ↓

Promoting Thinking and Working Mathematically

During this unit learners have opportunities to identify mathematical properties of objects (TWM.05) and organise them according to their properties (TWM.06). For example, when exploring coins, they will be sorting them against a variety of criteria. They will also compare and evaluate mathematical ideas (TWM.07) refine their ideas to develop more effective approaches (TWM.08). They will have opportunities to solve problems and explain their thinking, which involves specialising (TWM.01), generalising (TWM.02), making conjectures (TWM.03) and convincing (TWM.04).

Success criteria

Learners can:
- recognise the value of coins
- understand equivalence between different notes and coins
- understand that dollars are positioned to the left of the decimal point and cents to the right
- find totals of amounts of money
- calculate change.

Number – Money

Unit  Money

Lesson 1: **Writing money**

Learning objective

Code	Learning objective
3Nm.01	Interpret money notation for currencies that use a decimal point.

Resources

selection of coins in your own monetary system or dollars and cents, for example: 1c, 5c, 10c, 25 and 50c coins, $1, $2, $5 and $10 dollar notes or other representations (per pair)

Revise

Use the activity *Counting money (1)* from Unit 13: *Money* in the Revise activities to review counting in ones and tens and steps of five.

Teach [SB] 🖵 [TWM.01]

- Display **Slide 1**. Ask learners to identify the coins and notes they can see. Ensure you discuss the value of each in terms of the number of cents.
- Ask: **How many cents are there in one dollar? What about in two dollars? Three dollars? Four dollars?** and so on.
- 🖵 Display **Slide 2**. Ask: **What is the total amount of money showing on the first row?** Agree one dollar 23 cents. Invite learners to share their methods and discuss the most efficient strategies. Establish that adding the 10 cent coins, then counting on the 1 cent coins might be the most efficient method.
- Ask: **How can we use numbers to write one dollar 23 cents?** Demonstrate how to write this, using a dot between the dollars and cents. Explain that the dot separates the dollars from the cents and that we call this dot a decimal point. Dollars are always placed to the left of this dot and cents to the right. Write $1.23 on the board.
- Repeat for the other two amounts: $2.32 and $1.19.
- You might wish to use the Interactive teaching tool Money. Select dollars. Add dollars and cents. Ask learners to write how much money is showing. Invite learners to make given amounts of money.
- **[T&T]** Ask pairs of learners to take different numbers of coins and notes and to find their value and then write these on paper in dollar and cent notation.
- **[TWM.01]** Once they have done this, ask them to explore how they could make the same amount, using the fewest notes and coins.
- Direct learners to the Student's Book. Reinforce the fact that if there are no tens of cents, we must use zero as a place holder in the ten cents position.
- Give learners the opportunity to work through the paired activity before discussing the Guided practice example as a class.

Practise [WB] [TWM.01]

- Workbook

Title: Writing money

Pages: 102–103

- Refer to Activity 1 from the Additional practice activities.

Apply 👥 🖵 [TWM.07/08]

- Display **Slide 3**. Learners work out, from the information given, how much of her $20 Summer put in two piles. They are told that she put $4 more into one pile.
- Learners are given a diagram to use to help them. From the diagram, they should be able to deduce that the two empty parts are worth $16 altogether, so each is worth $8. Therefore one pile has $12 and the other $8.

Review

- Discuss what has been learned during the lesson.
- On the board, write different numbers of cents, for example: 345c, 187c, 607c.
- Invite learners to write these amounts, using dollar and cent notation, with the decimal point separating the cents from dollars.
- 🖵 Ask: **What does the dot or decimal point do when we write money? What is on the right of the dot? What is on the left?**

Assessment for learning

- How many cents would you get for one dollar?
- How many dollars and cents could you exchange for 376 cents?
- Why do we need to use a dot or decimal point when we write money?
- What goes to the left of the decimal point?
- What goes to the right of the decimal point?

Same day intervention

Support

- Learners work with amounts that include one dollar and up to 99 cents. Provide coins and a one dollar note, so that they can work practically.

Lesson 2: **Finding totals**

Learning objective

Code	Learning objective
3Nm.02	Add [and subtract] amounts of money [to give change].

Resources

selection of coins in your own monetary system or dollars and cents, for example: 1c, 5c 10c, 25 and 50c coins, $1, $2, $5 and $10 dollar notes or other representations (per pair) selection of 1c, 5c, 10c, 25c and 50c coins (per learner) (for the Workbook)

Revise

Use the activity *All about money* from Unit 13: *Money* in the Revise activities to review and consolidate learners' understanding of finding totals.

Teach 🔲 💻 [TWM.07/08]

- Display **Slide 1**. Ask: **What can you see on the slide?** Establish that there are four toys with different price labels. Discuss the toys. Which ones are the learners familiar with? Are there any that are new to them?
- 🗣 Ask: **How much does each toy cost?** Expect learners to be able to read the prices.
- 🗣 Ask: **What are these prices in cents?**
- [T&T] [TWM.07/08] Ask learners to work with a partner, choosing pairs of toys and finding the total price for the two. Encourage them to use the methods they think are the most efficient. For example, if adding the item priced in dollars and cents (i.e. bulldozer at $12.40), find the total of the dollars first, and then add on the cents.
- [TWM.07/08] After a few minutes, take feedback, inviting learners to share their choices and how they found the totals. Encourage other learners to share how they would use different methods to find the totals. Ask: **Which methods are more efficient?**
- Direct learners to Let's learn in the Student's Book. Spend some time working through the examples, so that they understand how to carry them out. You could explore purchasing other combinations of items that either you or learners suggest.
- Give learners the opportunity to work through the paired activity. Discuss the problem and together work through the different mental calculation strategies that could be used to find the total.
- Direct learners to the Guided practice example.

Practise 📓

- Workbook

Title: Finding totals

Pages: 104–105

- Refer to the variation in Activity 1 from the Additional practice activities.

Apply 👥 💻 [TWM.01]

- Display **Slide 2**. Learners work out the possible combinations of large and small notebooks that Rosy could have bought for her school friends.
- There are several possibilities. Can learners find them all?

Review

- 🗣 Ask: **What have we been learning about during this lesson? What else?**
- Expect them to be able to tell you that they have been working with money and finding totals, using mental calculation strategies.
- Recap the mental calculations considered.
- Ask: **Why do you think calculating mentally is important? Why else?**

Assessment for learning

- What does it mean to find the total?
- Which strategies did we use to find totals?
- Which strategy was the most useful? Why?

Same day intervention

Support

- Ensure that learners use money or an alternative concrete representation.

Number – Money

Unit  Money

Lesson 3: **Finding change**

Learning objective

Code	Learning objective
3Nm.02	Add and subtract amounts of money to give change.

Resources

mini whiteboard and pen (per learner); selection of 1c, 5c, 10c, 25c and 50c coins, optional (per learner) (for the Workbook)

Revise

Use the activity *Counting money (2)* from Unit 13: *Money* in the Revise activities to familiarise learners with the concept of sequences.

Teach 🔲 💻 [TWM.04/08]

- **[T&T]** Display **Slide 1**. Ask: **Look at these gifts. What is the price of each one?** Discuss the price labels. Ask learners to tell you how much the two gifts would cost altogether. Discuss strategies for working this out.

- **[T&T] [TWM.04/08]** Ask: **If you paid with a $100 note, what change would you receive?** Learners discuss possible strategies that they have learned for subtraction, and apply this to the context of money. Take feedback. Agree all methods that work. Establish that counting on is a good strategy. They could then count on from $64 to $100 to find the change ($36).

- Display **Slide 2** and provide learners with paper or a whiteboard. Ask similar problems for learners to solve by asking them to add together two or more prices to find the total, then subtracting the totals to find the change.

- Direct learners to Let's learn in the Student's Book and discuss how much change they would get from $5, $10 or $20 after buying different items of food. Include buying single items of food and also buying different combinations of two or more items of food.

- Give learners a few minutes to work through the paired activity.

- As a class, discuss the Guided practice example.

Practise 📒

- Workbook

Title: Finding change

Pages: 106–107

- Refer to Activity 2 from the Additional practice activities.

Apply 👥 💻 [TWM.04]

- Display **Slide 3**. Learners are given a scenario where Jemma thinks she will get $10 change from $40 after buying a jacket for $25.

- Learners discuss whether they think she is correct or not.

Review

- Recap what has been learned during the lesson.

- 🗩 Ask: **What is a good strategy for finding change? Can you think of another?** Agree a good strategy is counting on from the cost of an item to the amount of money you have to find the difference between the two amounts.

Assessment for learning

- What do we mean by change?
- What do we do to find the change?
- What was a useful strategy for finding the change?

Same day intervention

Support

- Focus on finding the change from $10 and then $20.

Number – Money

Lesson 4: **Solving problems with money**

Learning objective

Code	Learning objective
3Nm.02	Add and subtract amounts of money to give change.

Resources

mini whiteboard and pen (per learner)

Revise

Use the activity *Counting money (2)* from Unit 13: *Money* in the Revise activities to reinforce these skills. Focus on the variation activity.

Teach [SB] 🖵

- 🗒 Provide learners with paper or a whiteboard. Begin the lesson by revising the work covered in Lesson 2 on finding change. Write an amount such as $25 on the board. Call out amounts you have spent, for example $12. Learners work out how much change you will receive. Then give two amounts you have spent. They find the total and the change. Let them make jottings to help them.

- [T&T] Display **Slide 1**. Ask: **What do you notice? What is the same? What is different?** Agree that there is a selection of fruit and vegetables. There are two or three of each. Each has a price label showing the cost for one kilogram. Ask the learners to identify as many of them as they can. Compare the prices per kilogram.

- [T&T] Ask: **What do we need to do if we want to work out how much 1/2 kg of tomatoes costs?** Agree: halve $6.

- [T&T] Ask: **What if we want 3 kilograms?** Discuss the fact that this is multiplying. Ask similar questions that involve learners finding the prices of the other fruits and vegetables by multiplying.

- Ask: **If I have $10 and buy 2 kg of peas, how much change will I have?** Continue asking similar questions. Learners use whatever strategy they wish. Again, let them make jottings so that they don't have to remember too many things.

- Direct learners to the Student's Book. Discuss the Let's learn section, then let learners work through the paired activity.

- Discuss the Guided practice example.

Practise 🖥

- Workbook

Title: Solving problems with money

Pages: 108–109

- Refer to the variation in Activity 2 from the Additional practice activities.

Apply 👥 🖵

- Display **Slide 2**. Learners make a list of their top 10 favourite foods and give each one a price under $10.

- They take turns to choose two foods. Their partner works out the total cost and the change from $20.

Review 🖵

- Discuss what has been learned during the lesson.

- Display **Slide 1** again and ask questions that involve adding two of the fruit and vegetables together.

- Give an amount and ask learners to find the change from the cost of the pairs they worked out.

Assessment for learning

- If I buy three pairs of trainers costing $23 each, how much will I spend? How did you work that out?

- If I have $40 and spend $26, how much will be left? How did you work that out?

- Potatoes cost $4 a kilogram. How much will half a kilogram cost? How did you work that out?

Same day intervention

Support

- Focus on adding to make totals less than $10 and working out change from $10.

Number – Money

Additional practice activities

Activity 1

Learning objectives
- Interpret money notation for currencies that use a decimal point
- Add amounts of money to find totals

Resources
collection of coins and notes (per pair); paper (per pair)

What to do [TWM.04/05/06]
- Ask learners, in pairs, to take some coins and notes. They first estimate how much they think they have and write this on paper. Encourage them to give a range and then narrow that range as they count out the money.
- They count the exact amount they have in an efficient way. This may be to count the 10 cent coins first and then in fives, to add any 5 cent coins, and the finally in ones. They add this amount to any notes.
- They write the amount they have on the paper, using the correct notation.
- Finally, they compare the actual amount with the estimate. For some, this might be to say they have more or less. Expect others to find the exact difference.

Variation
 Each learner takes a collection of coins (less than $1) and finds the total. Learners then share their totals with each other, and each learner adds the two amounts together using their preferred strategy. Learners then compare totals and discuss their strategies. They then check by counting all the coins. They repeat several times before doing the same using notes.

Activity 2

Learning objective
- Add amounts of money to find totals and subtract to give change

Resources
paper (per learner); coloured pencils (per learner)

What to do [TWM.04]
- Ask learners to make a shopping list of six things that they might buy in a shop of their choice. They can illustrate their list with appropriate pictures.
- They then put a price beside each item they have chosen. Prices need to be a whole number of dollars less than $10.
- They give their shopping list to a partner who chooses pairs of items to buy. Once they have chosen the items, they find the total cost.
- Ask them to explore how many different possibilities there are for buying pairs of items. They find all the totals. Which is the most expensive pair of items? Which is the cheapest?

Variation [TWM.04]
Repeat the activity but with learners working out how much change they would receive from $20.

Learners choose three or four items, work out the total and then the amount of change they would receive from $50.

Unit 14: Place value and ordering

Collins International Primary Maths
Recommended Teaching and
Learning Sequence: Term 1, Week 3

Learning objectives

Code	Learning objective
3Np.01	Understand and explain that the value of each digit is determined by its position in that number (up to 3-digit numbers).
3Np.03	Compose, decompose and regroup 3-digit numbers, using hundreds, tens and ones.
3Np.04	Understand the relative size of quantities to compare [and order] 3-digit positive numbers, using the symbols =, > and <.

Unit overview

By the end of Stage 2, learners know that the value of any digit in a number depends on its position in that number. This unit focuses on formalising this by introducing the concept of place value in three-digit numbers. Learners will explore the way a number is made up by looking at the values of the digits in a place value chart. They will see how the value of each digit is determined by its position in the chart (the positional aspect). They will use the place value column heading to derive the actual value of the digit (multiplying by the appropriate power of ten – the multiplicative aspect). Learners will also partition and recombine to understand that the whole number is made by adding together the values that the digits represent (the additive aspect). If any position has no numerical value, then zero is used as a place holder.

Learners will also consider regrouping three-digit numbers by expressing them in different ways, for example 712 can be expressed as 71 tens and 2 ones, 712 ones or 7 hundreds, 1 ten and 2 ones. This unit also focuses on using the terms and symbols 'greater than' (>) and 'less than' (<) to compare two three-digit numbers.

Prerequisites for learning

Learners need to:
- partition two-digit numbers into tens and ones
- recombine tens and ones to make two-digit numbers
- regroup two-digit numbers in different ways
- compare and order two-digit numbers.

Vocabulary

hundreds, tens, ones, position, place, add, zero, place holder, regroup, compare, greater than, less than

Common difficulties and remediation

Learners sometimes write three-digit numbers 'literally', for example writing one hundred and forty-two as 100402. Providing practice with appropriate equipment will help learners understand the place value of each digit in a number. Discussion based on their experience of three-digit numbers in the world around them is also beneficial. In the decimal number system, zero is used as a placeholder. It is important that learners are exposed to this idea. The 'literal' number difficulty, mentioned above, often stems from a lack of understanding of the role of zero and the fact that the position of a digit within a number determines its value. It is important that learners have access to place value charts with headings of Hundreds, Tens and Ones, or 100s, 10s and 1s rather than H, T and O. Some learners might interpret the heading O as zero, which is not helpful.

Supporting language awareness

Place value, when applied to the reading and writing of numbers, can be a language-intensive topic. Give learners opportunities to see the words they will be using, for example hundred, twenty, thirty, forty, in and around the classroom as well as to practise spelling these words.

The word 'position' has been used throughout this unit when explaining the value of each digit in a number (up to three digits). If learners struggle to understand 'position' in the context of place value, then use the word 'place' instead. However, be sure to stress the similarities in meaning between the two words so that learners gradually begin to use 'position' rather than 'place'.

Promoting Thinking and Working Mathematically

During this unit learners will choose examples of three-digit numbers that satisfy criteria related to place value (TWM.01). They will explore the underlying patterns of place value and identify examples that satisfy given criteria (TWM.02). They have opportunities to spot mistakes, which involves conjecturing (TWM.03), convincing (TWM.04), characterising (TWM.05) and classifying (TWM.06).

Success criteria

Learners can:
- explain the place value of each digit in a three-digit number
- partition and recombine numbers
- regroup a number to represent it in different ways
- compare numbers, using the language of and symbols for 'greater than' and 'less than'.

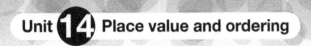

Unit 14 Place value and ordering

Number – Place value, ordering and rounding

Lesson 1: **Understanding place value (A)**

Learning objectives

Code	Learning objective
3Np.01	Understand and explain that the value of each digit is determined by its position in that number (up to 3-digit numbers).
3Np.03	Compose, decompose [and regroup] 3-digit numbers, using hundreds, tens and ones.

Resources

Resource sheet 3: Digit cards (per learner); Resource sheet 9: Place value chart (per learner); Base 10 equipment (per group)

Revise

Use the activity *Place value bingo* from Unit 14: *Place value and ordering* in the Revise activities to familiarise learners with the place values of three-digit numbers.

Teach 📖 💻 📊

- 📒 Display **Slide 1**. Ask: **What can you see on the slide?** Encourage learners to tell you all that they can see, including the digits, the chart, the headings for the columns. Ask: **Can you read the number in the chart? What does each digit represent?**

- Ask learners to use their digit cards and place value chart to make the number on the slide. Ask them to hold up the 3. Tell them that this is a digit because it is not part of a number so we don't know its value. Ask them to put it in the hundreds position. Encourage them to say: 'The 3 is in the hundreds position.' Repeat for 6 in the tens position and 4 in the ones position. Emphasise that whenever a numeral has no position, it is just a digit.

- **[T&T]** Ask: **What do we need to do to get the true values of the 3, 6 and 4?** Give learners a few minutes to discuss this. Take feedback. Establish that they need to look at the position of each digit in the place value chart. 3 is in the hundreds position so it must represent 300. Ask: **Six is in the tens position so what does that represent?** Establish that the 6 represents 60. Repeat for the 4 in the ones position representing 4.

- **[T&T]** Ask: **What do we need to do to make the whole number?** Agree that the three values need to be added together: 300 + 60 + 4 = 364. Ask learners to write this additive number statement.

- Ask groups of learners to make the number with their Base 10 equipment. Then confirm how to represent the number using the **Base 10 tool**.

- Direct learners to the Let's learn feature in the Student's Book to reinforce the concepts covered so far during the lesson.

- Display **Slide 2** and repeat the process above for 374, 384 and 394 (writing each number on the blank place value chart). Ensure learners make the numbers using their Base 10 equipment and that you confirm the representation using the **Base 10 tool**.

- Then write all four numbers on the board and ask: **What is the same about these numbers? What is**

different? Expect learners to tell you that the tens digit has increased by 1 each time; the other digits have stayed the same.

- Direct learners to the paired activity in the Student's Book and lead them through it. Learners work through the activity.

- Discuss the Guided practice example in the Student's Book.

Practise 📓 [TWM.04]

- Workbook

Title: Understanding place value (A)

Pages: 110–111

- Refer to Activity 1 from the Additional practice activities.

Apply 👥 💻 [TWM.07]

- Display **Slide 3**.

- Explain that Ahmed has described his number correctly but that he has made a mistake somewhere. Learners spot Ahmed's mistake and discuss what he has done wrong.

- They tell each other what Ahmed's number should be (437).

Review

- Review the fact that the value of each digit in a number depends on its position in the number.

- Write some three-digit numbers on the board. Ask learners to tell you the positions, the values and how they would be added to make the whole number.

Assessment for learning

- What is the value of the digit 3 in 365?
- What is the value of the digit 5 in 275?
- Which is greater, 423 or 453? How do you know?

Same day intervention

Support

- Some learners may not have fully understood how to find the values of digits in two-digit numbers. If this is the case, work on these before moving on to three-digit numbers.

Lesson 2: **Understanding place value (B)**

Learning objectives

Code	Learning objective
3Np.01	Understand and explain that the value of each digit is determined by its position in that number (up to 3-digit numbers).
3Np.03	Compose, decompose [and regroup] 3-digit numbers, using hundreds, tens and ones.

Resources

Resource sheet 10: Place value abacus (per learner); Resource sheet 3: Digit cards (per learner); Resource sheet 9: Place value chart (per learner); place value counters or Resource sheet 8: Place value counters (per pair)

Revise

Use the activity *Place that digit* from Unit 14: *Place value and ordering* in the Revise activities.

Teach 🔲 🖥 [TWM.01]

- 👥 Briefly recap learners' understanding of place value (up to three-digit numbers). Ask: **What is the value of the digit 3 in 436? How do you know?** Agree that it is 30 because the three is in the tens position. **What is the value of the digit 4 in 436? How do you know?** Agree that it is 400 because the four is in the hundreds position.

- [T&T] Display **Slide 1**. Ask: **What can you see? What number do you think this is supposed to represent?** Ask learners to talk in pairs about whether they think this is correct or if there is something missing. Establish that there is a digit missing and there should be a zero, which is a place holder, that goes in an 'empty' position. There are no tens so zero goes into the tens position. Without the place holder, the number looks like 34.

- Display **Slide 2**. Ask: **What do you notice this time?** Agree that the ones position is empty and should have a zero. Without the place holder the number looks like 85. Ask learners to draw some three-digit numbers on their abacuses from Resource sheet 10. They should include numbers with zero in the tens position or the ones position. Encourage learners to work in pairs and check each other's numbers.

- [TWM.01] [T&T] Display **Slide 3**. Ask: **What do you notice?** Ask learners to discuss what is the same and what is different about the representations. Agree that both show the same number, one as place value counters and the other as numerals in a place value chart.

- Ask learners to make the number with their place value counters and to make the number in their place value charts with digit cards.

- Say other numbers for learners to make in the same way. Include several with zero in the tens and ones positions. For each number, discuss the place values of the digits.

- Direct learners to the Let's learn feature in the Student's Book. Discuss the examples with zero shown. Ask: **Why is there a zero in the place value chart? Why is there no bead on the tens spike of**

the abacus? Remind learners that without zero as a place holder, the number would be 43.

- Direct learners to the paired activity in the Student's Book and lead them through it. Learners work through the activity.

- Discuss the Guided practice example in the Student's Book.

Practise 📘 [TWM.01/05]

- Workbook

Title: Understanding place value (B)

Pages: 112–113

- Refer to the first variation of Activity 1 from the Additional practice activities to include the use of zero.

Apply 👥 🖥 [TWM.01]

- Display **Slide 4**.

- The place value counters make up the number 143. Learners discuss whether Collette can make up other numbers using the place value counters.

- They should find that the place value counters all represent specific values and so there is only one number that can be made.

Review 📊

- Say some three-digit numbers with zeros in the tens or ones positions. Learners explain what position the zero is holding.

- Review the fact that the value of each digit in a number depends on its position in the number. Ask learners to provide their own examples to explain this. Finish the lesson by showing the **Place value tool**. Invite learners to create their own three-digit numbers. They explain what each digit is worth.

Assessment for learning

- What position is the zero holding in 304/870?

- What do we need to do to 400, 50 and 9 to make the whole number?

Same day intervention

Support

- Ensure learners have access to Base 10 equipment or place value counters.

Number – Place value, ordering and rounding

Unit 14 Place value and ordering

Lesson 3: Regrouping

Learning objectives

Code	Learning objective
3Np.01	Understand and explain that the value of each digit is determined by its position in that number (up to 3-digit numbers).
3Np.03	Compose, decompose and regroup 3-digit numbers, using hundreds, tens and ones.

Resources

place value counters or Resource sheet 8: Place value counters (per pair); Resource sheet 3: Digit cards (per learner); Resource sheet 9: Place value chart (per learner); place value counters or Resource sheet 8: Place value counters or Base 10 equipment (per learner) (for the Workbook)

Revise

Choose an activity from Unit 14: *Place value and ordering* in the Revise activities.

Teach 🔲 🖥 📊

- 🔲 Display **Slide 1**. Ask: **What number do the place value counters show?** (132) Invite a learner to demonstrate how this can be shown as an additive number statement, decomposing it as 100 + 30 + 2.

- Invite other learners to describe how 1 represents 100, 3 represents 30 and 2 represents 2. Then confirm how to represent the number with place value cards using the **Place value tool**.

- Ask pairs of learners to make this number with their place value counters and show it in their place value charts. Recap on the idea that the 1 is one unit of one hundred, the 3 is three tens and the 2 is two ones.

- [T&T] Ask: **How can we represent this number in ones?** Give learners a few minutes to discuss this with their partners bearing in mind that the concept of regrouping may not be immediately obvious to some learners. Prompt with questions such as: **How many ones are there altogether?** Establish that the number can be regrouped as 132 ones. Explain that this is called 'regrouping'. Ask: **How can we represent this number in tens and ones?** Establish that the number can be regrouped as 13 tens and 2 ones.

- [T&T] Display **Slide 2** and work through it as for **Slide 1** including how to represent the number with place value cards using the **Place value tool**. Ensure you include 233 ones, 23 tens and 3 ones as representations and that learners make the number with place value counters. For other representations, encourage learners to partition in different ways. There are numerous ways this can be done. Learners use their place value counters to explore them.

- Display **Slide 3**. Ask: **How is this number the same as the previous numbers? How is it different?** Establish that there are hundreds and ones but no tens. Invite a learner to write this number on the board. Stress the importance of zero as the place holder.

- Ask: **How many ones are there in this number?** (202) **How many tens could we make?** (20) Ask learners to work through some other ways to regroup 202.

- Direct learners to the Student's Book. Discuss the examples and let them complete the paired activity.
- Discuss the Guided practice example.

Practise 🔲

- Workbook

Title: Regrouping

Pages: 114–115

- Refer to Activity 2 from the Additional practice activities.

Apply 👥 🖥 [TWM.01]

- Display **Slide 4**.
- Learners show six more ways to regroup 706.
- Encourage them to talk in their pairs about the different ways they think of regrouping, such as 70 tens and 6 ones, 706 ones, and to write their suggestions on paper or in their books.

Review 📊

- Write some numbers on the board and ask learners to regroup them in different ways. Invite learners to demonstrate their ideas.

- Review what regrouping means, ensuring learners appreciate that regrouping can be done in various ways. Use the example 356 and ask learners to share how they could regroup the number. Confirm the different ways learners regroup 365 using the **Base 10 tool** and/or the **Place value counters tool**.

Assessment for learning

- How many ones are there in 456?
- How many tens are there in 456?
- How else can you regroup 456? Can you think of any more ways?

Same day intervention

Support

- If learners are not confident regrouping tens and ones, begin with two-digit numbers and gradually build up to three-digit numbers.

Lesson 4: **Comparing numbers**

Learning objectives

Code	Learning objective
3Np.01	Understand and explain that the value of each digit is determined by its position in that number (up to 3-digit numbers).
3Np.04	Understand the relative size of quantities to compare [and order] 3-digit positive numbers, using the symbols =, > and <.

Resources

14 cubes (per learner); Resource sheet 11: Comparison symbols (one of each symbol per learner); Resource sheet 3: Digit cards (per learner); Resource sheet 9: Place value chart (per learner); place value counters or Resource sheet 8: Place value counters (per pair)

Revise

Use the activity *Rearrange those digits* from Unit 14: *Place value and ordering* in the Revise activities to introduce comparing numbers.

Teach 🆂🅱 🖵 [TWM.01]

- **[T&T]** Display **Slide 1**. Ask: **What do you notice?** Ask learners to discuss what they think these symbols show. Expect them to know what the 'equal to' (=) symbol means. Emphasise that what is on one side must be the same as (equal to / equivalent to) what is on the other side.
- Discuss the other two symbols. Establish that one shows 'greater than' (>) and the other shows 'less than' (<). Give each learner a set of the three comparison symbols from Resource sheet 11.
- **[TWM.01]** 🖉 Ask: **What other numbers of cubes could go on either side of the greater than symbol? What about on either side of the less than symbol?** Take suggestions and model them on the board, using the comparison symbols and numbers, for example, 27 > 12, 12 < 27. Ask learners to write their own examples, using numbers, in their books, including at least one with the = symbol.
- Display **Slide 2**. Ask learners to make the two numbers on the slide, first with their digit cards in their place value charts and then with place value counters. Ask: **Which of these numbers is larger? How do you know? Which number is smaller? How do you know? How can we show this using the 'greater than' and 'less than' symbols?** Learners write two number statements to show 253 < 353 and 353 > 253.
- Display **Slide 3** and repeat for other numbers (writing each number on a blank place value chart). Keep the tens and ones digits the same and vary the hundreds. Next keep the hundreds and ones digits the same and vary the tens. Each time, ask: **Which digits must you use to compare?** Repeat by varying the ones only.

- Direct learners to the Student's Book. Discuss the examples and let them complete the paired activity.
- Discuss the Guided practice examples.

Practise 🆆🅱 [TWM.01]

- Workbook

Title: Comparing numbers

Pages: 116–117

- Refer to the first variation of Activity 2 from the Additional practice activities where after using place value counters they record, using the written method.

Apply 👥 🖵 [TWM.03/04]

- Display **Slide 4**.
- Learners explain why Seth is incorrect in his thinking that 897 is larger than 900.
- Expect learners to recognise both 900 and 874 as whole numbers and also to identify that the most significant digit in each number is the hundreds digit and as 9 hundreds is larger than 8 hundreds then 900 is larger than 874.

Review

- Give learners a number and ask them to give you some numbers that are greater than and less than the number you say.
- Draw < on the board. Invite a learner to place numbers on either side of the symbol to make a true statement. Repeat several times, then change to >.

Assessment for learning

- What does 'greater than' mean?
- What does 'less than' mean?
- Why are the greater than and less than symbols useful?

Support

- Focus on comparing three-digit numbers where the hundreds digit is the only digit that varies.

Additional practice activities

Activity 1

Learning objective
• Understand the value of the digits in a three-digit number

Resources

Resource sheet 3: Digit cards (per learner); Resource sheet 9: Place value chart (per learner); place value counters or Resource sheet 8: Place value counters (per pair)

What to do [TWM.04]

• Give each learner a set of digit cards and a place value chart.
• They make a three-digit number by placing any three digit cards, except zero, into the chart.
• They explain to a partner how their number is made up. They should discuss the position of the digits, what they represent and how the values are added together to make the whole number.

• Then they use place value counters to make their numbers.
• They each draw a picture of their number and write the additive number statement.

Variations

△2 Adapt the activity to include zero as a placeholder in the tens or ones positions.

①1 Learners who need to consolidate understanding of the value of the digits in a two-digit number can focus on making these.

Activity 2

Learning objectives
• Understand the value of the digits in a three-digit number
• Compose, decompose and regroup 3-digit numbers, using hundreds, tens and ones
• Compare 3-digit positive numbers, using the symbols > and <

Resources

Resource sheet 3: Digit cards (per learner); Resource sheet 9: Place value chart (per learner); place value counters or Resource sheet 8: Place value counters (per pair); mini whiteboard and pen (per pair) (for variation)

What to do [TWM.04/05]

• Give each learner a set of digit cards and a place value chart.
• They choose any three digit cards, which may include zero, to make a three-digit number. They do this by placing cards on the chart.
• They explain to a partner how their number is made up. They should should discuss the positions of the digits, what they represent and how the values are added together to make the whole number.
• Then they use place value counters to make their numbers.

• Learners then work out different ways in which they can be regrouped. Ensure they include tens and ones and just ones, and write these down.
• Next, learners work out other ways to regroup by partitioning their number in different ways. They write additive number statements to show how they regrouped, for example: 245 = 100 + 140 + 5.

Variations

△2 As above, but before making their numbers with place value counters pairs, write their numbers next to each other on a mini whiteboard and draw the correct symbol between them to show greater than or less than.

①1 Learners who need to consolidate understanding of the value of the digits in a two-digit number can focus on regrouping these, for example: 67 = 60 + 7, 67 = 50 + 17, 67 = 40 + 27.

Unit 15: Place value, ordering and rounding

Collins International Primary Maths Recommended Teaching and Learning Sequence: Term 2, Week 2

Learning objectives

Code	Learning objective
3Np.01	Understand and explain that the value of each digit is determined by its position in that number (up to 3-digit numbers).
3Np.02	Use knowledge of place value to multiply whole numbers by 10.
3Np.04	Understand the relative size of quantities to compare and order 3-digit positive numbers, using the symbols =, > and <.
3Np.05	Round 3-digit numbers to the nearest 10 or 100.

Unit overview

In Unit 14: Place value and ordering, learners focused on the positional, multiplicative and additive aspects of place value. In this unit, they will focus on a further aspect of place value – the base 10 element or property. This requires an understanding that the value of a digit in our number system increases or decreases by a power of 10 from one place value column to the next. Stage 3 learners will consider multiplying by 10.

In Unit 14, learners also compared numbers. In this unit they extend their understanding of comparison to order three-digit numbers in both increasing and decreasing order.

There is also a focus on rounding, which is new for Stage 3 learners. Rounding enables us to make a number simpler to use but keep it close to its value. We often use rounding to estimate what an answer to a calculation might be. For example, if we wanted to add 178 and 123, we could round the numbers to 180 and 120 and add these to give an approximate total of 300. It is less accurate but gives a reasonable estimate of the total. To be able to round, learners need a good basic understanding of place value.

Prerequisites for learning

Learners need to:
* partition three-digit numbers into hundreds, tens and ones
* understand the value of each digit in a three-digit number
* understand the role of zero as a place holder.

Vocabulary

hundreds, tens, ones, position, multiply, add, zero, place holder, order, ascending, descending, round, estimate, round up, round down

Common difficulties and remediation

It is common for learners to think that if they multiply a number by 10, they simply 'add a zero'. This clearly is not true. Multiplying a number by 10 makes it 10 times greater and all the digits move one place to the left. The zero is written in the ones position as a place holder. It is important that learners understand this. If they get into the habit of thinking that they simple 'add a zero', it will be problematic in later years. For example, by using the 'add a zero' approach to multiplying 1·6 by 10, the product effectively remains as 1·6 since placing a zero at the end gives 1·60. It is vital to ensure that learners don't develop misconceptions. Providing place value charts and asking learners to make a number with digit cards, physically moving them one place to the left and putting zero in the ones position, is helpful.

Supporting language awareness

When rounding numbers, it is important to have visual representations on display to illustrate any instructions that you give learners. It is common to tell learners that they should round up numbers ending with 5, without ever really explaining why.

There are ten digits from 0 to 9. The first half (0–4) are rounded down and the second half (5–9) are rounded up. This can usefully be displayed visually:

down < 0 1 2 3 4 | 5 6 7 8 9 > up

Promoting Thinking and Working Mathematically

During this unit learners will have opportunities to think and work mathematically, for example when exploring place value, ordering and rounding learners specialise (TWM.01) and make generalisations (TWM.02). When spotting mistakes and discussing what is the same and what is different about numbers, learners make conjectures (TWM.03), convince (TWM.04), characterise (TWM.05) and classify (TWM.06).

Success criteria

Learners can:
* order numbers in ascending and descending order
* understand what happens when a number is multiplied by 10
* round numbers to the nearest 10 and 100.

Lesson 1: **Ordering numbers**

Number – Place value, ordering and rounding

Learning objectives

Code	Learning objective
3Np.01	Understand and explain that the value of each digit is determined by its position in that number (up to 3-digit numbers).
3Np.04	Understand the relative size of quantities to compare and order 3-digit positive numbers, using the symbols =, > and <.

Revise

Use the activity *Pat, click, tap* from Unit 15: *Place value, ordering and rounding* in the Revise activities to reinforce place value in three-digit numbers.

Teach [SB] 🖳 [TWM.01/02]

- � Display **Slide 1**. Ask: **What can you tell me about the numbers on the slide?** Encourage them to tell you the digits are in place value charts. Work through the positional, and additive aspects of place value.

- [T&T] Ask: **What is the same about these three numbers? What is different?** Establish that the hundreds digits are all different, but the tens digits are the same and the ones digits are the same. Ask: **Which is the largest number? Which is the smallest number?** Invite learners to compare pairs of these numbers. Demonstrate using the greater than and less than symbols, for example 852 > 652, 652 < 852, ensuring that learners understand what they mean. There are six possibilities. Ensure the class finds them all.

- [T&T] Ask learners to work with a partner to make up pairs of three-digit numbers to compare in this way.

- [TWM.01/02] Referring back to **Slide 1**, ask: **How can we order these, from largest to smallest?** Agree that the order is 852, 652, 152. Ask: **How do you know?** Establish that they look at the hundreds digits and order by these. **What about from smallest to largest?**

- Display **Slide 2**. Ask: **What do you notice about these numbers?** Agree that these numbers have the same hundreds digits and the same ones digits. Ask: **How can we compare these numbers?** As above, ask learners to take pairs of the numbers and compare by using the greater than and less than symbols.

- [TWM.01/02] Ask: **How could we order these numbers?** Agree that this time learners need to look at the tens digits.

- Introduce the word 'ascending' for when the numbers increase or become larger in size and the word 'descending' for when the numbers decrease or get smaller in size. Ask learners to write the numbers on the slide in ascending order and then in descending order.

- [TWM.01/02] Display **Slide 3**. Ask: **What do you notice this time?** Agree that now the ones digits are different but the hundreds and tens are the same. Agree that this time, to order the numbers, learners need to look at the ones digits. Ask learners

to order the numbers in both ascending and descending order.

- [T&T] [TWM.01/02] Display **Slide 4**. Ask: **How can we order these numbers?** Ask learners to discuss what they need to do to order the four numbers. Elicit that they need to look at the hundreds digits, then the tens, then the ones. Ask learners to write the numbers in ascending and descending order.

- Direct the learners to the Student's Book and discuss the Let's learn section.

- Learners work through the paired activity.

- Discuss the Guided practice example in the Student's Book.

Practise [WB] [TWM.01/04]

- Workbook

Title: Ordering numbers

Pages: 118–119

- Refer to Activity 1 from the Additional practice activities.

Apply 👥 🖳 [TWM.07/08]

- Display **Slide 5**. Learners spot Saffir's mistake and discuss where he has gone wrong. They then order the numbers correctly.

Review

- Recap how to order numbers. Emphasise looking at the digits in the hundreds position first, then the tens and finally the ones.

- Write some three-digit numbers on the board. Ask � Ask: **How would you order these numbers, from greatest to least?**

Assessment for learning

- What is the first number when ordering these in ascending order: 324, 432, 423, 242? How do you know?
- What is the last number? How do you know?
- How can you compare pairs of these numbers: 783, 378, 873, 738? Is there another way?

Same day intervention
Support

- For those learners who may not be fully confident working with three-digit numbers, begin by ordering two-digit numbers in both ascending and descending order.

Lesson 2: **Multiplying by 10**

Learning objective

Code	Learning objective
3Np.02	Use knowledge of place value to multiply whole numbers by 10.

Resources

Resource sheet 3: Digit cards (per learner); Resource sheet 9: Place value chart (per learner); Base 10 equipment: ones, tens and hundreds (per pair)

Revise

Use the activity *Rearrange those digits* from Unit 15: *Place value, ordering and rounding* in the Revise activities to review comparing and ordering numbers.

Teach 🖿 💻

- 🄟 Display **Slide 1**. Ask: **What can you see?** Expect learners to be able to read the two numbers and to tell you how they are made, referring to the positional, multiplicative and additive elements of place value. Focus on the two representations made by Base 10 material. Ask: **How does the Base 10 material represent in 17? What about 170?**

- **[T&T]** Ask: **What is the same about 17 and 170? What is different?** Elicit that 170 is 10 times larger than 17. Ask learners to use digit cards to make 17 in their place value charts. Then ask them to make a representation of 17 with Base 10 material. They place these in their charts with the digit cards.

- Tell learners that to make 170, 17 is multiplied by 10. The value of each digit becomes 10 times larger, so 1 ten becomes 1 hundred and 7 ones becomes 7 tens. Ask: **What needs to be placed in the ones position?** Agree zero as a place holder. Ask learners to move their digit cards one place to the left and to add zero to the ones position. They then make the new number by adding the 0 digit card as the place holder. They place these in their charts.

- Display **Slide 2**. Repeat this process for multiplying 25 by 10. They make the number with digit cards in their place value charts, move the cards one place to the left and position the zero in the ones position. They also make the numbers with Base 10 material and the 0 digit card as the place holder.

- Display **Slide 3**. Repeat this process for multiplying other two-digit numbers by 10.

- Direct learners to the Student's Book. Discuss the example in the Let's learn section.

- Learners work through the paired activity.

- Discuss the Guided practice example in the Student's Book.

Practise 🆆 [TWM.02]

- Workbook

Title: Multiplying by 10

Pages: 120–121

- Refer to the second variation in Activity 1 from the Additional practice activities.

Apply 👥 💻 [TWM.04]

- Display **Slide 4**.

- Learners make a convincing statement to show why $35 \times 10 = 350$.

- Discuss the activity. Invite learners to share their explanations of how they convinced their partner.

Review

- Call out some two-digit numbers. Ask learners to tell you the products when the number is multiplied by 10.

- Call out some three-digit multiples of 10 and 100, for example 450, 310, 600 and 200, and ask learners to tell you what the numbers were before they were multiplied by 10.

- 🄟 Ask: **What happens when we multiply a number by 10?**

Assessment for learning

- What is the product when we multiply 54 by 10?
- What happens to the digits in 54?
- What do we need to put in the ones place when we multiply a two-digit number by 10? Why?

Same day intervention

Support

- Focus on multiplying two-digit numbers by 10. Ensure learners make the numbers in their place value charts and move each digit in turn to the left as they multiply.

Lesson 3: **Rounding to the nearest 10**

Learning objective

Code	Learning objective
3Np.05	Round 3-digit numbers to the nearest 10 [or 100].

Resources

mini whiteboard and pen (per learner); ruler (per learner); coloured pencil (per learner) (for the Workbook)

Revise

Use *Pat, click, tap* or *Rearrange those digits* from Unit 15: *Place value, ordering and rounding* in the Revise activities to rehearse place value. Include opportunities for multiplying by 10.

Teach [SB] 🖥 📊 [TWM.01/02]

- 👥 Display **Slide 1**. Ask: **What can you tell me about the numbers on the number line?** Expect learners to identify the numbers, tell you which are odd and even, that all the hundreds digits are 2 and most of the tens digits are 3. Accept any other observations they suggest that are correct.

- [T&T] [TWM.01/02] Ask: **What do you notice about the numbers that are circled?** Give learners a few minutes to discuss this with their partner. Remind them that in Stage 2, they rounded two-digit numbers to the nearest 10. Can they apply their prior knowledge to rounding three-digit numbers?

- [T&T] Display **Slide 2**. Ask: **What would these numbers be if we rounded them to the nearest 10?**

- Then ask: **Which other numbers would be rounded to 170?** Ask learners to use paper or their whiteboard and a ruler to draw their own number lines from 160 to 180. Agree that 166, 167, 168 and 169 round up to 170 and 171, 172, 173 and 174 round down to 170.

- Display **Slide 3**. Use this to discuss why numbers ending with 5 are rounded up. Emphasise the fact that the first half of the group of digits 0–9 in our number system are rounded down and the second half are rounded up.

- Using the **Number line tool** model examples of rounding three-digit numbers to the nearest 10 as above, including those with 5 tens and 2 ones.

- Remind learners about previous work on estimates, from Unit 6: Addition and subtraction (B). Ask: **What would be a good estimate for 32 + 39?** Elicit that learners could add 30 and 40, giving an estimate of 70. Review that the result is not accurate but it is close. Tell learners that to make this estimate they rounded 32 and 39 to the nearest 10. Apply this to 232 + 239. Agree that rounding numbers is a very useful strategy for estimating answers.

- Direct learners the Student's Book. Discuss the Let's learn section.
- Learners work through the paired activity.
- Discuss the Guided practice example in the Student's Book.

Practise [WB] [TWM.03/04]

- Workbook

Title: Rounding to the nearest 10

Pages: 122–123

- Refer to Activity 2 from the Additional practice activities.

Apply 🖥 👥👥 [TWM.04]

- Display **Slide 4**. Learners consider why rounding is a useful skill to develop and when they would use it in real life.

Review 📊

- Using the **Number line tool** set Number Line 1 to start at 340, end at 360. Set Increment and Sub-divide both to 1. Say: **350**. 👥 Ask: **What number would round to 350? Can you think of any others?** Repeat changing the numbers on the **Number line tool** accordingly.

Assessment for learning

- What is rounding?
- Which multiple of 10 would you round 168 to? Why?
- Which multiple of 10 would you round 765 to? Why?

Same day intervention
Support

- Begin by consolidating learners' understanding of rounding two-digit numbers to the nearest 10.

Lesson 4: **Rounding to the nearest 100**

Learning objective

Code	Learning objective
3Np.05	Round 3-digit numbers to the nearest [10 or] 100.

Resources

mini whiteboard and pen (per learner); ruler (per learner)

Revise

Use *Mathletics* from the Unit 15: *Place value, ordering and rounding* in the Revise activities to rehearse rounding to the nearest 10.

Teach 🆂🅱 🖥 📊

- Recap rounding three-digit numbers to the nearest 10, from Lesson 3.

- 📃 Display **Slide 1**. Ask: **What can you tell me about the numbers on the number line?** Expect learners to identify the numbers, tell you that they are three-digit numbers, multiples of ten, that most of the hundreds digits are 2, so they have a value of 200. Accept any other observations they suggest that are correct.

- [T&T] Ask: **What can you say about the numbers that are circled?** Give learners a few minutes to discuss this with their partner. Expect them to tell you that 210 is close to 200 and 280 is close to 300. Ask: **We have been thinking about rounding to the nearest 10. What number are we rounding to today?** Agree: rounding to the nearest 100.

- Ask learners to think back to Lesson 3: **Why is rounding useful?** Agree that it is a useful strategy to use when estimating. Give learners two numbers from the number line, for example 220 and 290. Ask them to round these numbers to the nearest 100 to give an estimate of the sum, i.e. 500 (200 + 300).

- [T&T] Display and discuss **Slide 2**. Agree that any three-digit number with 0, 1, 2, 3 and 4 tens is rounded down and any three-digit number with 5, 6, 7, 8, and 9 tens is rounded up. Ask: **What would 837 be rounded to? What about 557?** Agree 800 and 600 respectively. Ask: **Can you give me some three-digit numbers that are rounded to 300?** Accept any number in the range 250 to 349.

- Display **Slide 3**. Ask: **What numbers do you think are shown by the arrows?** Aim for reasonable estimates, for example, 517, 544, 569 and 591. Work together to round each number to the nearest 100.

- Ask learners to use paper or their whiteboard and a ruler to draw a number line marked in tens from 200 to 300. Tell them to draw arrows at certain points. They then give these to a partner who then estimates the numbers and then rounds them to the nearest 10 and 100. Using the **Number line tool** set Number Line 1 to start at 200, end at 300. Set Increment and Sub-divide both to 10 and model learners' examples.

- Write a pair of three-digit numbers on the board, for example, 243 and 487, and ask learners to estimate the sum by rounding to the nearest 100 (200 + 500 = 700). Then ask the learners to estimate the sum by rounding to the nearest 10 (240 + 490 = 730). Discuss which estimate is more accurate and why.

- Direct learners to the Student's Book. Discuss the Let's learn section. Learners work through the paired activity. Discuss the Guided practice example in the Student's Book.

Practise 🆆🅱 [TWM.04]

- Workbook

Title: Rounding to the nearest 100

Pages: 124–125

- Refer to the second variation in Activity 2 from the Additional practice activities.

Apply 🖥 👥 [TWM.03/04]

- Display **Slide 4**. Learners need to consider the estimates of Leo and Sophie and decide which one they agree with. They should explain their thinking to their partner

Review 📊

- Using the **Number line tool** set Number Line 1 to start at 400, end at 600. Set Increment and Sub-divide both to 10. Say: **500**. 📃 Ask: **What number would round to 500? Can you think of any others?** Repeat changing the numbers on the **Number line tool** accordingly.

Assessment for learning

- To which multiple of 100 would you round 368? Why?
- To which multiple of 100 would you round 809? Why?
- Why is rounding a useful skill?

Same day intervention

Support

- Focus on rounding three-digit numbers to the nearest 10 with learners who have not mastered this yet.

Additional practice activities

Activity 1

Learning objective
- Order three-digit numbers

Resources
Resource sheet 3: Digit cards (per learner)

What to do
- Give each learner a set of digit cards.
- They each make three three-digit numbers.
- Learners give these numbers to their partner who orders them in both ascending and descending order. They write down the numbers in order.
- Once they have done this, they rearrange the digits in their numbers to make new ones and repeat.
- They continue until they have made all the new numbers that they can.

Variations

1 For learners who need consolidation of comparing and ordering two-digit numbers, focus on ordering these in both ascending and descending order.

2 Learners make two-digit numbers with their digit cards, multiply them by 10 and then order them in both ascending and descending order.

Activity 2

Learning objective
- Round three-digit numbers to the nearest 10 and 100

Resources
Resource sheet 3: Digit cards (per learner)

What to do
- Give each learner a set of digit cards.
- They make a three-digit number. They write their number on paper or in their books.
- They then round it to the nearest 10 and explain to a partner how they did this.
- Next, they write down the other numbers that also round to that tens number.
- They repeat this as many times as time allows.

Variations

2 Working with their partner, one learner gives the other a three-digit multiple of 10 for example, 430. Their partner then writes down all the numbers that will round to it.

2 As well as rounding to the nearest 10, learners also round their three-digit numbers to the nearest 100.

They write down five more numbers that round to that multiple of 100.

Unit 16: Fractions (A)

Collins International Primary Maths
Recommended Teaching and
Learning Sequence: Term 1, Week 6

Learning objectives

Code	Learning objective
3Nf.01	Understand and explain that fractions are several equal parts of an object or shape and all the parts, taken together, equal one whole.
3Nf.02	Understand that the relationship between the whole and the parts depends on the relative size of each, regardless of their shape or orientation.
3Nf.03	Understand and explain that fractions can describe equal parts of a quantity or set of objects.

Unit overview

In this unit, learners focus on fractions of shapes and quantities. In Stages 1 and 2, learners explored unit fractions (numerator of 1), specifically $\frac{1}{2}$ and $\frac{1}{4}$. In Stage 2, learners explored three quarters. In this unit, they will consider other unit and non-unit fractions (numerators that are not 1). One of the key models for fractions, explored in this unit, is the part–whole model. Learners will explore fractions as being made up of equal parts of one whole. The number of parts into which the whole has been divided doesn't matter: provided that the parts are equal, they are fractions. A fraction is always one or more parts of the same whole, which can be anything. Learners need to be able to describe the whole as well as the parts. The vinculum is the line that separates the denominator and numerator. It is good practice to draw the vinculum first to show we are breaking the whole into parts; then write the denominator below it, to show the number of equal parts there are in the whole, and finally write the numerator, which shows how many parts we are considering. It is important that learners develop an understanding that a fraction of a whole can be the same fraction regardless of the size of the whole.

Prerequisites for learning

Learners need to:
- understand that half is one out of two equal parts
- understand that one quarter is one out of four equal parts
- be able to find halves and quarters of quantities and shapes.

Vocabulary

fraction, unit fraction, non-unit fraction, whole, part, equal, denominator, numerator

Common difficulties and remediation

Learners need to understand that fractions are made up of equal parts of a whole. It is important to display visual representations of different fractions. To enable learners to gain a good understanding of fractions, it is vital to use the part–whole model and representations such as double-sided counters or coloured cubes.

Supporting language awareness

Learners should be introduced to the words 'denominator' and 'numerator'. It would be helpful to display the image of a fraction, with each part labelled. The vinculum should always be drawn as a horizontal line. The terms 'fourth' and 'quarter' may be used interchangeably, although 'quarter' is the more common term. Within the context of fractions, using fourths can be helpful because it indicates four parts.

Promoting Thinking and Working Mathematically

During this unit, learners will have opportunities to think and work mathematically, for example specialising when they choose different fractions that fit certain criteria (TWM.01) and making generalisations about fractions (TWM.02). They discuss similarities and differences between fractions, which involves making conjectures (TWM.03), convincing (TWM.04), characterising (TWM.05) and classifying (TWM.06).

Success criteria

Learners can:
- understand that a fraction is one or more equal parts of a whole
- understand that fractions can describe a number of equal parts of a quantity or set of objects
- understand that a fraction of any whole is the same fraction, no matter the size of the whole.

Number – Fractions, decimals, percentages, ratio and proportion

Unit 16 Fractions (A)

Number – Fractions, decimals, percentages, ratio and proportion

Lesson 1: **Equal parts: quantities**

Learning objectives

Code	Learning objective
3Nf.01	Understand and explain that fractions are several equal parts of an object or shape and all the parts, taken together, equal one whole.
3Nf.03	Understand and explain that fractions can describe equal parts of a quantity or set of objects.

Resources

counters (per learner)

Revise

Use the activity *Five-minute halving* from Unit 16: *Fractions (A)* in the Revise activities.

Teach 🔲 💻 [TWM.01]

- ▶ **[TWM.01]** Display **Slide 1**. Ask: **What do you notice?** Encourage learners to talk about the different coloured sets of bars. Establish that these diagrams show the part–whole relationship for fractions.

- **[T&T] [TWM.01]** Ask: **What relationships can you see?** Give learners time to discuss. Ask: **What do the blue bars show?** Agree that the top bar is one whole and the second shows two halves. Repeat for the other bars. Establish that these diagrams all show wholes divided into equal parts/fractions and together the fractions make a whole. Explain that the top bar in each diagram shows the whole, the blue bars show halves, the green bars thirds, the red bars quarters and the yellow bars fifths.

- Demonstrate how to write these fractions correctly, using the correct words: draw the vinculum first to show the whole has been broken into parts, then write the denominator to show the number of parts, finally the numerator to show how many parts are being considered. As you draw each part of the fraction, say the word.

- Ask: **How many of each part make up one whole?** Agree that one whole is equivalent to two halves, three thirds, four quarters (or fourths) and five fifths. Elicit that when the numerator and denominator are the same, this represents one whole.

- Display **Slide 2**. Say: **Each whole represents a number. We know what each whole represents, but what are the parts?** Explain that knowing multiplication and division facts helps find fractions of numbers. For example, if we know our facts for 2, we know that $16 \div 2 = 8$ and if we know our facts for 5, we know that $45 \div 5 = 9$. If we need to find a quarter, we can halve and halve again. Provide counters, so that learners can practically halve and halve again. To find one third of 9, learners need to count on in steps of 3 (3, 6, 9), using a number line or counters if necessary, and then count the steps.

- Tell learners that a fraction with a numerator of 1 is called a 'unit fraction' and that fractions with any other numerator are called 'non-unit fractions'.

Establish that if learners can work out unit fractions of one whole, they can work out non-unit fractions. For example, one quarter of 16 is 4 so three quarters is $4 + 4 + 4 =$ or $4 \times 3 =$ which is 12. Show how three quarters is written as a fraction $\left(\frac{3}{4}\right)$. Ask them to work out three quarters of different numbers.

- Ask: **What is $\frac{1}{5}$ of 10?** Agree 2. Ask learners to tell you what $\frac{2}{5}$, $\frac{3}{5}$ and $\frac{4}{5}$ of 10 will be. Substitute different values for the whole (within the range of the tables learners know) and ask what the parts will be. Reverse this, by giving them the value of one part and asking them to work out the whole.

- Discuss Let's learn in Student's Book before asking learners to work on the paired activity.

- Discuss the Guided practice example.

Practise 🔲

- Workbook

Title: Equal parts: quantities

Pages: 126–127

- Refer to Activity 1 from the Additional practice activities.

Apply 👥 💻 [TWM.07]

- Display **Slide 3**. Learners discuss, and show, whether Tilly is correct.

Review

- Recap the key learning in this lesson, that a fraction is made up of equal parts of a whole.

- Say some multiples of 4 and ask learners to find one half, one quarter and three-quarters of them.

Assessment for learning

- What do you need to do to find half of 20?
- If you know one quarter of 16, how can you find three-quarters of 16?
- What is the relationship between one half and one quarter?

Same day intervention
Support

- Focus on halves and quarters.

Lesson 2: **Equal parts: shapes**

Learning objectives

Code	Learning objective
3Nf.01	Understand and explain that fractions are several equal parts of an object or shape and all the parts, taken together, equal one whole.
3Nf.02	Understand that the relationship between the whole and the parts depends on the relative size of each, regardless of their shape or orientation.

Resources

three squares of paper (per pair); scissors (per pair); coloured paper (per pair); ruler (per pair)

Revise

Use the activity *Five-minute halving* from Unit 16: *Fractions (A)* in the Revise activities.

Teach [SB] 🖵 [TWM.01]

- 🖏 Display **Slide 1**. Ask: **What do you notice?** Expect learners to be able to tell you that each shape has been divided into four equal parts and one part has been shaded. The fraction shaded is $\frac{1}{4}$.

- **[T&T] [TWM.01]** Display **Slide 2**. Ask: **What is the same about these shapes? What is different?** Agree that each shape has been divided into four parts. Ask: **Do these all show quarters?** Establish that A, C, D and E do but B and F don't. Ask learners to discuss why each part on shapes A, C, D and E is $\frac{1}{4}$ and why those on the other two are not. Shape A is the most unusual as the parts are different shapes. Learners should be able to see that the rectangle has been divided into half and each half has been halved, so each part must be $\frac{1}{4}$.

- Use **Slides 3** and **4** to reinforce the idea of equal and unequal parts.

- **[T&T]** Display **Slide 5**. Give each pair of learners three squares of paper and scissors. They fold the pieces of paper to show quarters in the three different ways shown on the slide. They then cut $\frac{1}{4}$ from each square and prove that all four of these quarters are the same size. Expect them to cut the rectangular quarter in half and arrange it so that it is the same as the square quarter. If necessary, help them to cut the triangular quarter in half so that it can be rearranged to match the square quarter. Ask: **What does this tell us about fractions of a shape?** Establish that fractions of shapes are not about their shape, they are about the amount of space they take up.

- Discuss Let's learn in Student's Book before asking learners to work on the paired activity.

- Discuss the Guided practice example.

Practise [WB]

- Workbook

Title: Equal parts: shapes

Pages: 128–129

- Refer to Activity 1 from the Additional practice activities.

Apply 👥 🖵 [TWM.01/02]

- Display **Slide 6**. Learners are shown an L shape and told it is a fraction of a shape. They establish how many of these shapes there will be in the whole and show different possibilities for the whole shape. Expect learners to put their L shapes together in different ways and to draw the new shapes that they can make.

- If necessary, before they begin, review **Slide 1**, where they were discussing three different whole shapes made from the same shaped quarter.

Review

- Recap the fact that, when finding fractions of shapes, the amount of space they take up is the fraction, irrespective of the shape it makes. Therefore, the same fraction can be represented in different ways.

- Discuss the fact that the whole is made up of all the parts. For example, if a shape or quantity is divided into thirds, three thirds make the whole.

- Discuss the generalisation that can be made: the denominator and numerator must be the same for the fraction to represent the whole.

Assessment for learning

- How many thirds/fifths make one whole?
- How many quarters are equivalent to one half?

Same day intervention

Support

- For **Apply**, tell learners that it is half of the shape. They explore different ways to make the whole using two L shapes only.

Number – Fractions, decimals, percentages, ratio and proportion

Unit **16** Fractions (A)

Lesson 3: **Same fraction, different whole**

Learning objective

Code	Learning objective
3Nf.02	Understand that the relationship between the whole and the parts depends on the relative size of each, regardless of their shape or orientation.

Revise

Use the activity *Fraction doodles* from Unit 16: *Fractions (A)* in the Revise activities to review and consolidate finding fractions of shapes.

Teach 🔲 💻 [TWM.03/04]

- 🔲 Display **Slide 1**. Ask: **What do you notice about these shapes?** Expect learners to tell you that the shapes have all been divided into five equal parts and that one of the five parts has been shaded in a darker colour. Ask: **What fraction of each shape has been shaded?** Agree $\frac{1}{5}$.

- Emphasise that the shapes are different sizes and in different orientations but, in each case, $\frac{1}{5}$ of the whole has been shaded. Each shape has been divided into five equal parts and each part is $\frac{1}{5}$, although the shapes are different sizes.

- [T&T] Display **Slide 2**. Ask: **Do you agree with Ami?** Take feedback after a few minutes. Establish that Ami is incorrect and that in each shape $\frac{1}{4}$ is shaded. Each quarter is $\frac{1}{4}$ of the whole of the shape. Stress that it doesn't matter that the shapes are different or in different positions.

- [T&T] [TWM.03/04] Display **Slide 3**. Ask learners to describe what they can see on the slide and to discuss with a partner which whole shape they think is longer, the red or the yellow. Ask: **Which shape do you think is the longer of the two? How can you prove it?** Agree that the part of the red rectangle labelled $\frac{1}{2}$ needs one more part just like the part showing to give the whole. Elicit that the part of the yellow rectangle labelled $\frac{1}{3}$ needs two more parts just like the part showing to make the whole. Therefore, the whole yellow rectangle must be longer than the whole red rectangle.

- Direct learners to the Student's Book and let them work through Let's learn.

- [TWM.04] Learners work through the paired activity. Can they identify the odd one out and explain why?

- Discuss the Guided practice example in the Student's Book.

Practise 🔲 [TWM.04]

- Workbook

Title: Same fraction, different whole

Pages: 130–131

- Refer to Activity 2 from the Additional practice activities.

Apply 👥 💻 [TWM.01/02]

- Display **Slide 4**. Learners are shown a square divided in such a way that a rectangle representing one third is marked and labelled $\frac{1}{3}$. Then the remainder of the shape has been divided into halves by a diagonal line, and one half is shaded.

- They need to consider the fraction that has been shaded and say what fraction this is.

- Encourage them to discuss what fraction the shaded part is.

- If necessary, help learners to identify that the whole must be $\frac{3}{3}$, so the rectangle enclosing the shaded part must be $\frac{2}{3}$ of the whole shape. Then the shaded part must be $\frac{1}{3}$ as it is is half of the remaining $\frac{2}{3}$.

Review

- Discuss the fact that, when looking at fractions of a shape, the shape is the whole and the parts must be equal.

- Recap how the shape, size and position of a shape don't matter; the parts representing the fractions relate to that shape and not any other shapes that may be smaller, larger or positioned differently.

Assessment for learning

- How can we find a fraction of a shape?
- Can a square and a rectangle both be divided into thirds? How many equal parts would there be in each shape?
- Can two shapes of different sizes both show one third? How?
- Can two shapes in different positions both show one third? How?

Same day intervention
Support

When looking at fractions of shapes, focus on finding halves and quarters.

Lesson 4: **Fractions of quantities and sets**

Learning objective

Code	Learning objective
3Nf.03	Understand and explain that fractions can describe equal parts of a quantity or set of objects.

Resources
cubes (per pair)

Revise

Use any activity from Unit 16: *Fractions (A)* in the Revise activities.

Teach [SB] [💻] [TWM.03/04]

- 🄿 Display **Slide 1**. Ask: **How many apples would make up one half of them?** Expect learners to tell you that there are 12 and half of them would be 6.

- **[T&T] [TWM.03/04]** Ask learners to work with a partner to answer the question on the slide. After a few minutes, ask: **How do you know that she puts three apples in the basket?** Give learners a few minutes to discuss this with a partner. Expect them to be able to tell you that $\frac{1}{4}$ of 12 is 3, so she puts three apples in the basket. Discuss the part–whole model and how this helps them visualise the whole divided into four equal parts.

- Ask: **What if she puts $\frac{3}{4}$ in the basket? What about $\frac{1}{3}$? $\frac{2}{3}$? $\frac{1}{6}$?** Continue asking about other fractions, including sixths.

- **[T&T]** Ask: **What do you notice about $\frac{1}{2}$, $\frac{2}{4}$ and $\frac{3}{6}$?** Agree that they all represent the same number of apples.

- **[T&T]** Display **Slide 2**. Ask learners to look at the question and to talk to their partners about how to find the number of ducks on the farm. After a few minutes take feedback. Establish that the whole is five fifths. They know $\frac{1}{5}$ is five, so they need to multiply 5 by 5 to find $\frac{5}{5}$ or the whole. Again, discuss how the part–whole model helps.

- **[T&T] [TWM.03/04]** Display **Slide 3**. Ask learners to work with a partner to find out how many toffees Maddie started with. Take feedback. Establish that if she ate $\frac{2}{3}$, the number she has left must represent $\frac{1}{3}$. $\frac{1}{3}$ of the amount is therefore 4, so $\frac{2}{3}$ must be 8 and the whole must be 12. Discuss how the part–whole model helps.

- Direct learners to the Student's Book. Discuss the Let's learn section.

- Learners work through the paired activity.
- Discuss the Guided practice example in the Student's Book.

Practise [WB] [TWM.01]

- Workbook

Title: Fractions of quantities and sets

Pages: 132–133

- Refer to the variation in Activity 2 from the Additional practice activities.

Apply 👥 💻

- Display **Slide 4**. Learners work out how many children there are in the school Maths club. They are given some information that will help them solve the problem.
- Learners draw a part–whole model and add the information given in order to find the answer.

Review

- As a class, discuss the activity from **Apply**. Invite learners to draw the part–whole model for the problem and to explain how many learners were in the school Maths club.
- Discuss the value of part–whole models in solving problems involving fractions.
- Set problems for learners to solve by using a part–whole model, similar to those considered in this lesson. Invite learners to demonstrate how to solve them.

Assessment for learning

- If 12 represents half, what is the whole? How do you know?
- If 8 represents $\frac{1}{4}$, what is $\frac{3}{4}$? What is the whole?
- If 6 represents $\frac{2}{5}$, what is $\frac{1}{5}$? What is the whole?

Same day intervention

Support

- Focus on finding halves, thirds and quarters of quantities.

Number – Fractions, decimals, percentages, ratio and proportion

Additional practice activities

Activity 1

Learning objective
- Understand and explain that fractions are several equal parts of an object and all the parts, taken together, equal one whole

Resources
40 small objects such as counters, cubes or beads (per pair)

What to do
- Provide each pair with 40 small objects.
- The first learner takes a handful of objects, counts them, ensures there is an even number and then makes up a number story based on finding half of the total.

- The second learner uses the strategy of their choice to work out what half of the first learner's number is.
- They split the objects into two equal piles to check the answer is correct.

Variation
2 The first learner collects an even number, from 10 to 40, of objects. Ensure they collect an even number. They then ask the second learner to find half of those objects.

Activity 2 **2**

Learning objective
- Understand and explain that fractions are several equal parts of an object and all the parts, taken together, equal one whole

Resources
A4 paper cut across the width into strips (per learner)

What to do
- Give each learner four strips of paper.
- They keep one strip whole and write 1 in the middle.
- They fold a second in half. What fraction have they made? They label each part $\frac{1}{2}$.
- They fold the third in half and half again. What fraction have they made now? They label each part $\frac{1}{4}$.
- They fold the fourth strip in half, half again and half once more. What fraction have they made this time? They label each part $\frac{1}{8}$.
- Ask them to write a statement to show how many parts make the whole, for example $\frac{2}{2} = 1$, $\frac{4}{4} = 1$.

Variation
2 Give learners a value for the whole. They then work out the values of the fractional parts. For example, the whole is 24, $\frac{1}{2} = 12$, $\frac{1}{4} = 6$ and $\frac{1}{6} = 3$.

Unit 17: Fractions (B)

Collins International Primary Maths
Recommended Teaching and
Learning Sequence: Term 3, Week 4

Learning objectives

Code	Learning objective
3Nf.06	Recognise that two fractions can have an equivalent value (halves, quarters, fifths and tenths).
3Nf.08	Use knowledge of equivalence to compare and order unit fractions and fractions with the same denominator, using the symbols =, > and <.

Unit overview

In this unit, learners focus on equivalent fractions. They will have already started to notice equivalence when they explored halves and quarters (or fourths). This unit looks at this in more depth, as learners begin to understand that fractions can have the same value despite looking different. They will then use their understanding of fractions and equivalences to compare and order fractions. As they learn about equivalence and start comparing fractions, learners will have the opportunity to reinforce their knowledge of the symbols for 'greater than', 'less than' and 'equal to'.

Prerequisites for learning

Learners need to:

- understand the equivalence of $\frac{1}{2}$ and $\frac{2}{4}$

- understand the equivalence of $\frac{2}{2}$, $\frac{4}{4}$ and 1

- understand that $\frac{1}{2}$ is greater than $\frac{1}{4}$ when the whole is the same.

Vocabulary

fraction, whole, part, equal, denominator, numerator, equivalent, compare, greater than, less than, order

Common difficulties and remediation

Fractions are different from whole numbers in that there are infinitely many fractions that have the same value. With whole numbers, each number has a one particular value. It is important to display visual representations to show fraction equivalences.

Supporting language awareness

Learners should be reminded of the words 'denominator' and 'numerator' as they are introduced to equivalent fractions. It would be helpful to have the image of a fraction displayed with each part labelled. It would also be helpful to display bar models showing the equivalent fractions studied during this unit.

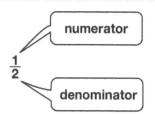

Promoting Thinking and Working Mathematically

During this unit, learners will have opportunities to think and work mathematically; for example, when they look at equivalent fractions that have the same value but look different, this involves specialising (TWM.01), generalising (TWM.02) and critiquing (TWM.07). They explore fractions of different wholes and make conjectures (TWM.03) about them and present evidence to convince others about their mathematical thinking (TWM.04). They also identify the properties of fractions and organise them into groups, which involves characterising (TWM.05) and classifying (TWM.06).

Success criteria

Learners can:

- understand and explain that equivalent fractions have the same value but look different

- understand and explain that the greater the denominator, the smaller the fraction of the same whole

- order unit fractions with different denominators and non-unit fractions with the same denominator.

Number – Fractions, decimals, percentages, ratio and proportion

Lesson 1: **Same value, different appearance**

Learning objective

Code	Learning objective
3Nf.06	Recognise that two fractions can have an equivalent value (halves, quarters, fifths and tenths).

Resources

four strips of paper the width of an A4 sheet (per learner)

Revise

Use the activity *Fractions of numbers (1)* from Unit 17: *Fractions (B)* in the Revise activities to reinforce and consolidate finding fractions of numbers.

Teach 🟦 💻 [TWM.01/02]

- 🔲 Display **Slide 1**. Discuss what it shows. Discuss the appearance of the diagram and establish that it looks like a wall, with equal rows but bricks of different sizes. Ask: **What do you notice?** Encourage learners to talk about the different coloured bars in the fraction wall. Ask them to tell you what each bar represents. Agree that the first shows one whole, the second shows halves, the third shows quarters (or fourths) and the fourth shows eighths.
- Ask learners to use their strips of paper to fold and make bars similar to those on the slide. They label the first strip as 1 to show that it is one whole. They label each part of the other strips with the appropriate fraction.
- [T&T] Ask: **What else do these bars show?** Give learners a few minutes to discuss this. Take feedback and, if necessary, ask: **Can you see any equivalent fractions?** Make sure that learners understand or recall the term 'equivalent' meaning 'having the same or equal value'. Agree that the whole bar is equivalent to one whole in each case, then elicit that one half is equivalent to two quarters and also to four eighths. Can anyone tell you that one quarter is equivalent to two eighths?
- Ask learners to look at their strips and identify the equivalent fractions. Ask them to write statements that show these, for example: $\frac{1}{2} = \frac{2}{4} = \frac{4}{8}$.
- [TWM.01/02] [T&T] Display **Slide 2**. Ask: **Why does it say 'Same value, different appearance!' on the slide?** Establish that each fraction has the same value: they are all showing one half. Each fraction looks different. Ask learners to explain how each looks different.
- Repeat for **Slide 3**.
- Direct learners to the Student's Book. Discuss the Let's learn section before learners work on the paired activity. Then discuss the Guided practice example.

Practise 🆆🅱 [TWM.02]

- Workbook

Title: Same value, different appearance

Pages: 134–135

- Refer to Activity 1 from the Additional practice activities.

Apply 👥 💻 [TWM.03/04]

- Display **Slide 4**. Learners decide if Nina is correct in thinking that the two fractions show that $\frac{1}{2}$ is equivalent to $\frac{2}{4}$. They should discuss this with their partner and explain why they agree or disagree. Encourage them to write a written explanation.
- Discuss the activity. Invite learners to share their thinking. Establish that the diagrams do not show that $\frac{1}{2}$ is equivalent to $\frac{2}{4}$ because the two diagrams are different sizes, so the wholes are not equivalent. The diagram with one half shaded is much smaller than the diagram with $\frac{2}{4}$ shaded. Fractions can be equivalent only if the wholes they are part of are the same size.

Review

- Invite learners who completed Challenge 3 in the Workbook to explain what they noticed about the fractions that are equivalent to one half. Establish that the numerator is half of the denominator and the denominator is double the numerator.
- Write some numerators on the board for learners to double to give their denominators and so make fractions that are equivalent to one half.

Assessment for learning

- What do we mean by 'same value, different appearance'? Why is this a good way to describe equivalent fractions?
- Can you tell me two equivalent fractions? Can you tell me two more?

Same day intervention

Support

- Focus on equivalences between halves, quarters and eighths. Always provide visual representations to aid learners understanding.

Number – Fractions, decimals, percentages, ratio and proportion

Lesson 2: **Equivalent fractions**

Learning objective

Code	Learning objective
3Nf.06	Recognise that two fractions can have an equivalent value (halves, quarters, fifths and tenths).

Resources

mini whiteboard and pen (per learner)

Revise

Use the activity *Fractions of numbers (1)* from Unit 17: *Fractions (B)* in the Revise activities to reinforce and consolidate finding fractions of numbers.

Teach 📖 🖥 [TWM.01]

- 📖 [TWM.01] Display **Slide 1**. Ask: **What do you notice?** Expect learners to be able to tell you different equivalent fractions. Ask them to write the equivalent statements for the fraction wall, for example, $\frac{1}{2} = \frac{2}{4} = \frac{4}{8} = \frac{5}{10}$ and $\frac{1}{5} = \frac{2}{10}$.

- Recap what they know about the numerators and denominators for equivalent fractions for $\frac{1}{2}$ from the Lesson 1.

- [T&T] Display **Slide 2**. Ask: **What is the same about the shaded fractions? What is different?** Give learners the opportunity to talk to a partner. Take feedback. Establish that the fractions all have the same value. The difference is that they are made up of different numbers of parts.

- [T&T] Ask: **What do you notice about the numerators?** Agree they increase by 1 each time (one quarter, two eighths and three twelfths). Ask: **What do you notice about the denominators?** Agree that they increase by 4 each time. Establish that these numbers are all products in the multiplication facts for 4. Ask: **What do you think the next equivalent fraction will be?** Allow time for learners to discuss with a partner, then take feedback. Agree $\frac{4}{16}$. You could ask learners to predict what the next five or six equivalent fractions would be.

- Display **Slide 3**. Ask: **What do you notice? Does anything look familiar?** Discuss how the number lines show the 2, 4, 5 and 10 times tables facts. Draw the link between the multiplication facts and equivalent fractions, and the patterns or sequences made by the numerators and denominators. Referring to each number line in turn, ask learners to predict what the missing equivalent fractions are and write these on the number lines.

- [T&T] Display **Slide 4**. Ask: **What is the same about the shaded fractions? What is different?** Establish that the shaded fractions are all equivalent to $\frac{1}{5}$. The value is the same but they all look

different. Ask learners to describe how they are different.

- Direct learners to the Student's Book. Discuss the Let's learn section before learners work on the paired activity. Then discuss the Guided practice example.

Practise 📓 [TWM.01/04]

- Workbook

Title: Equivalent fractions

Pages: 136–137

- Refer to the variation in Activity 1 from the Additional practice activities.

Apply 👥🖥 [TWM.01]

- Display **Slide 5**. The diagram shows that $\frac{1}{5}$ is equivalent to $\frac{2}{10}$. Learners work with a partner to write down other equivalences shown by the diagram, for example: $\frac{2}{5} = \frac{4}{10}$, $\frac{3}{5} = \frac{6}{10}$.

Review

- Write some fractions (halves, quarters, fifths and tenths) on the board. Ask learners to tell you some equivalent fractions for each of them.

Assessment for learning

- What is an equivalent fraction?
- Can you give me a fraction that is equivalent to $\frac{1}{2}$?
- Can you give me a fraction that is equivalent to $\frac{1}{5}$?
- What about $\frac{1}{4}$?

Same day intervention

Support

- Focus on equivalences between fifths and tenths. Always provide visual representations, such as diagrams, to aid learners understanding.

Unit **17** Fractions (B)

Lesson 3: **Comparing fractions**

Learning objective

Code	Learning objective
3Nf.08	Use knowledge of equivalence to compare [and order] unit fractions and fractions with the same denominator, using the symbols =, > and <.

Resources

mini whiteboard and pen (per learner); squared paper (per learner)

Revise

Use the activity *Equivalent fractions* from Unit 17: *Fractions (B)* in the Revise activities to reinforce and consolidate equivalent fractions.

Teach 📖 🖥 [TWM. 03/04]

- **[T&T]** Display **Slide 1**. Ask: **What fractions can you see? Which pentagon has more parts shaded? Which is the larger fraction? Which is smaller?** Give learners the opportunity to discuss with a partner. Agree that the fractions are $\frac{2}{5}$ and $\frac{3}{5}$, and that $\frac{3}{5}$ is the larger. Establish that when the denominators of the fractions are the same, you can compare them by looking at the numerator.

- Ask: **What symbols do we use to compare numbers?** Learners should recall that we use greater than (>) and less than (<). Ask: **How can we use those symbols to compare these two fractions?** Ask learners to write two statements, then check that they have written $\frac{2}{5} < \frac{3}{5}$ and $\frac{3}{5} > \frac{2}{5}$.

- Give learners squared paper. Ask them to draw two bars that each enclose five squares. They shade each bar to show a different fraction, then they write two statements to compare their fractions.

- Display **Slide 2**. Ask: **What can you tell me about the fractions in this fraction wall?** Expect learners to be able to tell you that it shows halves, quarters and eighths. Ask: **What does it tell us about the sizes of the fractions?** Agree that it shows that some fractions are equivalent to others. Ask: **What else does it tell us?** Establish that, in this wall, half is the largest fraction and one eighth is the smallest.

- **[TWM.03/04] [T&T]** Ask: **Two is a smaller number than eight, so why is a half a larger fraction than an eighth?** Give learners a few minutes to discuss this with their partner. Establish that when we cut up a pizza, cutting it into two equal parts gives larger slices than cutting it into eight. The more slices we cut, the smaller they will be. So the largest unit fraction of any whole is one half.

- Ask learners to write two statements to compare halves and eighths. Check that they have written: $\frac{1}{2} > \frac{1}{8}$ and $\frac{1}{8} < \frac{1}{2}$. Ask: **Is a half larger or smaller than one quarter?** Agree that one half of a whole

is always the largest fraction. Also agree that the larger the denominator, the smaller the fraction.

- **[T&T]** Display **Slide 3**. Ask: **What do these pictures tell us?** Take feedback after a few minutes. Agree that one half of the small pizza is smaller than one quarter of the larger one. Establish that we can only compare fractions if they are part of the same whole, or the wholes are the same size.

- Direct learners to the Student's Book. Discuss the Let's learn section before learners work on the paired activity. Then discuss the Guided practice example.

Practise 📝 [TWM.04]

- Workbook

Title: Comparing fractions

Pages: 138–139

- Refer to Activity 2 from the Additional practice activities.

Apply 👥 🖥 [TWM.03/04]

- Display **Slide 4**. Learners are given an explanation of how to compare fractions. They discuss with a partner whether it is a good explanation and why.

- Discuss the activity as a class. Invite learners to explain their reasoning about Bhavana's explanation.

Review

- 🗣 Ask: **Does the larger the denominator mean that the fraction is larger? Why?**

- Recap the fact that if the denominators are the same, the numerators need to be considered. The largest numerator will be the largest fraction.

Assessment for learning

- Which is the larger fraction, $\frac{1}{3}$ or $\frac{1}{5}$? How do you know?

- Which is the smaller fraction, $\frac{1}{6}$ or $\frac{1}{8}$? How do you know?

- Which is the larger fraction, $\frac{3}{10}$ or $\frac{7}{10}$? How do you know?

Same day intervention
Support

- Focus on comparing fractions with the same denominator.

Number – Fractions, decimals, percentages, ratio and proportion

Lesson 4: **Ordering fractions**

Learning objective

Code	Learning objective
3Nf.08	Use knowledge of equivalence to [compare and] order unit fractions and fractions with the same denominator, using the symbols =, > and <.

Resources

two sheets of squared paper and pen (per learner); coloured pencils (per learner)

Revise

Use the activity *Comparing fractions* from Unit 17: *Fractions (B)* in the Revise activities to reinforce and consolidate comparing fractions.

Teach 🖪 💻 [TWM.03/04]

- 🖐 **[TWM.03/04]** Display **Slide 1**. Give learners two sheets of squared paper and a pen. Ask learners to take one sheet and to draw five identical rectangles like the one shown on the slide and then to divide them to show the fractions listed. Ask: **Which is the smallest fraction that you made? How do you know?** Expect learners to tell you that one tenth is the smallest because that rectangle has been divided into the most parts, so tenths must be smaller than the other fractions.

- **[T&T]** Ask: **Can you order the fractions on the slide from smallest to greatest?** Give learners a few minutes to discuss this with their partner. Expect them to be able to tell you the order: $\frac{1}{10}, \frac{1}{8}, \frac{1}{5}, \frac{1}{4}, \frac{1}{3}$.

- Display **Slide 2**. Ask: **What do you notice on this slide?** Agree that the diagrams show different amounts of tenths shaded. Using their second sheet of squared paper, ask learners to draw three bars, each enclosing ten squares. Using coloured pencils they shade them to show different numbers of tenths. Then they order their fractions, from smallest to largest. Repeat the activity for fifths, but ask them to order from largest to smallest.

- **[T&T]** Display **Slide 3**. Ask: **What do you notice on this slide?** Agree that there are number lines that show different fractions, in order. Ask: **How does this help us to order fractions?** Allow time for discussion, in pairs. After a few minutes, take feedback. Establish that this is a useful representation to show how to order unit fractions. Agree that the slide shows that $\frac{1}{8}$ is the smallest fraction and $\frac{1}{2}$ is the largest.

- Ask learners to draw their own number line from 0 to 1 and plot the fractions from $\frac{1}{10}$ to $\frac{8}{10}$.

- Direct learners to the Student's Book. Discuss the Let's learn section before learners work on the paired activity. Then discuss the Guided practice example.

Practise 🔳 [TWM.01/07]

- Workbook

Title: Ordering fractions

Pages: 140–141

- Refer to the variation in Activity 2 from the Additional practice activities.

Apply 👥 💻 [TWM.04]

- Display **Slide 4**.

- Learners discuss Barnie's way of ordering fractions and think about what he should have done.

Review

- Recap the learning from this lesson, particularly that the larger the denominator, the smaller the fraction. Therefore, unit fractions with different denominators can be ordered according to the size of the denominator.

- 🖐 Ask: **If the denominators of fractions are the same, what do we look at to order them?**

Assessment for learning

- How would you order $\frac{1}{3}, \frac{1}{6}, \frac{1}{4}, \frac{1}{5}, \frac{1}{10}$ and $\frac{1}{2}$? Is there another way?

- Can you explain what we must do to order fractions with different denominators?

- How would you order $\frac{2}{5}, \frac{4}{5}$ and $\frac{1}{5}$? Is there another way?

- Can you explain what we must do to order fractions with the same denominator?

Same day intervention

Support

- Focus initially on ordering fractions with the same denominator, then look at different denominators. Ensure learners always have visual representations to help aid their understanding.

Number – Fractions, decimals, percentages, ratio and proportion

Additional practice activities

Activity 1

Learning objective
- Recognise that two fractions can have an equivalent value (halves, quarters, fifths and tenths)

Resources

plain sheet of paper (per pair); set of 0 to 10 number cards (per learner)

What to do [TWM.01]

- Give each pair of learners a sheet of plain paper and each learner a set of 0 to 10 number cards. Ask learners to draw the following on their sheet of paper:

$$\underline{\qquad} = \underline{\qquad}$$

- Learners work together using their two sets of number cards to make some equivalent fractions. One card represents the numerator, and another card (or two other cards) represents the denominator, for example:

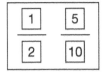 or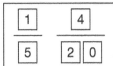

- They should make the unit and non-unit fractions that they have been considering in this unit: in particular, halves, quarters, fifths and tenths.
- How many can they make using two sets of cards?
- Ask learners to record each of their fractions, for example: $\frac{1}{2}$ and $\frac{5}{10}$.

Variations

2 Learners make a set of fraction cards for all the fractions they have recorded. They then play 'Concentration'. Learners spread the fraction cards face down on the table and take turns to choose two cards. If the fractions are equivalent, they keep the two cards. If not, they put them back face down on the table. The game continues until all the cards have been taken. The winner is the player with the most cards.

3 Encourage learners to use their digit cards to create other fractions considered in this unit, i.e. eighths.

Activity 2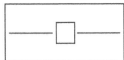

Learning objective
- Use knowledge of equivalence to compare and order unit fractions and fractions with the same denominator, using the symbols =, > and <

Resources

plain sheet of paper (per pair); Resource sheet 11: Comparison symbols (one of each symbol per pair); set of 0 to 10 number cards (per learner)

What to do

- Give each pair of learners a sheet of plain paper and the =, > and < symbol cards from Resource sheet 11, and each learner a set of 0–10 number cards. Ask learners to draw the following on their sheet of paper:

$$\underline{\qquad} \; \square \; \underline{\qquad}$$

- One learner uses the number cards to make a fraction (half, quarters, fifths or tenths), placing the cards on the sheet of paper. One card represents the numerator, and another card (or two other cards) represents the denominator. They then place one of the comparison cards in the box, for example:

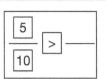

- The other learner uses the number cards to complete a true statement, for example:

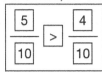

- Learners record both fractions, i.e. $\frac{5}{10}$, $\frac{4}{10}$
- Learners swap roles and repeat several times, each time recording each pair of fractions.

Variation

From the list of fractions they recorded, learners sort the fractions into four groups: unit fractions, quarters, fifths and tenths. They then order each set of fractions, from smallest to largest.

Number – Fractions, decimals, percentages, ratio and proportion

Unit 18: Fractions (C)

Collins International Primary Maths
Recommended Teaching and
Learning Sequence: Term 3, Week 5

Learning objectives

Code	Learning objective
3Nf.04	Understand that a fraction can be represented as a division of the numerator by the denominator (half, quarter and three-quarters).
3Nf.05	Understand that fractions (half, quarter, three-quarters, third and tenth) can act as operators.
3Nf.07	Estimate, add and subtract fractions with the same denominator (within one whole).

Unit overview

In this unit, learners will discover that a fraction can be represented as a division of the numerator by the denominator, for example $\frac{1}{2}$ is 1 divided by 2. They will also consider fractions as operators: quantities that can be divided by the denominator and multiplied by the numerator. This builds on work previously covered in Stages 1 and 2 on finding fractions of quantities.

They will also use what they already know about fractions to add and subtract them. The fractions they add and subtract will have the same denominator. The maximum sum will be one whole. Learners will use their knowledge of the law of commutativity and their understanding of inverse with whole numbers and apply this to fractions.

Prerequisites for learning

Learners need to:

• understand the link between fractions and division

• be able to find $\frac{1}{2}$ and $\frac{1}{4}$ of different quantities

• understand that $\frac{1}{2}$ and $\frac{1}{4}$ can be combined to form another fraction.

Vocabulary

fraction, unit fraction, non-unit fraction, whole, part, equal, denominator, numerator, equivalent, commutative, inverse, add, subtract, divide, multiply

Common difficulties and remediation

As with all work on fractions, visual representations are vital. The bar model is a key representation that shows the parts and the whole. These help learners' understanding of fractions. It is important to have these on display. Fraction walls are also useful. Ensure that these show the fractions you are focusing on during this unit.

Supporting language awareness

Learners should continue to use the vocabulary of 'denominator', 'numerator' and 'equivalent fractions'. It may be be helpful to have an image of a fraction displayed, with each part labelled.

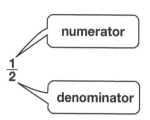

Promoting Thinking and Working Mathematically

During this unit, learners will have opportunities to think and work mathematically; for example, when they explore fractions with the same denominator that make one whole, they specialise (TWM.01) and make generalisations (TWM.02). They have opportunities to sort fractions according to their own criteria, which involves making conjectures (TWM.03), convincing (TWM.04), characterising (TWM.05) and classifying (TWM.06). They also compare and evaluate their mathematical ideas through problem solving with fractions, including making up their own problems (TWM.07).

Success criteria

Learners can:

• understand that a fraction can be represented as a division of the numerator by the denominator

• understand that fractions can act as operators

• add fractions with the same denominator

• subtract fractions with the same denominator.

Unit 18 Fractions (C)

Lesson 1: Fractions and division (A)

Learning objective

Code	Learning objective
3Nf.04	Understand that a fraction can be represented as a division of the numerator by the denominator (half, quarter and three-quarters).

Resources

two strips of paper (per pair); scissors (per pair)

Revise

Use the activity *Fractions of numbers (2)* from Unit 18: *Fractions (C)* in the Revise activities to reinforce and consolidate finding fractions of numbers.

Teach 🖼️ 🖵 [TWM.01/02]

- **[T&T]** 🗣️ Display **Slide 1**. Ask: **How is the fraction one half related to 1 divided by 2?** Give learners time to talk to their partner about what they can see on the slide. Take feedback. Discuss the idea that $\frac{1}{2}$ is equivalent to $1 \div 2$. Ask pairs of learners to take a strip of paper, fold it in half and cut or tear the strip to show half. Establish that they have divided one whole into two parts; each part is worth half.

- Repeat with $\frac{1}{4}$ as being equivalent to $1 \div 4$. Expect learners to fold a strip of paper into half and then half again. They open it up and cut or tear each part to show that one whole divided by 4 will give one quarter.

- **[T&T] [TWM.01/02]** Display **Slide 2**. Invite an individual to read the problem to the class before asking learners to discuss it in pairs. Agree that the two boys will need to share the cake between the two of them. Elicit that this is the same idea as on the previous slide: $\frac{1}{2}$ is equivalent to $1 \div 2$ or one shared between two. The division model for fractions is sharing. Ask: **How much of the cake would each person receive?** Agree: half before displaying **Slide 3**.

- Ask a learner to read out the second part of the problem on **Slide 3**. Ask: **What should the four children do?** Agree that the cake needs to be divided into 4, so each person will receive $\frac{1}{4}$.

- Direct learners to the Student's Book. Discuss the Let's learn section and then the paired activity and Guided practice example.

Practise 📒

- Workbook

Title: Fractions and division (A)

Pages: 142–143

- Refer to Activity 1 from the Additional practice activities.

Apply 👥 🖵

- Display **Slide 4**.

- Learners are asked to make up a word problem for $3 \div 4$ to show that this is equivalent to $\frac{3}{4}$.

Review

- Recap that fractions can be represented as division. Stress the fact that the numerator is divided by the denominator.

- On the board, write some fractions that learners are familiar with. Ask them to tell you what the division would be, for example: $\frac{1}{4}$ is equivalent to $1 \div 4$.

Assessment for learning

- How can we represent $\frac{1}{2}$ as a division?
- How can we represent $1 \div 4$ as a fraction?
- How can we represent $\frac{3}{4}$ as a division?
- How can we represent $2 \div 4$ as a fraction?

Same day intervention

Support

- Focus on dividing by 2 and 4 to find halves and quarters, using the sharing model for division. Ensure learners have counters or similar resources to help them.

Number – Fractions, decimals, percentages, ratio and proportion

Lesson 2: **Fractions and division (B)**

Learning objectives

Code	Learning objective
3Nf.04	Understand that a fraction can be represented as a division of the numerator by the denominator (half, quarter and three-quarters).
3Nf.05	Understand that fractions (half, quarter, three-quarters, third and tenth) can act as operators.

Resources

mini whiteboard and pen (per learner)

Revise

Use the activity *Fractions of numbers (2)* from Unit 18: *Fractions (C)* in the Revise activities to reinforce and consolidate finding fractions of numbers and link to fractions as operators, which will be covered later in this unit.

Teach 🔲 💻 [TWM.01/03/04/07]

- 🔲 **[TWM.01]** Display **Slide 1**. Ask: **What can you tell me about these fractions?** Expect learners to tell you the different fractions that are on the slide. They should be able to describe them as equal parts of one whole.

- 🔲 Ask: **If the whole is the same, which is the largest fraction? Which is the smallest? Is $\frac{1}{4}$ larger than or smaller than $\frac{1}{3}$? How do you know?**

- **[TWM.03/04] [T&T]** Ask: **How could we sort these fractions?** Give learners some time to discuss with their partner ways to sort the fractions. Take feedback. What different ways did they think of? (for example: tenths and not tenths, unit fractions and non-unit fractions, equivalent fractions) There are lots of possible ways. Discuss some of them.

- **[T&T]** Display **Slide 2**. Ask: **How does this model help us to find $\frac{1}{4}$ of 12?** Give learners the opportunity to talk to a partner. Take feedback. Establish that finding $\frac{1}{4}$ is the same as dividing by 4, and so you need to share 12 equally into 4 parts and as we are only considering 1 part, the value is 3.

- **[T&T]** Display **Slide 3**. Ask: **How does this model help us to find $\frac{3}{4}$ of 12?** Give learners the opportunity to talk to a partner. Take feedback. Establish that they need to find $\frac{1}{4}$ first, by sharing 12 equally into 4 parts. They then need to multiply by the numerator to find the value of three parts (9).

- Call out multiples of 4. Ask learners to find $\frac{1}{2}$, $\frac{1}{4}$, then $\frac{3}{4}$ of each number.

- Repeat for multiples of 3, asking learners to find $\frac{1}{3}$ and for multiples of 10, asking learners to find $\frac{1}{10}$.

- **[TWM.07]** Direct learners to the Student's Book. Discuss the Let's learn section with learners before they work on the paired activity. Discuss the Guided practice example.

Practise 📒

- Workbook

Title: Fractions and division (B)

Pages: 144–145

- Refer to the variation in Activity 1 from the Additional practice activities.

Apply 👥 💻 [TWM.04/05/06]

- Display **Slide 4**. Learners are given four fractions. They think about the fractions that could be the odd one out.

- They explain their thinking to their partner.

- Once they have found one option, they consider other possibilities.

Review

- Write the fractions $\frac{1}{2}$, $\frac{1}{4}$, $\frac{3}{4}$, $\frac{1}{3}$ and $\frac{1}{0}$ on the board and underneath some numbers that are multiples of some of these fractions, for example, 6, 8, 10, 12, 15, 24, 30 and 40. Point to a fraction and a number and ask learners to work out the fraction of the number. Encourage them to divide by the denominator and multiply by numerator.

Assessment for learning

- Can you describe how you would find $\frac{3}{4}$ of 36? Is there another way?

- Can you describe how to find $\frac{5}{10}$ of a shape?

- If you find $\frac{5}{10}$ of a shape, what other fraction have you made?

Same day intervention

Support

- Focus on finding halves and quarters of numbers. Ensure that you give learners counters or a similar resource to help them share.

Number – Fractions, decimals, percentages, ratio and proportion

Lesson 3: **Adding and subtracting fractions (A)**

Learning objective

Code	Learning objective
3Nf.07	Estimate, add and subtract fractions with the same denominator (within one whole).

Resources

mini whiteboard and pen (per learner)

Revise

Use the activity *Adding and subtracting fractions* from Unit 18: *Fractions (C)* in the Revise activities to review and consolidate how many of a fraction makes one whole.

Teach 📖 🖥 [TWM.01/02]

- 🅟 Display **Slide 1**. Ask: **What fractions can you see?** Point to the top row of the bar model and ask: **How many of these fractions make one whole?** Expect learners to be able to tell you two halves. Repeat for the other two rows and elicit that four quarters make one whole and eight eighths make one whole. Ask: **What rule have we found for fractions to be one whole?** Agree that the numerator and denominator of a fraction must be the same to make one whole.

- [T&T] Return to the halves bar model on the slide and say: **We know that two halves make one whole.** Ask: **Can we make an addition statement for halves?** On the board write: $\frac{1}{2} + \frac{1}{2} = 1$. Then ask: **Can you make addition statements for the other fractions that you can see?** Give learners a few minutes to write down some addition calculations with their partner. Take feedback and discuss examples given, for example: $\frac{1}{4} + \frac{1}{4} = \frac{2}{4}$, $\frac{1}{4} + \frac{2}{4} = \frac{3}{4}$. Continue asking for addition statements, encouraging learners to extend, for example, to $\frac{1}{8} + \frac{5}{8} = \frac{6}{8}$. Reinforce equivalent fractions, for example: $\frac{6}{8} = \frac{3}{4}$.

- [TWM.01/02] Ask: **If we know that $\frac{1}{8} + \frac{5}{8} = \frac{6}{8}$, what else do we know?** Agree that, just as with whole numbers, addition of fractions is commutative, so $\frac{5}{8} + \frac{1}{8} = \frac{6}{8}$. Ask learners to write the commutative facts for their earlier addition statements.

- Ask: **If we have an addition statement, what else do we know?** Agree that addition and subtraction are inverse operations. Knowing that $\frac{5}{8} + \frac{1}{8} = \frac{6}{8}$ means that learners also know that $\frac{6}{8} - \frac{1}{8} = \frac{5}{8}$ and $\frac{6}{8} - \frac{5}{8} = \frac{1}{8}$. Ask them to write the two inverse facts for each of their addition statements.

- [T&T] Display **Slide 2**. Ask: **How can the bars we have seen help us to solve this problem?** Allow discussion then, after a few minutes, take feedback.

Agree that we could cross out three $\frac{1}{8}$ sections of the bar to show the fraction eaten. Model this on the board. Learners then count the parts left, to find $\frac{5}{8}$.

Discuss the calculation that shows the problem: $\frac{8}{8} - \frac{3}{8} = \frac{5}{8}$.

- Display **Slide 3** and discuss how the diagrams show $\frac{3}{5} + \frac{1}{5} = \frac{4}{5}$ and $\frac{3}{5} - \frac{1}{5} = \frac{2}{5}$.

- Direct learners to Let's learn in the Student's Book. Discuss the paired activity and Guided practice example.

Practise 📒

- Workbook

Title: Adding and subtracting fractions (A)

Pages: 146–147

- Refer to Activity 2 from the Additional practice activities. Focus on halves, quarters and eighths.

Apply 👥 🖥 [TWM.03/04]

- Display **Slide 4**. Learners are given an explanation of how to add and subtract fractions. They discuss with a partner if it is a good explanation and why.

Review

- Call out some fractions with denominators of 3, 4, 5 and 8 for learners to add. Ensure that the total is not more than 1.

- Call out some fractions with denominators of 3, 4, 5 and 8 for learners to subtract.

Assessment for learning

- What is the total of $\frac{1}{4}$ and $\frac{1}{4}$? How do you know?
- What is the total of $\frac{3}{8}$ and $\frac{1}{8}$? How do you know?
- What is the difference between $\frac{3}{8}$ and $\frac{7}{8}$? How do you know?

Same day intervention

Support

- Focus on adding and subtracting thirds, quarters and fifths, for example $\frac{1}{4} + \frac{2}{4} = \frac{3}{4}$. Ensure that you include commutative and inverse facts.

Number – Fractions, decimals, percentages, ratio and proportion

Lesson 4: **Adding and subtracting fractions (B)**

Learning objective

Code	Learning objective
3Nf.07	Estimate, add and subtract fractions with the same denominator (within one whole).

Revise

Use *Adding and subtracting fractions* from Unit 18: *Fractions (C)* in the Revise activities to reinforce what was studied in Lesson 3. Focus on the variation.

Teach 🟦 🖥

- 🔲 Display **Slide 1**. Ask: **What fractions can you see?** Point to the top row of the bar model and ask: **How many of these fractions make one whole?** Expect learners to be able to tell you five fifths. Repeat for the other row and elicit that ten tenths make one whole. Ask: **What rule have we found for a fraction to be one whole?** Agree that the numerator and denominator of a fraction must be the same to make one whole.

- **[T&T]** Return to the fifths bar model on the slide and say: **Can we make any addition statements for fifths?** Give learners a few minutes to write down some addition statements with their partner. Take feedback and discuss examples given, for example: $\frac{1}{5} + \frac{2}{5} = \frac{3}{5}$. Repeat for tenths, looking for examples such as $\frac{1}{10} + \frac{5}{10} = \frac{6}{10}$. Reinforce equivalent fractions, for example: $\frac{6}{10} = \frac{3}{5}$.

- Ask: **If we know that $\frac{1}{10} + \frac{5}{10} = \frac{6}{10}$, what else do we know?** Learners should recall from the previous lesson that addition of fractions is commutative, so $\frac{5}{10} + \frac{1}{10} = \frac{6}{10}$. Ask learners to write the commutative facts for their earlier addition statements.

- Ask: **If we have an addition statement, what else do we know?** Since addition and subtraction are inverse operations, $\frac{6}{10} - \frac{1}{10} = \frac{5}{10}$ and $\frac{6}{10} - \frac{5}{10} = \frac{1}{10}$. Ask learners to write the two inverse facts for each of their addition statements.

- **[T&T]** Display **Slide 2** and discuss the problem. Ask: **How can the bar help us to solve this problem?** Allow discussion then, after a few minutes, take feedback. Agree that crossing out three $\frac{1}{10}$ sections and one $\frac{1}{10}$ section would show the fractions that have been used. Learners then count the parts left to make six tenths, which is what is left. Discuss the calculations that can be made to show the problem: $\frac{3}{10} + \frac{1}{10} = \frac{4}{10}$ and then $\frac{10}{10} - \frac{4}{10} = \frac{6}{10}$.

- Direct learners to Let's learn in the Student's Book. Discuss the paired activity and the Guided practice example.

Practise 🟦 [TWM.04]

- Workbook

Title: Adding and subtracting fractions (B)

Pages: 148–149

- Refer to the variation in Activity 2 from the Additional practice activities. Ensure you include fifths and tenths as well as halves, quarters and eighths.

Apply 👥 🖥 [TWM.04]

- Display **Slide 3**.
- Learners are given a problem to solve. They find the solution and discuss with their partner how the visual representation can help them.

Review

- Call out some fractions with denominators of 6, 7, 9 and 10 for learners to add.
- Call out some fractions with denominators of 6, 7, 9 and 10 for learners to subtract.
- Recap the fact that the denominators always stay the same. Ask learners to explain why.

Assessment for learning

- What is the total of $\frac{1}{6}$ and $\frac{1}{6}$? How do you know?
- What is the total of $\frac{3}{10}$ and $\frac{1}{10}$? How do you know?
- What is the difference between $\frac{3}{10}$ and $\frac{7}{10}$? How do you know?

Same day intervention

Support

- Focus on adding and subtracting different numbers of fifths and tenths, for example: $\frac{1}{5} + \frac{2}{5} = \frac{3}{5}$. Ensure that you include commutative and inverse facts.

Number – Fractions, decimals, percentages, ratio and proportion

Additional practice activities

Activity 1

> **Learning objectives**
> - Understand that a fraction can be represented as a division
> - Understand that fractions (half, quarter, three-quarters, third and tenth) can act as operators.

Resources

five strips of paper (per learner); scissors (per learner); ruler (per learner)

What to do

- Give each learner five strips of paper.
- Ask them to use three of the strips of paper and fold the strips to show $\frac{1}{2}$, $\frac{1}{4}$ and $\frac{1}{8}$.
- For each strip, they cut off the parts to demonstrate that $\frac{1}{2}$ is one strip divided by 2, $\frac{1}{4}$ is one strip divided by 4 and $\frac{1}{8}$ is one strip divided by 8.

- Repeat for fifths and tenths using the remaining two strips of paper. For this, learners should measure strips of paper so that they are 20 cm in length. They could measure each part to show that $\frac{1}{5}$ is one strip divided by 5 and each part is 4 cm.

Variation

2 Use the strips of paper again, but give values to the whole. For example, if the whole is 80 what is $\frac{1}{2}$, $\frac{1}{4}$, $\frac{3}{4}$, $\frac{1}{5}$, $\frac{1}{10}$?

Activity 2

> **Learning objective**
> - Add and subtract fractions with the same denominator (within one whole)

Resources:

mini whiteboard and pen (per learner)

What to do

- On the board, write two fractions with the same denominator. Initially begin with thirds, quarters, fifths and eighths.
- Ask learners to make up the two addition and two subtraction calculations and to write these down.
- Ask them to work with a partner to make up word problems to go with their calculations and to solve them.
- Repeat these several times with different fractions.

Variation

2 Repeat the activity but focus on sixths, sevenths, ninths and tenths.

Unit 19: Time

Collins International Primary Maths
Recommended Teaching and
Learning Sequence: Term 1, Week 7

Learning objectives

Code	Learning objective
3Gt.01	Choose the appropriate unit of time for familiar activities.
3Gt.02	Read and record time accurately in digital notation (12-hour) and on analogue clocks.
3Gt.03	Interpret and use the information in timetables (12-hour clock).
3Gt.04	Understand the difference between a time and a time interval. Find time intervals between the same units in days, weeks, months and years.

Unit overview

In Stage 2, learners were taught to read and record to five minutes on analogue and digital clocks. In this unit, they learn to tell the time to the nearest minute on analogue and 12-hour digital clocks. They start by looking at minutes past the hour and linking that to digital time, for example 40 minutes past 2 is 2:40. They will consider how this could be a morning, afternoon or evening time. Learners will suggest suitable unit of time for measuring familiar activities. They will also explore timetables and calculate time intervals between the same units in years, months, weeks and days.

Prerequisites for learning

Learners need to:

- understand the relationships between units of time, for example, a second is shorter than one minute
- be able to measure time taken for activities
- know how to tell the time to the nearest five minutes on analogue and 12-hour digital clocks
- be able to interpret calendars.

Vocabulary

time, analogue, digital, minutes past, minutes to, time interval, second, minute, hour, day, week, month, year, timetable

Common difficulties and remediation

Learners will usually understand the concept of time because of its relevance in everyday situations. This unit focuses on 'real-life' situations for telling the time. Finding time durations and intervals can be problematic for some learners, who may think that they can simply use the methods for whole numbers to add for durations or subtract for differences. The relationship between the units used means that this is not always straightforward. Time is a measurement that does not use only metric units.

There are 60 seconds in a minute, 60 minutes in an hour, 24 hours in a day, and so on. Some learners find working with these units of measure challenging.

Supporting language awareness

In this unit, vocabulary will include: 'days of the week', 'months of the year', 'day', 'week', 'fortnight', 'month', 'year', 'weekend', 'calendar', 'date', 'timetable', 'arrive', 'depart', 'hour', 'minute', 'second', 'digital/ analogue clock'. You may find it useful to refer to the audio glossary on Collins Connect. It is important to display the words, as they are introduced into the lesson, on cards in the classroom with visual clues to help learners recognise and understand what each one means. Use question starters such as: **How long ago was…? How long will it be until…? How long will it take to…? How often do we…?**

Promoting Thinking and Working Mathematically

During this unit learners will have opportunities to think and work mathematically, for example, when exploring instruments to tell time, and when they explore equivalent units of time they specialise (TWM.01) and make generalisations (TWM.02). They have opportunities to make conjectures (TWM.03), convince (TWM.04), characterise (TWM.05) and classify (TWM.06) when looking at different types of clock.

Success criteria

Learners can:

- choose and use units of time for activities
- read and record time in digital notation (12-hour) and on analogue clocks
- interpret and use the information in timetables
- understand the difference between a time and a time interval
- find time intervals between the same units in days, weeks, months and years.

Unit 19 Time

Geometry and Measure – Time

Lesson 1: **Units of time**

Learning objectives

Code	Learning objective
3Gt.01	Choose the appropriate unit of time for familiar activities.
3Gt.04	Understand the difference between a time and a time interval. Find time intervals between the same units in days, weeks, months and years.

Resources

stopwatch (per group/pair); minute timer (per pair); interlocking cubes (per group/pair); selection of 3D shapes (per group); counters (per pair)

Revise

Use the activity *Units of time* from Unit 19: *Time* in the Revise activities.

Teach 📖 🖥

- ▶ Display **Slide 1**. Ask: **What is this? When might we use one?** Agree that it is a stopwatch and that we could use it to measure short lengths of time. Give each group a stopwatch to examine. Ask: **What unit would we use to measure a very short period of time?** Agree seconds.

- **[T&T]** Say: **Describe what you can see on the stopwatch.** Ask: **How do you think you would use it?** Take feedback. Explain how to use the stopwatch. Time learners as they carry out activities, for example making a tower of ten cubes, sorting 3D shapes into those with curved surfaces and those with no curved surfaces.

- Ask: **What other instruments can we use to measure time?**

- **[T&T]** Display **Slide 2**. Ask: **What can you see? What's the same? What's different?** Agree there is an analogue clock and a digital clock.

- **[T&T]** Ask: **What units do we use to measure time?** Expect learners to be able to tell you seconds, minutes, hours, days, weeks, months and years. Discuss familiar activities and ask learners to tell you what units they would use to measure them, for example reading a page of a book would take minutes, a day at school would last hours, a weekend would be days, a school holiday might be weeks, time between birthdays or anniversaries or special festivals might be months or years.

- Ask questions related to finding time differences between familiar events, for example:
  - **How long is it from your birthday last year until your next birthday?**
  - **Sophie was born in 2011 and Amy was born on exactly the same date as Sophie but in 2009. How many years older is Amy than Sophie?**
  - **Freddie was born in 2013. His brother was born in 2007. How many years younger is Freddie?**
  - **Anya arrived at her granmother's house on Monday and stayed until Saturday. How many days was she with her grandmother?**

- **How many months are there from 1st February to 1st October?**

- Establish that learners have been finding time intervals to answer these questions. They are working out the difference between units of time. Ensure that learners understand the difference between actual times, which are read from a clock, and time intervals, which are the lengths of time that pass between two given times.

- **[T&T]** Ask learners to work with a partner to create similar scenarios to ask the class.

- Direct learners to the Student's Book. Discuss the Let's learn section.

- Learners work through the paired activity.

- Discuss the Guided practice example in the Student's Book.

Practise 📕 [TWM.04]

- Workbook

Title: Units of time

Pages: 150–151

- Refer to Activity 1 from the Additional practice activities.

Apply 👥 🖥 [TWM.01]

- Display **Slide 3**. Learners discuss how many counters they think they could place into piles of 10 in one minute.

- They write down both their estimates. They then test it out by taking turns to see how many counters they can put into piles of ten in one minute.

- They compare their results with their estimates.

Review

- Review the lesson by recapping the different units and instruments used to measure time.

Assessment for learning

- Would it take minutes, hours or days to read a page of your book? Why do you think that?
- When would we use a stopwatch?

Same day intervention
Support

- Focus on equivalent units of minutes and hours, and measuring on minutes.

Lesson 2: **Telling the time (A)**

Learning objective

Code	Learning objective
3Gt.02	Read and record time accurately in digital notation (12-hour) and on analogue clocks.

Resources

analogue clock (per learner); mini whiteboard and pen (per learner/pair); two copies of Resource sheet 12: Clock faces (per learner)

Revise

Use the activity *Units of time* from Unit 19: *Time* in the Revise activities.

Teach 🔲 🖥 📊 [TWM.07]

- 📖 Display **Slide 1**. Say: **This is the world's biggest clock face. It is in Mecca, Saudi Arabia.** Ask: **What do you notice? Can you describe the clock face? How is it the same as many other clocks? How is it different?** Agree it is the same because of the hands but different because there are no hour numbers.
- [T&T] Ask: **What time is showing on the clock? How do you know?** Agree approximately 5 minutes to 6 because the hour hand is pointing towards the 6 and the minute hand is just after 5 minutes to the hour.
- [T&T] Display the analogue clock on the **Clock tool**. Ask: **How can you read the time on an analogue clock?** Encourage learners to discuss this with their partner and use their clocks to help them. Take feedback. Tell learners that focusing on 'minutes past' helps to make the link between reading and recording time on analogue and 12-hour digital clocks.
- Display **Slide 2**. Count around the hour numbers on the clockface in fives and link to the multiplication table for 5. Stress that these are minutes past the hour. Using the **Clock tool** show clock 1 as an analogue clock and clock 2 as a digital (12-hour) clock. Show 50 minutes past 6 on the analogue clock and discuss the digital time of 6:50.
- Ask: **If the time is 50 minutes past 6 or 6:50, how many minutes will it be until 7 o'clock?** Together count up from the 10 to 12 in fives and agree 10 minutes. Say: **So this time also shows 10 minutes to 7.**
- Repeat for other times to five minutes to and past the hour. Include showing times only on the analogue clock and asking learners to write the equivalent 12-hour digital time on their whiteboards. Where appropriate, ask learners to calculate the number to minutes to the next hour and say the corresponding time.
- [TWM.07] Direct learners to the Student's Book. Discuss the Let's learn section with learners.

- Learners work through the paired activity.
- Discuss the Guided practice example in the Student's Book.

Practise 📒 [TWM.04]

- Workbook

Title: Telling the time (A)

Pages: 152–153

- Refer to the first variation in Activity 2 from the Additional practice activities.

Apply 👥 🖥

- Display **Slide 3**. Learners draw a total of 12 clock times on their two copies of Resource sheet 12.
- For each time the minute hand must be on a different *hour* number (i.e. in order to show a time to the nearest five minutes past or to the hour). They can choose where to put the hour hand.
- Under each clock, they write the equivalent digital time.
- They check what they have done with their partner, to see if they agree.
- Invite learners to share the times they made up.

Review 📊

- Using the **Clock tool** show clock 1 as an analogue clock and set the time to 5 minutes to 5. Ask: 📖 **What would this time show on a digital clock?** Agree 04:55 and write it on the whiteboard. **How many minutes is it until 5 o'clock?** Agree that 55 minutes past 4 can also be read as 5 minutes to 5.
- Repeat with other times.

Assessment for learning

- What other way can you say 35 minutes past 8? Is there another? And another?
- What is the digital time for 15 minutes past 3?

Same day intervention

Support

Focus on reading and recording time to the hour, half hour and quarter hour before moving onto five-minute intervals.

Geometry and Measure – Time

Lesson 3: **Telling the time (B)**

Geometry and Measure – Time

Learning objective

Code	Learning objective
3Gt.02	Read and record time accurately in digital notation (12-hour) and on analogue clocks.

Resources

analogue clock (per learner); mini whiteboard and pen (per pair)

Revise

Use the activity *Reading the time* from Unit 19: *Time* in the Revise activities to assess learners' understanding of telling the time to the nearest minute. You may wish to change the times so that they are in five-minute intervals.

Teach 🆂🅱 🖥

- ▶ Display **Slide 1**. [T&T] One by one ask: **Where would the minute hand go on this clock to show 6 minutes past the hour?**
  - ... to show **18 minutes** ...?
  - ... to show **29 minutes** ...?
  - ... to show **37 minutes** ...?
  - ... to show **43 minutes** ...?
  - ... to show **54 minutes** ...?

- Give learners time to discuss where on the clock the different minutes would be marked. After each time, take feedback, agree for example, that 6 minutes past would be positioned between 5 and 10, but closer to 5 and it would point to the minute mark on the clock after the hour number 1.

- Ask learners to show 51 minutes past 5 on their analogue clock. Suggest that they first find 50 minutes past and then count on three more minutes. Discuss the digital time of 5:51. Ask them to show similar times on their clocks and to write the digital equivalent on their mini whiteboards.

- Ask: **If the time is 51 minutes past 5, or 5:51, how many minutes will it be until 6 o'clock?** Together, count in ones from the minute hand and agree 9 minutes. Say: **So this time also shows 9 minutes to 6.** Look at the other times learners recorded and work out how many minutes until the next o'clock time.

- Direct learners to the Student's Book. Discuss the Let's learn section with them.

- Learners work through the paired activity.

- Discuss the Guided practice example in the Student's Book.

Practise 🆆🅱 [TWM.01/03]

- Workbook

Title: Telling the time (B)

Pages: 154–155

- Refer to Activity 2 from the Additional practice activities.

Apply 👥 🖥 [TWM.07]

- Display **Slide 2**. Jodie has made a mistake in reading the clock. Learners discuss what Jodie has done wrong.

- They then write the actual time shown on the clock.

Review 📊

- Using the **Clock tool**, show clock 1 as an analogue clock and set the time to 7 minutes to 9. ▶ Ask: **What would this time show on a digital clock?** Agree 08:53 and write it on whiteboard. **How many minutes is it until 9 o'clock?** Agree that 53 minutes past 8 can also be read as 7 minutes to 9.

- Repeat several times.

Assessment for learning

- What other way can you say 43 minutes past 5? Is there another? And another?
- What is the digital time for 17 minutes past 9?

Same day intervention

Support

Focus on reading and recording time to five-minute intervals before moving onto one-minute intervals.

Lesson 4: **Timetables**

Learning objectives

Code	Learning objective
3Gt.03	Interpret and use the information in timetables (12-hour clock).
3Gt.04	Understand the difference between a time and a time interval. Find time intervals between the same units [in days, weeks, months and years].

Resources

analogue clock (per learner)

Revise

Use the activity *Reading the time* from Unit 19: *Time* in the Revise activities to reinforce and consolidate telling the time.

Teach 🆂🖥

- **[T&T]** 🗣 Display **Slide 1**. Ask: **What do you think this is? What does it tell us?** After a few minutes, take feedback. Expect learners to be able to tell you it is a school timetable and it shows activities carried out during the day. Ask: **Are the times analogue or digital? How do you know?** Agree they are digital because the times are written in numbers.

- Ask questions about the timetable, for example: **What happens on a Monday/Tuesday? What does Kim do on Monday between half past 1 and 3 o'clock? What does Kim do after school on Friday?** Expect learners to find the information from the timetable.

- **[T&T]** Ask: **What time does Science begin on Wednesday? What time does it end?** Agree that it begins at 11:00 and ends at 12:00. Establish that these are clock times.

- Ask: **For how long does that Science lesson last?** Agree that they need to find out how long it is between 11:00 and 12:00. They can count on one hour and then 30 minutes. Write 1:00 on the board. Ask learners why this way of recording might be confusing. Establish that this looks like a digital clock time. They need to know the time interval between the two times. Ask: **What might be a clearer way of recording the length of the lesson?** Agree, writing 1 hour.

- Ask learners to tell you the starting times for all the lessons and show you on a clock.

- **[T&T]** Display **Slide 2**. Ask: **What can you tell me about this timetable?** Give learners a few minutes to discuss with a partner. Take feedback. Establish that it shows the departure times of buses from different bus stops. As before, ask questions, such as: **What time does the first bus leave Town Station? What time is it due to leave Shopping Centre?**

- Direct learners to Let's learn in the Student's Book. Discuss the timetable, and the departure and arrival times of different trains from Ayetown to Beeville.

- Learners work through the paired activity.

- Discuss the Guided practice example in the Student's Book.

Practise 🆆🅱

- Workbook

Title: Timetables

Pages: 156–157

- Refer to the second variation in Activity 2, focusing on the variation, from the Additional practice activities.

Apply 👥🖥 [TWM.01]

- Display **Slide 3**. Learners make up their own timetable about an imaginary activity. It can show anything – school, a bus route or a train timetable.

- When the timetable is complete, they make up their own questions to ask the class.

Review

- Recap the work covered in this unit. Ensure you include finding 'past' times and 'to' times.

- Discuss why timetables are useful and how they can be used.

- Say some pairs of times and ask learners to find their differences, for example: 3:45 and 4:05, 20 minutes past 11 and 10 minutes to 12.

Assessment for learning

- If a bus leaves the bus stop at 2:35 and take 25 minutes to get to town, what time does it arrive in town?

- What is the time difference between 10:55 and 12: 30? How do you know?

- If a train leaves the station at 10:30 and arrives at its destination at 2:45, how long is the journey time?

Same day intervention

Support

Encourage learners to work in pairs or groups to use and interpret the different timetables and to make time interval statements involving times to the nearest 30-, 15-, 10-, or 5-minutes.

Enrichment

Encourage learners to make time interval statements involving times to the nearest minute.

Geometry and Measure – Time

Additional practice activities

Activity 1 ▪▪ ▲2

> **Learning objective**
> • Choose the appropriate unit of time for familiar activities

Resources

stopwatch (per pair); Resource sheet 13: Tangram set (per pair); scissors (per pair); A4 paper and pen (per pair)

What to do

• Ask learners to cut the tangram into seven pieces. Let them play with the pieces and see what irregular shapes they can make.
• After they have explored, one learner times the other as that learner remakes the original square. They record the length of time shown on the stopwatch. They then swap roles.
• They continue in this way, making other irregular shapes.
• Once they have done these, they compare times taken to make the shapes, working out who was quicker.

Variation

1 Learners put the tangram pieces together to make any different patterns, shapes, objects or animals. that they want to make. Their partner times them as above and records the time.

Activity 2 ▪▪ ▲2

> **Learning objective**
> • Read and record time accurately in digital notation (12-hour) and on analogue clocks.
> • Find time intervals between the same units

Resources

Resource sheet 3: Digit cards (per pair); Resource sheet 12: Clock faces (per pair); paper (per pair) (for the second variation)

What to do

• Learners choose three digit cards (per pair). They use these to make all the digital times they can, for example,
  - if they choose 3, 6 and 8, they could make 6:38 and 8:36
  - if they choose 1, 2 and 4, they could make 1:24, 1:42, 2:14, 2:41, 4:12 and 4:21.
• For each time they make, they draw it on an analogue clock face and write the digital time underneath.
• They pick two times and find the time difference between them.
• They then select another three cards and make up more times to draw and label.

Variation

1 For reading and recording time to five-minute intervals, learners must choose the 0 or 5 digit cards and three other cards. For example, if they choose 5, 2, 3 and 7, they could make 3:25, 7:25, 2:35 and 7:35.

 Learners make a timetable to show their week at school. Write the information they need on the board, for example, subjects they study and the times they begin. Learners draw their own timetable and add the information you have given.

Unit 20: 2D shapes, symmetry and angles

Collins International Primary Maths
Recommended Teaching and
Learning Sequence: Term 1, Week 8

Learning objectives

Code	Learning objective
3Gg.01	Identify, describe, classify, name and sketch 2D shapes by their properties. Differentiate between regular and irregular polygons.
3Gg.09	Identify both horizontal and vertical lines of symmetry on 2D shapes and patterns.
3Gg.10	Compare angles with a right angle. Recognise that a straight line is equivalent to two right angles or a half turn.

Unit overview

This unit focuses on regular and irregular 2D shapes. Learners will explore the properties of shapes including regular and irregular polygons. The properties include symmetry, the numbers of sides and vertices, and whether the vertices are right angles or not. Learners best understand shape when they cut them out, fold them, turn them, rotate them, fit them together, make patterns, match them, sort and classify them. This unit provides plenty of opportunities for them to do these things.

Learners identify 2D shapes in pictures and patterns. They explore these shapes, lines of symmetry and right angles where they occur in the school environment and also the wider world in, for example, building structures, the animal kingdom and the art world.

Note: For those teachers/schools following the CIPM Recommended Teaching and Learning Sequence this unit is taught prior to Unit 21: 3D shapes. However, some schools may prefer to change the order of these two units and teach Unit 21 before teaching this unit.

Prerequisites for learning

Learners need to:
* recognise regular polygons
* identify symmetry on 2D shapes and patterns
* predict and check how many times a shape looks identical as it completes a full turn
* understand that an angle is a description of a turn
* identify right angles in shapes and recognise that a right angle is a quarter turn.

Vocabulary

polygon, non-polygon, pentagon, hexagon, heptagon, octagon, nonagon, circle, semi circle, quadrilateral, regular, irregular, side, curved, straight, vertices, right angle, quarter turn, half turn, symmetry, symmetrical, line of symmetry, vertical, horizontal

Common difficulties and remediation

Learners are often introduced to regular shapes in one particular orientation. They often then associate the name of the shape with the regular shape and orientation, so it is important that learners see the same shape in different positions. The name of a shape depends on the number of sides and vertices it has, so it is important to emphasise that, for example, a pentagon is any shape with five straight sides and five vertices. Give learners opportunities to draw lots of different pentagons, introducing the word 'regular' when the shape has sides that are equal and vertices that are equal and 'irregular' otherwise.

Some learners may be confused by the concept of reflective symmetry. It is a good idea to use mirrors so that learners can understand that a reflected shape, or part of a shape, is one that is identical but reversed.

Supporting language awareness

Recap the shape words that the class have already learned and introduce new vocabulary with words written on cards and displayed in the classroom, with pictures or actual objects beside them. You may find it useful to refer to the audio glossary on Collins Connect. Tell learners that the shapes with straight sides are 'polygons'. The names of most shapes end with 'gon'. Ask them to say 'gon'. Break up the words so that learners can hear how they are made up.

Ask questions such as: **What is the same?** and **What is different?** when describing properties of shape and expect learners to answer in complete sentences.

Promoting Thinking and Working Mathematically

During this unit learners will be making conjectures (TWM.03), convincing (TWM.04), characterising (TWM.05) and classifying (TWM.06) when they sort shapes according to different criteria and explain their thinking. They also have opportunities to specialise (TWM.01) and generalise (TWM.02) when identifying shapes in the real world.

Success criteria

Learners can:
* identify, describe, classify, name and sketch 2D shapes by their properties
* differentiate between regular and irregular polygons
* identify horizontal and vertical lines of symmetry
* compare angles with a right angle
* recognise that a straight line is two right angles or a half turn.

Geometry and Measure – Geometrical reasoning, shapes and measurements

Lesson 1: **2D shapes**

Learning objective

Code	Learning objective
3Gg.01	Identify, describe, classify, name and sketch 2D shapes by their properties. Differentiate between regular and irregular polygons.

Resources

card (per learner); ruler (per learner); scissors (per pair); paper (per learner); class geometric set of shapes optional (per class); coloured pencils (per learner) (for the Workbook)

Revise

Use the activity *Shapes around us* from Unit 20: *2D shapes, symmetry and angles* in the Revise activities to assess learners' knowledge of 2D shape.

Teach [SB] 🖥

- Display **Slide 1**. Ask: **Do you recognise these buildings?** The first is The Pentagon, which is the headquarters of the United States Department of Defence in Washington, USA. Ask: **Why is it called The Pentagon?** Agree that the building has five sides and five-sided shapes are all called pentagons. Inform learners that this is a regular pentagon because all the sides are the same length and the angles the same size. Repeat for the hexagonal town square in the second image.

- 🖥 Display **Slide 2**. Ask: **What shapes can you see in the top row?** Agree triangle, square, pentagon, hexagon, heptagon and octagon. Stress that they all have lengths that are the same and vertices that are equal, therefore they are regular shapes. The ones in the second row are irregular. Tell learners that a polygon is any 2D shape with three or more straight sides.

- Ask: **Are the colours important?** Agree the colours are irrelevant. Sides and vertices are properties of shape, but colours are not.

- Ask: **What can you tell me about 2D shapes?** Agree that 2D shapes are identified by the number of sides and vertices (corners) they have. Ask learners to identify the number of sides and vertices for each shape. Tell learners that some shapes have right angles. Remind them that a right angle is a quarter turn. Point to the square and the irregular triangle and pentagon to show what right angles look like. Ask: **Where can we see right angles in the classroom?**

- Display **Slide 3**. Ask: **What can you tell me about these shapes?** Agree that some have straight sides and some have curved sides. Agree that the circle and semicircle are not polygons because they have curved sides.

- Direct the learners to the Student's Book. Discuss the Let's learn section.

- Learners work through the paired activity.

- Discuss the Guided practice example in the Student's Book.

Practise [WB] [TWM.05]

- Workbook

Title: 2D shapes

Pages: 158–159

- Refer to Activity 1 from the Additional practice activities.

Apply 👥 🖥 [TWM.01]

- Display **Slide 4**.

- Learners draw two regular triangles on card and cut them out. If they have difficulty drawing the shapes, they can draw around a regular triangle from a class geometric set of shapes. They arrange the two triangles to make other shapes. They can place them side by side or overlap them.

- They work with a partner and make sketches of the new shapes.

Review 📊

- Display the **Geoboard tool**. Invite learners to make any polygons that you name, including those covered in previous stages, i.e. pentagons, hexagons and octagons.

- 🖥 Ask: **Where might we see 2D shapes in real life?** Encourage learners to identify shapes around the classroom, in picture books and in the environment.

Assessment for learning

- Where can you see right angles in the classroom? Where else?

- Describe the properties of a regular shape. What about an irregular shape?

- What is the same about a regular octagon and an irregular octagon? What is different?

Same day intervention

Support

- Focus on regular shapes. Gradually introduce irregular versions of the shapes when learners are confident at naming and describing the properties of regular shapes.

Geometry and Measure – Geometrical reasoning, shapes and measurements

Lesson 2: **Sorting 2D shapes**

Learning objective

Code	Learning objective
3Gg.01	Identify, describe, classify, name and sketch 2D shapes by their properties. Differentiate between regular and irregular polygons.

Resources

A4 paper cut into six equal pieces (per learner); ruler (per learner) (for the Workbook)

Revise

Use the activity *Naming shapes* from Unit 20: *2D shapes, symmetry and angles* in the Revise activities.

Teach 🆂🅱 💻 📊 [TWM.03/04/06]

- **[TWM.04]** 🔁 Display **Slide 1**. Ask: **What type of pattern can you see?** Explain that it is an example of Islamic art. Say: **Identify the hexagons that you can see in the pattern. How do you know they are hexagons? Are they regular or irregular? Why?** Encourage learners to use the correct mathematical vocabulary when they explain.

- Ask: **What other shapes can you see?** Agree that there are some irregular quadrilaterals, irregular octagons as well as some irregular ten-sided shapes (decagons). You could inform learners that the 24 sided-shape (star) is called an icositetragon.

- **[T&T]** Ask: **What are the properties of regular and irregular shapes?** Remind learners that regular 2D shapes have equal sides and equal vertices. Irregular shapes do not.

- Display the **Geoboard tool**. Introduce seven-, nine- and ten-sided shapes by name (heptagon, nonagon, decagon). Ask learners to describe them.

- **[TWM. 03/04] [T&T]** Display **Slide 2**. Ask: **What shapes can you see?** Agree that there is a mixture of polygons and shapes that are not polygons because their sides are curved, not straight. Ask: **What would be a good way to sort these shapes?** After a few minutes take feedback, for example: according to the individual shapes, whether they are polygons or not polygons, whether they have right angles. Accept all suggestions that work.

- Give each learner six pieces of paper and ask them to sketch six irregular shapes, one on each piece of paper. They should be different from those drawn by their partner. **[T&T]** Working in pairs, learners sort the shapes.

- Direct learners to the Student's Book. Discuss the Let's learn section with learners.

- **[TWM.06]** Learners work through the paired activity.

- Discuss the Guided practice example in the Student's Book.

Practise 🆆🅱 [TWM.06]

- Workbook

Title: Sorting 2D shapes

Pages: 160–161

- Refer to Activity 1 from the Additional practice activities.

Apply 👥 💻 [TWM.05]

- Display **Slide 3**. Learners need to prove that Gwen's shape is a hexagon.

- Encourage them to write an explanation to help Sam understand.

- It is important that learners understand that any shape with six straight sides is a hexagon.

- Invite learners to share what they think about Sam's thoughts. Agree that he is wrong and that any shape with six straight sides is a hexagon.

Review

- Recap that a polygon is any 2D shape made entirely of straight sides and that a shape with at least one curved side is not a polygon.

- 🔁 Ask: **What do we look for when we name a shape?** Agree number of sides and vertices.

- Describe different 2D shapes and ask learners to identify them.

- Ask learners to describe different 2D shapes for the class to identify.

Assessment for learning

- What can you tell me about a semi circle?
- What can you tell me about regular shapes?
- How are irregular shapes different from regular shapes?

Same day intervention

Support

- When sorting 2D shapes, provide criteria for learners to sort against, for example: regular and not regular, polygons and not polygons.

Geometry and Measure – Geometrical reasoning, shapes and measurements

Lesson 3: **Symmetry**

Learning objective

Code	Learning objective
3Gg.09	Identify both horizontal and vertical lines of symmetry on 2D shapes and patterns.

Resources

squared paper (per learner and per pair); ruler (per learner); mirror (per pair)

Revise

Use the activity *Symmetry* from Unit 20: *2D shapes, symmetry and angles* in the Revise activities to assess learners' knowledge of symmetry.

Teach 🔲 🖥 ⏸

- **[T&T]** 🔁 Display **Slide 1**. Discuss the countries shown by the flags. Can learners tell you where the different countries are in the world? If appropriate and possible, find these on a map. Ask: **What do you notice about the flags? What do they all have in common?** Give learners a few minutes to discuss with their partner. Take feedback. Agree that they are all symmetrical.

- Ask: **What do you know about symmetry?** Establish that a shape or pattern is symmetrical if a straight line (a line of symmetry) can be drawn that divides the shape into two identical halves that mirror each other. Refer to the flags again and ask learners to show you where the lines of symmetry are. Some have one line of symmetry; some have two. Ensure they find both the horizontal and vertical lines of symmetry on those that have two.

- Display the **Symmetry tool** showing the square. Invite learners to draw lines of symmetry onto it. At this stage, only focus on horizontal and vertical lines of symmetry. However, learners may also notice that the square has diagonal lines of symmetry. Repeat with other shapes.

- Give learners squared paper and a ruler. Ask them to draw their own symmetrical shapes and mark the horizontal and/or vertical lines of symmetry. They may need to use a mirror to check.

- Direct learners to the Student's Book. Discuss the Let's learn section with learners.

- Learners work through the paired activity. Guide them towards folding the paper along the line of symmetry to prove their shape is symmetrical.

- Discuss the Guided practice example in the Student's Book.

Practise 🔲 [TWM.05]

- Workbook

Title: Symmetry

Pages: 162–163

- Refer to Activity 2 from the Additional practice activities. Ensure learners include symmetry as a property of the pentominoes.

Apply 👥 🖥 [TWM.06]

- Display **Slide 2**. Learners spot items in the classroom that they think have symmetry.

- They share ideas with their partner and make a list or draw pictures.

- Invite learners to share the symmetrical items they noticed in the classroom, for example the top of a table, the cover of a book.

Review ⏸

- Recap what symmetry is, include the vocabulary of 'line of symmetry', 'horizontal' and 'vertical'.

- Display the **Symmetry tool** and show a variety of shapes for learners to identify those with one, two and no lines of symmetry.

Assessment for learning

- What does the word 'symmetry' mean?
- How many lines of symmetry does a rectangle have?
- What about a regular triangle?
- Show / Draw me a shape that has no line of symmetry.

Same day intervention

Support

- Focus on vertical lines of symmetry until learners have a sound grasp of these and then introduce horizontal lines of symmetry.

Lesson 4: **Angles**

Learning objective

Code	Learning objective
3Gg.10	Compare angles with a right angle. Recognise that a straight line is equivalent to two right angles or a half turn.

Resources

two strips of card (per learner); paper fastener (per learner); paper (per learner); ruler (per learner)

Revise

Use the activity *Angles* from Unit 20: *2D shapes, symmetry and angles* in the Revise activities.

Teach 🆂🅱 🖥

- **[T&T]** 🔖 Ask: **What is a right angle? What shapes have right angles?** Take feedback, asking learners to explain their reasoning.
- Display **Slide 1**. Ask: **What is the same about these shapes and what is different?** Agree that they all have right angles. They have different numbers of sides and vertices so they are different shapes.
- Explain that right angles can be positioned in different orientations. Point this out on the pentagon on the slide. It has three right angles: at bottom left and right and at the top.
- Display **Slide 2**. Explain that a right angle is a quarter of a turn. Ask learners to stand up and make clockwise and anticlockwise quarter turns. Ask learners to show you right angles around the classroom.
- Give each learner two strips of card and a paper fastener. They fix the strips together with the paper fastener to make a right angle measurer. Ask: **Can you show me a right angle?** They use it to locate right angles in the classroom or outside. Ask: **Can you use your angle measurer to show angles that are less than a right angle? What about greater than a right angle?**
- Display **Slide 3**. It shows right angles in different orientations. Ask: **What do you notice about the two red right angles at the bottom of the slide?** Agree that they form a straight line. Say: **We already know that one right angle is the same as a quarter of a turn.** Ask: **What would two right angles be the same as?** Ask learners to make two quarter turns. Establish that they make a half turn. Link to fractions work that two quarters are equal to one half.
- Direct learners to the Student's Book. Discuss the Let's learn section.
- Learners work through the paired activity. They could use their right angle measurer to check whether the angles are greater or less than a right angle.

- Discuss the Guided practice example in the Student's Book.

Practise 🆆🅱

- Workbook

Title: Angles

Pages: 164–165

- Refer to the variation in Activity 2 from the Additional practice activities. Ask learners to identify and count how many right angles there are in each pentomino.

Apply 👥 🖥 [TWM.04]

- Display **Slide 4**. Learners explain how two right angles make a straight line, to help Isaac understand.
- They need to include 'half', 'quarter' and 'turn' in their definition. When they are happy with their explanation, they write it down.
- Invite learners to share their explanations. For each, ask: **Is this a good explanation? Why?** Make links to fractions – two quarters is the same as a half.

Review 📊

- Display the **Geoboard tool** with a variety of 2D shapes. Ask: **Can you identify the shapes with right angles?** Ensure to include shapes in different orientations.
- Discuss with the learners how right angles are right angles no matter which position they are in.

Assessment for learning

- What is a right angle?
- Where would you see right angles in real life?
- What size of turn is a right angle?
- Is this angle smaller or larger than a right angle? How do you know?

Same day intervention
Support

- Give learners a small piece of paper with a corner that is a right angle and other corners that are not. Ask them to use this to compare different angles with a right angle.

Additional practice activities

Activity 1 👥 ⚠️2

Learning objective
• identify, describe, classify, name and sketch 2D shapes by their properties

Resources
squared paper (per learner); scissors (per pair); glue (per pair); Resource sheet 14: Carroll diagram template (per pair); large piece of paper (per pair) (for variation)

What to do [TWM.03/04/05/06]
• Learners draw a variety of regular and irregular shapes on squared paper and cut them out.
• They label their shapes with the correct names and give a detailed description of each shape's properties to their partner.

• Remind them about Carroll diagrams and how they sort things according to one criterion, for example: four sides / not four sides; symmetrical / not symmetrical; right angle / no right angle.
• The pairs choose a criterion and sort the shapes accordingly and stick them onto the Carroll diagram on Resource sheet 14.

Variation
1 Learners draw another set of shapes and stick these on a large piece of paper to make an information poster. They list the properties of each shape, including right angles and symmetry.

Activity 2 👥 ⚠️2

Learning objectives
• identify, describe, classify, name and sketch 2D shapes by their properties
• Identify horizontal and vertical lines of symmetry
• Recognise right angles and investigate right angles in shapes and patterns

Resources
squared paper (per learner); plain paper (per pair); scissors (per pair); small mirror (per pair)

What to do [TWM.03/04/05]
• Learners work in pairs to make pentominoes (shapes made from five squares placed side by side) on squared paper.
• There are 12 possible pentominoes (see below) – can the class find all of them? They should not be reflections or rotations of shapes previously drawn. Ask learners to name the shape each pentomino makes and then to find something they have in common. Agree that they all have straight edges.

• Learners then identify the pentominoes that are symmetrical. They can use a small mirror to help them.
• Learners each make a copy of all 12 pentominoes and cut them out. In pairs, they take turns to place a pentomino on a piece of plain paper. Their partner matches it to develop a symmetrical pattern.

Variation
2 Ask learners to explore the angles in each pentomino. What do they notice? Agree all angles are right angles. How many does each pentomino have? Ensure they investigate the internal angles of each one. Can they see any shapes that have two or three right angle turns?

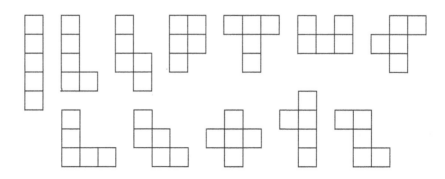

Unit 21: 3D shapes

Collins International Primary Maths Recommended Teaching and Learning Sequence: Term 1, Week 9

Learning objectives

Code	Learning objectives
3Gg.05	Identify, describe, sort, name and sketch 3D shapes by their properties.
3Gg.08	Recognise pictures, drawings and diagrams of 3D shapes.

Unit overview

This unit focuses on 3D shapes. It includes work on pyramids and prisms. Learners will be taught to classify 3D shapes into those that have flat surfaces, such as cubes, and those that have curved surfaces, such as spheres. The mathematical name for the former is polyhedra, although learners are not expected to know this term at this stage. Two **faces** that meet form an **edge** and the corner where three or more edges meet is called a **vertex**. Regular polyhedra (such as cubes and tetrahedra) have faces that are exactly the same shape and edges that are the same length. Some polyhedra are prisms. A prism has two identical 2D polygons at its opposite ends, joined by parallel edges that, together with the parallel edges of the end faces, form rectangular faces. A pyramid is made from a polygonal base with edges from each vertex of the base, meeting at an apex. The faces from the base to the apex are triangular.

Note: For those teachers/schools following the CIPM Recommended Teaching and Learning Sequence, this unit is taught after Unit 20: 2D shapes, symmetry and angles. However, some schools may prefer to change the order of these two units and teach this unit prior to teaching Unit 20.

Prerequisites for learning

Learners need to:
• recognise different 3D shapes
• recognise the 2D faces on 3D shapes
• sort 3D shapes according to their properties
• use 3D shapes to make patterns.

Vocabulary

cube, cuboid, square-based pyramid, tetrahedron, triangular prism, sphere, hemisphere, cylinder, tetrahedron, polygonal, edge, face, vertex, vertices

Common difficulties and remediation

Shape is a strand of mathematics in which learners who are less confident often excel, so have high expectations of what these learners, in particular, can achieve! Some learners may confuse names of 2D shapes with names of 3D shapes, particularly since prisms and pyramids are identified by the shape of their polygonal faces. Provide activities that encourage learners to name 3D shapes and describe their properties in terms of the numbers and shapes of their faces and the numbers of edges and vertices. Ask them to tell you the similarities and differences between different 3D shapes.

Another aspect learners may find difficult is that of visualisation. Provide plenty of opportunities for learners to do this by, for example, asking them to visualise a 3D shape (e.g. a triangular pyramid) and imagine cutting along some of its edges and laying it out flat (this is its net, but the term is not introduced until Stage 4). Then ask them to draw this flat shape on paper and visualise the 3D shape it would make if it were folded up. Ask them to describe what they would see after cutting the top off a squared-based pyramid.

Make clear in practical demonstrations that the cross-section of a prism is the constant shape seen when the prism is cut across its length, parallel to its polygonal ends and perpendicular to the base, so that learners can see that it is the same shape from one end of the shape to the other.

Supporting language awareness

Ask questions such as: **What is the same?** and **What is different?** when classifying shapes and expect earners to answer in complete sentences.

Promoting Thinking and Working Mathematically

During this unit learners will be making conjectures (TWM.03), convincing (TWM.04), characterising (TWM.05) and classifying (TWM.06) when they sort 3D shapes according to different criteria and explain their thinking. They also have opportunities to specialise (TWM.01) and generalise (TWM.02) when identifying the 2D faces of 3D shapes and 3D shapes in the real world.

Success criteria

Learners can:
• identify, describe, classify, name and sketch 3D shapes by their properties
• recognise pictures, drawings and diagrams of 3D shapes.

Unit 21 3D shapes

Lesson 1: **Identifying 3D shapes**

Learning objectives

Code	Learning objective
3Gg.05	Identify, describe, classify, name and sketch 3D shapes by their properties.
3Gg.08	Recognise pictures, drawings and diagrams of 3D shapes.

Resources

small piece of modelling clay (per learner); cuboid and triangular prism (per pair); pencil and paper (per learner)

Revise

Use the activity *Naming shapes* from Unit 21: *3D shapes* in the Revise activities to assess learners' knowledge of 3D shapes.

Teach 🔲 🖥 [TWM.05]

- 📺 Display **Slide 1**. Ask: **What do all these shapes have in common?** Elicit that they are all 3D shapes with flat faces and the faces are polygons. Ask: **What do we need to think about when we describe 3D shapes?** Elicit that we can look at the shapes of their faces. We can also count their faces, edges and vertices. If necessary, use a 3D shape to explain that edges occur when faces meet and vertices occur when edges meet. Tell learners that the shapes of the faces and the numbers of faces, edges and vertices are the properties of the shapes.

- Ask: **What 3D shapes can you see? What are their properties?** Discuss the similarities and differences between them. Ask learners to name them and describe their properties. Elicit the names of the shapes and agree there is a cube, a cuboid, a triangular pyramid (tetrahedron) and a square-based pyramid. Introduce learners to the triangular prism.

- [T&T] Ask: **Where might you see objects like these 3D shapes in the classroom and in your house?** Take feedback, for example a cereal box is a cuboid. Do the class agree?

- Elicit that all the shapes they have looked at so far have flat faces, straight edges and vertices.

- [T&T] Ask: **Are all 3D shapes like this?** Agree that they are not. Display **Slide 2**. Ask learners to name the familiar shapes and describe their properties. Agree that there is a sphere and a cylinder. Introduce learners to the hemisphere. Point out that each of these has a curved surface. [T&T] Ask: **Where might you see objects like these 3D shapes in the classroom or in your house?**

- Give learners modelling clay. Ask them to make a sphere, then a cube, a cuboid, a square-based pyramid and a triangular pyramid. Question learners about the properties of each shape they make.

- Direct learners to the Student's Book. Discuss the Let's learn section.

- [TWM.05] Learners work through the paired activity.

- Discuss the Guided practice example in the Student's Book.

Practise 🔲 [TWM.01/06]

- Workbook

Title: Identifying 3D shapes

Pages: 166–167

- Refer to Activity 1 from the Additional practice activities.

Apply 👥 🖥 [TWM.01]

- Display **Slide 3**. Learners take a cuboid and triangular prism. Make sure that the rectangular faces of the triangular prism are the same as at least two faces of the cuboid.

- They put the prism on top of the cuboid. Then they each sketch the new shape and talk to their partner about its properties.

- Invite learners to show their sketches of the new shape.

- Invite one or two to describe the properties of this shape. Some may compare it to a house.

Review

- Recap the differences between the various 3D shapes. Ask learners to give you examples of each.

- Finish by drawing, for example, a square on the board and ask learners to tell you what 3D shape may have a face this shape. Repeat for triangle, rectangle and circle.

Assessment for learning

- What is a 3D shape?
- Tell me the names of some 3D shapes that you know.
- What properties do we need to think about when describing 3D shapes?
- What 3D shape has six faces? Is there another one?

Same day intervention
Support

- Focus on prisms, particularly cubes and cuboids. Ensure you have a supply of real objects that match these geometric shapes so learners can make links between 3D shapes and those seen in the real world.

Lesson 2: **Prisms**

Learning objectives

Code	Learning objective
3Gg.05	Identify, describe, classify, name and sketch 3D shapes by their properties.
3Gg.08	Recognise pictures, drawings and diagrams of 3D shapes.

Resources

variety of prisms (per group)

Revise

Use the activity *What's my shape?* from Unit 21: *3D shapes* in the Revise activities.

Teach [SB] 🖵 [TWM.05]

- Invite learners to choose a 3D shape and describe it for the rest of the class to identify.
- **[T&T]** Display **Slide 1**. Ask: **What can you see in the picture?** Agree a tent. Ask: **What shape is the tent?** Agree that it is the shape of a triangular prism.
- **[T&T]** Ask: **How would you describe a prism?** Agree a prism is a 3D shape that has two identical ends and rectangular faces joining the ends. Inform learners that the rectangular faces may be either squares or rectangles. Explain that most prisms are named by the shape of the two identical polygons.
- Give learners a selection of prisms to examine. They sketch and label them.
- Display **Slide 2**. Ask: **What do these shapes have in common?** Agree that they are all prisms because the two identical polygon faces are joined by rectangles. Ask them to name the prisms. Agree triangular prism, two rectangular prisms (often known as cubes and cuboids), pentagonal prism, hexagonal prism and octagonal prism.
- Direct learners to the Student's Book. Discuss the Let's learn section.
- **[TWM.05]** Learners work through the paired activity.
- Discuss the Guided practice example in the Student's Book.

Practise 📖 [TWM.05]

- Workbook

Title: Prisms

Pages: 168–169

- Refer to Activity 1 from the Additional practice activities.

Apply 👥 🖵 [TWM.03/04]

- Display **Slide 3**. Ameera thinks a cylinder is a prism. Adam thinks it is not. Learners decide who they agree with and explain why.

- Establish that Adam is correct because the end faces are not polygons and the circular faces of the cylinder are not joined by rectangles.

Review 🖵

- Display **Slide 2** again. Recap that a prism is a 3D shape that has two polygonal end faces and all other faces are rectangles. Ask learners to tell you the names of some prisms and to identify their end shapes.
- Describe different 3D shapes and ask learners to identify them.
- Ask learners to describe the properties of different 3D shapes for the class to identify. Finish the lesson by playing Falling Shapes 3D on the Interactive tool. Learners need to move the net to catch a specific shape. Ask learners to name those that should not be caught and state whether they are prisms or not prisms.

Assessment for learning

- What is a prism?
- Can you tell me some 3D shapes that are prisms?
- Describe the properties of a hexagonal prism? How many edges? How many vertices?
- What about an octagonal prism?

Same day intervention

Support

- Initially explore cubes and cuboids using examples in the classroom, for example, boxes. When learners are confident with the properties of these and what a prism is, introduce triangular prisms.

Enrichment

- Ask learners to examine a cuboid and to try to make one from card, based on their observations. They could draw around the faces of the cuboid, cut them out and stick them together.

Geometry and Measure – Geometrical reasoning, shapes and measurements

Unit 21 3D shapes

Geometry and Measure – Geometrical reasoning, shapes and measurements

Lesson 3: **Pyramids**

Learning objectives

Code	Learning objective
3Gg.05	Identify, describe, classify, name and sketch 3D shapes by their properties.
3Gg.08	Recognise pictures, drawings and diagrams of 3D shapes.

Resources

modelling clay (per learner); pencil and paper (per learner); scissors (per learner); access to two identical square-based pyramids (per pair)

Revise

Use the activity *Odd one out* from Unit 21: *3D shapes* in the Revise activities.

Teach SB 🖥 [TWM.05]

- **[T&T]** � Display **Slide 1**. Ask: **What are the two shapes in the photograph?** Agree that they are pyramids. These ones are in Egypt.
- Discuss the pyramids in the photograph. Elicit that each has a square base and sloping triangular faces. Discuss the properties of a pyramid, in terms of numbers and shapes of faces and numbers of edges and vertices.
- **[T&T]** Display **Slide 2**. Ask: **What do all pyramids have in common?** Agree pyramids have sides made from triangular faces. Discuss the different pyramids on the slide. Inform learners that the 2D shape of the base gives the pyramid its name.
- Can learners name them all? Give them a few minutes to identify the different pyramids. Encourage them to explain why they made their decisions. Take feedback. Agree that the:
  - blue pyramid is a square-based pyramid because its base is a square
  - red pyramid is a pentagonal pyramid because the base is a pentagon
  - orange pyramid is a triangular pyramid because its base is a triangle. Tell learners that its proper mathematical name is 'tetrahedron', and that 'tetra' means four, which is the number of faces a tetrahedron has
  - green pyramid is a hexagonal based pyramid because its base is a hexagon.
- Ask: **What do you think a pyramid with a base of a heptagon is called?** (heptagonal prism) **What about an octagon?** (octagonal prism)
- Direct learners to the Student's Book. Discuss the Let's learn section.
- **[TWM.05]** Learners work through the paired activity. (The shape created is an octahedron.)
- Discuss the Guided practice example in the Student's Book.

Practise WB [TWM.01]

- Workbook

Title: Pyramids

Pages: 170–171

- Refer to Activity 2 from the Additional practice activities.

Apply 👥 🖥 [TWM.05/06]

- Display **Slide 3**. Learners make a square-based pyramid out of modelling clay.
- Each learner then visualises and sketches what it would look like if it were 'opened up'.
- They then cut it out, fold it and see if they have a square-based pyramid before discussing with their partner what they could do to make it more accurate.
- Invite learners to share their pyramids. Ask them to reflect on how easy or difficult they found the visualising. Agree that to get a more accurate pyramid, learners need to measure and be sure that they make a square and that all four triangles are identical.

Review 🖥

- Display **Slide 2** again. Recap what a pyramid is. Ensure learners understand that it is a 3D shape with the base of a named shape and triangular faces. Ask learners to tell you the names of some and to describe their properties in terms of numbers of faces, edges and vertices. Ask: 🖒 **What do you notice about all the triangular faces of a pyramid?** If not suggested, discuss with the class how all the triangular faces of a pyramid meet as a point called an apex.
- Finish by drawing, for example, a square and a triangle on the board. Ask learners to tell you what 3D shape these could be the faces of.

Assessment for learning

- What are the properties of a ...-based pyramid?
- What is the same about all pyramids? What is different?

Same day intervention

Support

- For those less confident with the properties of pyramids, focus on square-based pyramids. Explore a variety of different sizes, so learners can see that the properties remain the same, no matter the size.

Lesson 4: 3D shapes in real life

Learning objectives

Code	Learning objective
3Gg.05	Identify, describe, classify, name and sketch 3D shapes by their properties.
3Gg.08	Recognise pictures, drawings and diagrams of 3D shapes.

Resources

real-life 3D objects, e.g. can of drink, dice, box of cereal (eight per group); squared paper (per learner); ruler (per learner); scissors (per learner)

Revise

Use any activity from Unit 21: *3D shapes* in the Revise activities.

Teach [SB] 🖵 [TWM.03/04]

- **[T&T]** 🖐 Display **Slide 1**. Ask the learners to name each object and to tell you where the object might be found. Ask: **What 3D shapes can you see? What are their properties?** Discuss the similarities and differences between them. Ask learners to tell you other objects that are the same shapes as these.

- **[T&T]** Give each group of learners a selection of eight objects. Ask them to discuss what each object is and identify which type of 3D shape it is. Ask: **How do you know what these shapes are?** Expect them to talk about their properties in terms of numbers and shapes of faces, whether they are prisms or not and numbers of edges and vertices. Emphasise that most objects in real life are made up of one or more of the 3D shapes they know about.

- **[TWM.03/04]** Ask learners to sort their groups of items according to their own criteria. Invite groups to share how they sorted them. Ask the class: **Do you think this is a good way of sorting? Why?**

- Invite a learner to secretly choose an object and describe it to the class. Ensure they use mathematical language to describe the shape of the object. Give the class no more than five chances to name the shape and object.

- Direct learners to the Student's Book. Discuss the Let's learn section.

- Learners work through the paired activity.

- Discuss the Guided practice example in the Student's Book.

Practise [WB]

- Workbook

Title: 3D shapes in real life

Pages: 172–173

- Refer to Activity 2 from the Additional practice activities.

Apply 👥 🖵 [TWM.05]

- Display **Slide 2**. Learners look at the picture of a cube and then visualise what it would look like 'opened up'. They use what they know about the properties of cubes to do this.

- They draw what they think on squared paper (using a ruler). They cut it out and fold it to make a cube.

- Once they have done this, they use what they have done and their knowledge of properties of a cuboid to draw one of these.

- Invite learners to share their drawings. Ask them to reflect on the ease or difficulty they had in completing the activity.

Review

- Recap the work covered on 3D shapes in this unit.

- Discuss the various shapes that they have been learning about, especially prisms and pyramids. Ask learners to tell you the properties of these shapes and to identify them as real-life objects.

- Invite them to share how confident they are at identifying, describing and making 3D shapes.

Assessment for learning

- Where can you see cubes in real life?
- What shape is a cereal box?
- Where can we see spheres in real life?

Same day intervention

Support

- Work with learners to identify cubes, cuboids and cylinders in the classroom.

Enrichment

- Ask learners to create a poster that shows a selection of the shapes covered in this unit. They make a sketch of examples of prisms and pyramids and write a list of the properties of each one. Ensure these include faces, edges and vertices. They could also write where each of their examples might be seen in real life.

Geometry and Measure – Geometrical reasoning, shapes and measurements

Additional practice activities

Activity 1

Learning objective
- Identify, describe, classify, name and sketch 3D shapes by their properties

Resources
sphere, cylinder, cube, cuboid, square-based pyramid, triangular prism (per group); Resource sheet 14: Carroll diagram template, enlarged to A3 (per group); Resource sheet 15: Venn diagram template, enlarged to A3 (per group) (for the variation)

What to do [TWM.03/05/06]
- Give small groups of learners a selection of 3D shapes. They discuss their properties – numbers and shapes of faces, numbers of edges and vertices.

- As a group, they decide the criteria for a Carroll diagram, for example: prism/not prism. They draw their own Carroll diagram, or use the template from Resource sheet 14, and place their shapes in the appropriate sections.
- They then write the shape names in their diagram.

Variation
1 Learners repeat the activity but use a single circle Venn diagram instead of a Carroll diagram. Give them more shapes to sort. They think of a criterion to sort by. The shapes that fit the criterion are placed inside the oval and those that don't are placed outside it.

Activity 2

Learning objectives
- Identify, describe, classify, name and sketch 3D shapes by their properties
- Recognise pictures, drawings and diagrams of 3D shapes

Resources
newspapers, magazines, supermarket advertising flyers (per group); sheet of A3 or A2 paper (per group); scissors (per group); glue (per group)

What to do [TWM.04/05/06]
- Learners work together as a group to look through newspapers, magazines, supermarket advertising flyers, and so on, for real-world examples of familar 3D shapes.
- Groups create a poster by cutting out and sticking the images onto large paper and labelling each object with its 3D shape name.
- They then add the properties of each shape.
- Ask groups to make statements about pairs of shapes that give differences and similarities between them.

Variation
1 Give learners images and ask them to talk about the similarities and differences between them.

Geometry and Measure – Geometrical reasoning, shapes and measurements

Unit 22: Length, perimeter and area

Collins International Primary Maths
Recommended Teaching and
Learning Sequence: Term 3, Week 7

Learning objectives

Code	Learning objective
3Gg.02	Estimate and measure lengths in centimetres (cm), metres (m) and kilometres (km). Understand the relationship between units.
3Gg.03	Understand that perimeter is the total distance around a 2D shape and can be calculated by adding lengths, and area is how much space a 2D shape occupies within its boundary.
3Gg.04	Draw lines, rectangles and squares. Estimate, measure and calculate the perimeter of a shape, using appropriate metric units and area on a square grid.
3Gg.11	Use instruments that measure length[, mass, capacity and temperature].

Unit overview

Learners will begin by exploring the appropriate equipment and units to use when estimating, and then measuring and recording, different lengths. They will learn when and how to use a ruler, metre stick and trundle wheel and also whether to use centimetres, metre or kilometres. Learners will develop an understanding of the relationship between the units of length: centimetres, metres and kilometres. They will reinforce and consolidate their skills in using a ruler accurately.

Learners will also explore perimeter and area. It is important they understand that perimeter is a length: it is the distance around the outside of a shape or region. Area is the amount of flat space within the boundary or perimeter of a shape or region. The units for measuring perimeter are those used for length. Area is measured in square units. This unit is very practical, so it is important to have the appropriate equipment ready for learners to use.

Prerequisites for learning

Learners need to:
- understand that length is a fixed distance between two points
- understand that two lengths can be added or subtracted
- understand why it is better to measure length in standard units
- measure lengths in centimetres and metres
- understand that when using a ruler to take a measurement, they should start from zero.

Vocabulary

kilometre (km), metre (m), centimetre (cm), ruler, metre stick, tape measure, trundle wheel, length, height, width, distance, estimate, actual, perimeter, area, space, square units

Common difficulties and remediation

Some learners may forget equivalences between units, for example how many metres make one kilometre. It would be helpful to rehearse these facts regularly and to refer to them when covering fractions and mental and written calculations.

Some learners may use a ruler incorrectly. They may position the end of the object at the end of the ruler (even if there is a gap between the end and the mark for zero) or at the division that shows one centimetre. Provide opportunities for learners to practise placing the end of the object they are measuring so that it aligns exactly with the zero on the scale.

Supporting language awareness

Display key words and their meanings, in particular, length, height, width, perimeter and area. Also display the following units of measure alongside their abbreviations: centimetre (cm), metre (m) and kilometre (km).

Promoting Thinking and Working Mathematically

Learners are given facts, such as 1 km = 1000 m, and asked to work out other facts they know, which involves specialising (TWM.01) and making generalisations (TWM.02). They make decisions on equipment and units to use to measure, which involves making conjectures (TWM.03) and convincing (TWM.04). When working on perimeter and area they identify and describe mathematical properties, which involves characterising (TWM.05) and classifying (TWM.06).

Success criteria

Learners can:
- estimate and measure lengths in centimetres (cm), metres (m) and kilometres (km)
- understand the relationship between units
- understand that perimeter is the total distance around a 2D shape and can be calculated by adding lengths
- understand that area is how much space a 2D shape occupies within its perimeter or boundary.

Geometry and Measure – Geometrical reasoning, shapes and measurements

Lesson 1: **Units of length**

Geometry and Measure – Geometrical reasoning, shapes and measurements

Learning objective

Code	Learning objective
3Gg.02	Estimate and measure lengths in centimetres (cm), metres (m) and kilometres (km). Understand the relationship between units.

Resources

ruler (per learner); metre stick (per pair/group) four classroom objects (per pair); paper and pencil (per pair)

Revise

Use the activity *Ruler or metre stick* from Unit 22: *Length, perimeter and area* in the Revise activities to assess learners' understanding of metres and centimetres.

Teach 🅂🅱 🖥 [TWM.03/04]

• **[T&T]** Display **Slide 1**. Ask: **What can you see in the photograph?** Take feedback, accepting all their ideas. Explain that these people are building a road and that they are taking measurements as part of the road-building process.

• **[T&T] [TWM.03/04]** 🔀 Ask: **What unit would be used to measure the length of the section of road they are building?** Accept metres, then explain that longer distances, such as between two towns, are measured in kilometres. Remind learners that centimetres are used for small lengths, then metres are used for longer lengths, such as the width of a room or the length of a field. On the board write: 1 m = 100 cm and 1 km = 1000 km. Discuss the equivalences and also the abbreviations for centimetre, metre and kilometre as these abbreviations are new to learners.

• **[TWM.03/04]** Ask questions such as: **How many centimetres are there in 3 m/4 m/5 $\frac{1}{2}$ m? How many metres are there in 800 cm/900 cm/1000 cm?** Expect learners to consider whether their answer is sensible and to explain the strategy they used. Ask further questions for equivalences converting kilometres to metres and converting metres to kilometres.

• Give pairs or groups of learners rulers and metre sticks and let them measure the lengths of different items around the classroom. Expect them to estimate the lengths first and then record the actual measurements in their own way.

• Direct learners to the Let's learn section in the Student's Book. After discussion, allow them to complete the paired activity, giving support if necessary. Discuss the Guided practice example.

Practise 🆆🅱

• Workbook

Title: Units of length

Pages: 174–175

Apply 👥 🖥 [TWM.05/06]

• Display **Slide 2**. Pairs of learners find four objects in the classroom. They discuss the units they will use to measure each object. They also discuss what equipment they will use to measure them.

• They then estimate and measure the length of each of their items and copy and complete the table.

Review

• Recap the units used to measure different lengths heights, widths and distances.

• Recap the possible equipment that learners can use to measure.

Assessment for learning

• How many centimetres are there in 1 metre? If we know this, what else do we know?

• How many metres are there in 1 kilometre? If we know this, what else do we know?

• What unit would you use to measure the width of a book?

• What about the depth of a swimming pool?

• What about the distance from one country to another?

Same day intervention

Support

• Focus on measuring in metres. Ensure learners know how to use a metre stick. Guide them in this as necessary.

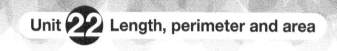
Lesson 2: **Measuring lines**

Learning objectives

Code	Learning objective
3Gg.04	Draw lines, [rectangles and squares. Estimate, measure and calculate the perimeter of a shape, using appropriate metric units and area on a square grid].
3Gg.11	Use instruments that measure length[, mass, capacity and temperature].

Resources

ruler (per learner); six strips of paper of different lengths (per learner); plain paper (per learner/pair); paper and scissors (per learner) (for the Workbook)

Revise

Use the activity *Units of length* from Unit 22: *Length, perimeter and area* in the Revise activities to reinforce and consolidate converting between centimetres, metres and kilometres.

Teach 🆂🅱 🖥 [TWM.03/04/07]

- 🗣 **[TWM.03/04]** Briefly recap units of length. For example, write 200 cm = 2 m on the board. Ask learners to use doubling, halving, adding or multiplying to generate as many new facts as they can. Take feedback after several minutes.
- **[T&T]** Display **Slide 1**. Say: **This photograph shows a pigmy mouse lemur. It is the smallest primate in the world. Its body measures 6 cm and its tail is about twice as long as its body.** You could explain that primates are a type of mammal, and they include humans, apes, monkeys and lemurs.
- **[TWM.03/04]** Ask learners to use their hands to estimate the length of the lemur's body, then its tail and finally the whole animal. They then use their rule to measure the length of its body, tail and the whole animal and compare with their estimates.
- **[T&T]** 🗣 Display **Slide 2**. Ask: **Can you explain to your partner how to use a ruler?** Agree that they must line up the zero on the ruler with one end of the object they are measuring. The length of the object is where the other end reaches, along the ruler.
- Give learners strips of paper to measure. Ask learners to estimate the length of each of their strips and to record their estimate on the strip of paper and to circle it. Remind learners not to use their ruler at this stage – they are only estimating, not measuring.
- Once learners have estimated, they use their ruler to measure the length of each of their strips, recording the lengths of the strips.
- Once learners have measured their strips, tell them to swap with a partner to measure and check that the measurements are correct.
- Next, provide learners with a sheet of plain paper and ask them to draw, and label, various lines between 1 cm and 25 cm. Once learners have drawn their lines, tell them to swap with a partner to measure and check that the measurements are correct.

- **[TWM.07]** Direct the class to the Let's learn section in the Student's Book. After discussion, allow learners to complete the paired activity, giving support if necessary. Discuss the Guided practice example.

Practise 🆆🅱

- Workbook

Title: Measuring lines

Pages: 176–177

- Refer to Activity 1 from the Additional practice activities.

Apply 👥 🖥 [TWM.03/04]

- Display **Slide 3**. Pairs decide which of the pictured items would be measured in centimetres. They then make a list of six objects in the classroom that they think would be measured in centimetres. They estimate their lengths and then check.
- They make a list of their objects and write their estimates and the actual lengths next to the name of each one.

Review

- Look back at **Slide 3**. Ask learners what units they would use to measure the tree and the book. What units would they use to describe how far they live from school?
- Ask them to describe how to measure with a ruler and then how to use one to draw lines.

Assessment for learning

- How would you measure a line to the nearest centimetre?
- How would you draw a line 12 cm long?

Same day intervention

Support

- Focus on measuring in centimetres. Ensure learners know how to hold and use a ruler. Guide them in this as necessary.

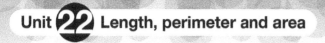

Unit **22** Length, perimeter and area

Lesson 3: **Perimeter**

Geometry and Measure – Geometrical reasoning, shapes and measurements

Learning objectives

Code	Learning objective
3Gg.03	Understand that perimeter is the total distance around a 2D shape and can be calculated by adding lengths[, and area is how much space a 2D shape occupies within its boundary].
3Gg.04	Draw [lines,] rectangles and squares. Estimate, measure and calculate the perimeter of a shape, using appropriate metric units [and area on a square grid].
3Gg.11	Use instruments that measure length[, mass, capacity and temperature].

Resources

plain paper (per learner); pencil (per learner); ruler (per learner); 1 cm squared paper (per pair); 1 cm squared paper (per learner) (for Workbook)

Revise

Use the activity *Using rulers* from Unit 22: *Length, perimeter and area* in the Revise activities to reinforce and consolidate learners' knowledge of drawing and measuring accurately with a ruler.

Teach [SB] 🖵

- **[T&T]** 🖑 Display **Slide 1**. Ask: **What is the length of the football pitch? What is its width?** Agree that both the lengths are 105 m and the widths are 65 m each. Ask: **What is the total distance around the outside of the football pitch?** Ask learners to discuss this with their partner. Encourage them to make an estimate first, for example 300 m, because 105 m is close to 100 m and 65 m to 50 m. After a few minutes take feedback. Establish that the total distance is 105 m + 105 m + 65 m + 65 m, which is 340 m. Ask: **What is the distance around the outside of a shape called?** If necessary, introduce the term 'perimeter'. Take time to ensure that learners understand that the perimeter is the total length around the outside of the shape.
- **[T&T]** Display **Slide 2**. Say: **These shapes have been drawn on centimetre squared paper.** Ask: **How can we find the perimeters of these shapes?** Give learners a few minutes to discuss this with their partner. Take feedback. Agree that they find the lengths and widths and add them together. Ask: **Can you think of a quicker way to do this?** Agree that for the rectangle, they could double the length and width and add them together or add the length and width and double. For the square they could multiply the length by 4 or double and double again.
- **[T&T]** Display **Slide 3**. Repeat the calculations for these shapes, which are not drawn on squared paper.
- Give learners plain paper, pencils and rulers. Ask them to choose the lengths and widths of three rectangles and draw them carefully to the exact lengths. Encourage learners to estimate the perimeters first. They then work out the perimeters of their rectangles.

- Direct the class to the Let's learn section in the Student's Book. After discussion, allow learners to complete the paired activity, giving support if necessary. Discuss the Guided practice example.

Practise [WB] [TWM.01/03]

- Workbook

Title: Perimeter

Pages: 178–179

- Refer to Activity 2 from the Additional practice activities.

Apply 👥 🖵 [TWM.03/04]

- Display **Slide 4**. Learners are shown 12 square paving stones. They need to design a patio for Bert. The stones need to be joined edge to edge to make rectangular shapes.
- Learners work with a partner and sketch different designs on squared paper.
- They choose one and find its perimeter.

Review

- Ask learners to explain what perimeter is and how it can be calculated.

Assessment for learning

- What do we mean by perimeter?
- How can we find the perimeter of a rectangle?
- Is there another way?
- How can we find the perimeter of a square?
- Is there another way?

Same day intervention

Support

- Provide drawings of rectangles on plain paper so that learners can focus on measuring their sides and finding sums to give their perimeters.

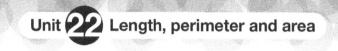

Lesson 4: **Area**

Learning objectives

Code	Learning objective
3Gg.03	Understand that [perimeter is the total distance around a 2D shape and can be calculated by adding lengths, and] area is how much space a 2D shape occupies within its boundary.
3Gg.04	Draw [lines,] rectangles and squares. Estimate, measure and calculate [the perimeter of a shape, using appropriate metric units and] area on a square grid.
3Gg.11	Use instruments that measure length[, mass, capacity and temperature].

Resources

1 cm squared paper (per learner); pencil (per learner); ruler (per learner) 1 cm squared paper (per pair)

Revise

Use the activity *Solving length problems* from Unit 22: *Length, perimeter and area* in the Revise activities to apply learners' knowledge of length. Include perimeter problems.

Teach 🔲 💻

- **[T&T]** 🔲 Display **Slide 1**. Ask: **What do you see?** After a few minutes take feedback. Explain that they are flowerbeds each enclosed by four paved sides. Ask: **How would you work out the perimeter of one of the flowerbeds?** Agree that you would need to measure the length and width of the bed where the flowers are growing and then add the measurements. Remind learners that the quickest way to work out the perimeter of a rectangle is to add the length and width and double it.

- Ask: **Does anyone know what we call the amount of space that is inside the paved sides, where the flowers are growing?** Establish that it is the area of the flowerbed. Take time to make clear the distance *around the outside* of the flowerbed is the perimeter, and the *amount of space inside* the paved sides is the area.

- Display **Slide 2**. Ask: **How could we find the areas of these rectangles?** Agree that they could count the squares.

- Ask learners to count the squares for each rectangle. Agree that in the first rectangle there are 5 rows of 8 squares, which gives an area of 40 squares. Agree that in the second rectangle there are 2 rows of 9 squares, which gives an area of 18 squares. Can you think of a quicker way to find these areas? Agree they could multiply, for example, 8 × 5 and 9 × 2.

- Give learners squared paper, pencils and rulers. Ask them to draw rectangles and estimate their areas. They then find the actual areas by counting the squares. Invite learners to share with the class what they did.

- Direct the class to the Let's learn section in the Student's Book. After discussion, allow learners to complete the paired activity, giving support if necessary. Discuss the Guided practice example.

Practise 🔲

- Workbook

Title: Area

Pages: 180–181

- Refer to the variation in Activity 2 from the Additional practice activities.

Apply 👥 💻 [TWM.01]

- Display **Slide 3**. Learners are given a rectangle and asked to find its area, then to draw another rectangle with the same area but different perimeter.

- Once they have found the area, they need to use knowledge of their tables to find another multiple that is equal to 24.

Review

- Discuss where, in the classroom, they can see different areas; for example, the area of the surface of a table, the boards, the door of a cupboard.

Assessment for learning

- Can you explain what is meant by area? Can you explain it in another way?
- What surrounds an area?
- I have a rectangle that is six rows of two squares, what is its area?

Same day intervention
Support

- Give learners prepared rectangles. Encourage them to count the squares to find the area. Once they have, discuss more efficient ways to find the area, for example finding the number of squares in one row and repeatedly adding the rows.

Additional practice activities

Activity 1

Learning objectives
- Estimate and measure lengths in centimetres
- Use instruments to draw and measure length

Resources
20 strips cut from A4 paper (per group); glue (per group); scissors (per learner); two rulers (per group); metre stick (per group)

What to do
- Ask learners to take strips of paper and to make different lengths from them by cutting and gluing.
- Learners should make approximately 10 strips in a variety of lengths: some should be shorter than a ruler, and some longer than a metre stick. They can glue different lengths together to make longer lengths.
- Learners order the strips, from shortest to longest. They estimate the lengths in centimetres, metres

and centimetres, then use rulers and metre sticks to measure each length and compare with their estimates.
- Finally, learners make a table of their results similar to that below.

Strip	Estimate	Actual
1		
2		
3		
4		
5		

Activity 2

Learning objectives
- Estimate, measure and calculate the perimeter of a shape, using appropriate metric units and area on a square grid
- Use instruments to draw and measure length

Resources
1–20 number cards from Resource sheet 16: 0–100 number cards (per pair); plain paper (per pair); ruler (per pair); 1 cm squared paper, for the variation (per pair)

What to do
- Learners shuffle the number cards and place them in a pile face down on the table.
- Learners take turns to pick a card. They use the number on the card as a length and use a ruler to draw it.
- They pick another card and use the number as the width, and use a ruler to draw this.

- They then draw the other two sides of the rectangle before finding the perimeter of their rectangle.
- Their partner checks to make sure it is correct.
- Repeat several times.

Variation
 Repeat the activity but give learners 1 cm squared paper. The number cards represent the number of squares for the lengths and widths of rectangles.

Once they have drawn a rectangle, they find its perimeter and area.

Geometry and Measure – Geometrical reasoning, shapes and measurements

Unit 23: Mass

Collins International Primary Maths
Recommended Teaching and
Learning Sequence: Term 3, Week 8

Learning objectives

Code	Learning objective
3Gg.06	Estimate and measure the mass of objects in grams (g) and kilograms (kg). Understand the relationship between units.
3Gg.11	Use instruments that measure [length,] mass[, capacity and temperature].

Unit overview

Learners will begin by exploring the appropriate equipment and units to use when estimating, measuring and recording different masses, including whether to use grams or kilograms. Learners will develop understanding of the relationship between the units of grams and kilograms. They will use various strategies to find equivalences between them; for example, using what they already know to generate other facts, and scaling up and down. They will find the mass of real items in order to learn how to read intervals on a scale. Learners will also solve real-life problems involving mass.

It is important to allow sufficient time for the Additional practice activities in this unit. The key to a rounded understanding of measurement is through repeated 'hands-on learning'; for example, being aware of how heavy 1 kg actually is, as well as familiarity with real units of measure rather than just the theory.

Prerequisites for learning

Learners need to:
• understand that mass is the quantity of matter in an object
• be able to use non-standard or standard units to estimate and measure familiar objects
• understand a measuring scale as a continuous number line where intermediate points have value.

Vocabulary

kilogram (kg), gram (g), mass, heavier, lighter

Common difficulties and remediation

As with length, some learners may forget equivalences between units, for example how many grams make one kilogram. It is helpful to rehearse these facts regularly and refer to them when covering fractions and mental and written calculation.

Some learners may confuse 'weight' and 'mass'. They often think weight is how heavy something is. This is, in fact, mass. Mass is a measurement of how much matter there is in an object. Weight is a measurement of how strongly gravity is pulling on that object. You could explain this, using the example of a person in space. Their weight would be nothing, because there is no pull of gravity. Their mass would be the same

as it is on Earth, because the amount of matter they have is still the same. It is important to use the correct vocabulary with learners.

Some learners may have problems with conservation of mass. If you give them, for example, a flat piece of malleable material to find the mass of, then roll it into a ball and give it back to them, they may measure it again, thinking that altering the appearance of the ball changes its mass. Learners need to understand that the same piece of material will have the same mass, no matter what it looks like.

Some learners have difficulty estimating mass. It is important to give them something to compare with the mass they are estimating. Let them feel, for example, the mass of a 1 kg bag of rice or a 2 kg bag of potatoes, in order to make a sensible estimate.

Supporting language awareness

In this unit, vocabulary includes: kilogram, gram, mass, heavier, lighter, scales, intervals. You may find it useful to refer to the audio glossary on Collins Connect. It is important to display on cards in the classroom the words used, with visual clues, to help learners recognise and understand what each one means. It is helpful to write the abbreviations used for gram (g) and kilogram (kg) beside these words.

Promoting Thinking and Working Mathematically

During this unit, learners will have opportunities to think and work mathematically. For example, they work out other facts that they know from being given a fact such as 1 kg = 1000 g, which involves specialising (TWM.01) and making generalisations (TWM.02). They look at similarities and differences between different instruments used to find the mass of objects, which involves making conjectures (TWM.03), convincing (TWM.04), characterising (TWM.05) and classifying (TWM.06).

Success criteria

Learners can:
• estimate and measure mass in kilograms (kg) and grams (g)
• understand the relationship between units
• use instruments to measure mass.

Unit **23** Mass

Lesson 1: **Units of mass**

Geometry and Measure – Geometrical reasoning, shapes and measurements

Learning objective

Code	Learning objective
3Gg.06	Estimate and measure the mass of objects in grams (g) and kilograms (kg). Understand the relationship between units.

Resources

selection of gram and kilogram weights (per group)

Revise

Use the activity *Kitchen scales or balance scales?* from Unit 23: *Mass* in the Revise activities to assess learners' understanding of measuring items to find their masses.

Teach 〔SB〕 ▯ [TWM.01/02]

- [T&T] Display **Slide 1**. Ask: **What can you see?** Agree the photographs show a meerkat and an antelope. Ask: **Can anyone tell me how big a meerkat is?** Elicit that it is quite small. Tell learners that a meerkat has a mass of about 700 grams. Repeat for the antelope, and establish that a meerkat is a lot bigger than a meerkat, and has a mass of about 40 kilograms
- [T&T] Show **Slide 2**. Ask: **What can you tell me about the relationship between grams and kilograms?** Agree that kilograms are the heavier of the two units for measuring mass, and grams are the lighter. Give groups a selection of weights and ask them to order them, from heaviest to lightest.
- [TWM.01/02] Establish that there are 1000 grams in one kilogram. Ask questions such as: **How many grams are there in 2 kg/4 kg/5 kg? How many kilograms are there in 2000 g/3000 g/8000 g?**
- Direct learners to the Student's Book. Discuss the Let's learn section before learners work on the paired activity. Then discuss the Guided practice example.

Practise 〔WB〕 [TWM.01/04]

- Workbook

Title: Units of mass

Pages: 182–183

- Refer to Activity 1 from the Additional practice activities.

Apply 👥 ▯ [TWM. 01/04]

- Display **Slide 3**. Learners are given the fact 1 kg = 1000 g. They work with a partner to make up new equivalent facts.
- They should challenge themselves by, for example, using halves and quarters of kilograms such as 1 kg and $\frac{1}{4}$ of a kg and use strategies such as doubling and halving.
- Expect them to explain their strategies to their partner and make a list to share with the class at the end of the lesson.

Review

- Invite pairs to suggest some equivalent masses. Make a list of these on the board. Then ask learners to find more by adding and subtracting those that have been suggested.
- Finish the lesson by giving a mass, for example 4 kg, and asking learners to give different ways to make it using a mixture of kilograms and grams.

Assessment for learning

- How many grams are there in one kilogram? What about 2 kg, 5 kg, 6 kg?
- How many kilograms are there in 5000 g? What about 2000 g, 9000 g, 3000 g?

Same day intervention

Support

- Focus on whole kilogram masses.

Lesson 2: **Measuring in kilograms**

Learning objectives

Code	Learning objective
3Gg.06	Estimate and measure the mass of objects in grams (g) and kilograms (kg). Understand the relationship between units.
3Gg.11	Use instruments that measure [length,] mass[, capacity and temperature].

Resources

kitchen and balance scales with a selection of known weights (per group); objects from around the classroom that have a mass of more than a kilogram (per group); 1 kg bag of rice or sugar or 1 litre bottle of water (per pair); access to 1 kg of rice or similar and kitchen scales (per learner) (for the Workbook)

Revise

Use the activity *Units of mass* from Unit 23: *Mass* in the Revise activities to reinforce and consolidate converting between grams and kilograms.

Teach 🔲 🖥

- **[T&T]** Display **Slide 1**. Ask: **What can you see? What is the same about them? What is different?** Agree that these are all weighing scales. Two have numeric scales, like number lines, that show the mass. One is a set of kitchen scales, for measuring food; the other is used to find the mass of people. Ask: **Have any of you used scales like these? Which scales would you use to find the mass of yourself / some flour / ...?**

- Point out the balance scale. **[T&T]** Ask: **How does this balance scale work?** Give learners a few minutes to discuss this. Take feedback. Establish that we put the item on one side and add known masses (often called weights) to the other. When the scales balance, the total mass of the weights gives the mass of the item being measured.

- Ask learners if they have any scales at home. Invite any who say they have used scales at home to share what they use them for and how they use them.

- **[T&T]** Ask: **What might you measure in kilograms?** Invite pairs to share their thoughts.

- Give groups kitchen and balance scales with a selection of known weights. Ask them to find objects from around the classroom that they think will have masses measured in kilograms. They check by putting them on the scales. They record the masses they find, to the nearest kilogram.

- Direct learners to the Student's Book. Discuss the Let's learn section before learners work on the paired activity. Then discuss the Guided practice example.

Practise 📖 [TWM.01]

- Workbook

Title: Measuring in kilograms

Pages: 184–185

- Refer to the variation in Activity 1 from the Additional practice activities.

Apply 👥 🖥 [TWM.03/04]

- Display **Slide 2**.

- Pairs need to feel a 1 kg weight or a real object, for example a 1 kg bag of rice or 1 litre bottle of water, to get a feel for the mass of 1 kg.

- They make a list of objects from around the classroom that they think will have a mass of more than 1 kg.

- They check by holding each item and comparing it with their kilogram weight.

Review

- Ask: **Can you describe how to find the mass of an object on kitchen scales? What about on balance scales?**

Assessment for learning

- What objects do you think have a mass of 1 kg or more?
- Can you think of anything else?
- How would you use balance scales to measure a mass in kilograms?

Same day intervention

Support

- Focus on finding the mass of items to the nearest whole kilogram.

Unit 23 Mass

Lesson 3: **Measuring in grams**

Geometry and Measure – Geometrical reasoning, shapes and measurements

Learning objectives

Code	Learning objective
3Gg.06	Estimate and measure the mass of objects in grams (g) and kilograms (kg). Understand the relationship between units.
3Gg.11	Use instruments that measure [length,] mass[, capacity and temperature].

Resources

set of standard kitchen scales (per group); scales with divisions (per pair); four small plastic (sandwich) bags (per pair); sand, rice or similar (per pair)

Revise

Use the activity *Using scales* from Unit 23: *Mass* in the Revise activities to rehearse and consolidate learners' knowledge of measuring accurately with scales.

Teach 🔲 🖥 [TWM.03]

- **[T&T]** Display **Slide 1**. Ask: **What can you see in the photograph?** Encourage pairs to discuss what they can see. After a few minutes take feedback. Establish that there is an apple on a digital scale.

- Ask: **What is a digital scale? How is it different from the kitchen scales you saw in previous lessons?** Agree that both use numbers but you need to read the mass from a numeric scale on the previous kitchen scales. The digital scale tells you the exact mass. Ask: **What is the mass of the apple? Is this in kilograms or grams? How do you know?** Agree 167 grams. An apple could not have a mass of 167 kilograms and there is a g above 167, which indicates grams.

- **[TWM.03]** Give groups of learners a set of standard kitchen scales. **[T&T]** Ask: **Describe the scales to each other. What do you notice?** Expect them to discuss the pointer that moves round the scale to show the mass of an object, the divisions and the whole numbers. They should be able to identify the divisions that indicate grams and those that indicate kilograms. Ask them to find objects from around the classroom that they think will have a mass of less than a kilogram. They put each object on the scale and read its mass, rounded to the nearest whole number. Ask them to make a table to record their results.

- Direct learners to the Student's Book. Discuss the paired activity and Guided practice example.

Practise 🔲 [TWM.01]

- Workbook

Title: Measuring in grams

Pages: 186–187

- Refer to Activity 2 from the Additional practice activities.

Apply 👥 🖥 [TWM.01]

- Display **Slide 2**.
- Pairs fill four sandwich bags with sand and then place them on a set of kitchen scales.
- They read the mass of each one, to the nearest division. They then order the bags, from lightest to heaviest.

Review

- Ask: 🗣 **What should we do if the needle on a set of weighing scales points between divisions?** Expect them to suggest rounding up or down to the nearest labelled division.

Assessment for learning

- How would you read a mass on a set of scales?
- What would you do if the pointer was between divisions?

Same day intervention

Support
- Focus on finding the mass of items to the nearest multiple of 100 g.

Lesson 4: **Measuring in kilograms and grams**

Learning objectives

Code	Learning objective
3Gg.06	Estimate and measure the mass of objects in grams (g) and kilograms (kg). Understand the relationship between units.
3Gg.11	Use instruments that measure [length,] mass[, capacity and temperature].

Revise

Use the activity *Solving mass problems* from Unit 23: *Mass* in the Revise activities.

Teach 🔲 💻 [TWM.01/02]

• **[T&T]** Display **Slide 1**. Ask: **What can you see in the photograph?** (a fennec fox, which lives in North Africa)

• **[T&T]** Say: **An average fennec fox has a mass of between 1 kg and 2 kg, about 1 kg 600 g. Which weights can you see on the slide that give a mass the same as the an average fennec fox?** Encourage them to think of different ways to make the mass. After a few minutes, take feedback, for example: 1 kg and three 200 g, three 500 g and 100 g.

• **[T&T]** 🗣 Display **Slide 2**. Ask: **What is the man doing? Why?** Agree that the man is finding the mass of fruits and vegetables on a pan balance. **[T&T]** Ask: **How do you think he does this?** Agree that he puts the vegetables on one pan and known weights on the other until the scales balance.

• **[T&T] [TWM.01/02]** 🗣 Ask: **What units of mass might this man use?** Agree grams and kilograms. Say: **Someone wants to buy 400 g of tomatoes. The man places known weights that total 400 g on one side of the balance and adds tomatoes until they balance. [T&T]** Ask: **If the man has 50 g, 100 g and 200 g weights, how can he make 400 g?** Think of all possibilities. Ask learners to discuss this with a partner. After a few minutes, invite learners to share their thoughts.

• Display **Slide 3**. Ask: **What is the mass of the pet food?** Agree that the scale reads 1200 g. Ask: **Can we convert this to kilograms and grams?** Learners know that there are 1000 g in 1 kg. Say: **The other way of reading a mass of 1200 g is 1 kg 200 g.** Establish that most masses can be given in terms of kilograms and grams.

• Ask learners problems, for example: **Rice costs $2 dollars a kilogram, Fahmida buys $\frac{1}{2}$ kg of rice. How much does she spend?** Expect learners to share how they solved this.

• Invite learners to work with a partner to make up similar problems that link money to mass. Ask them to share these so that the class can answer them.

• Direct learners to the Student's Book. Ask: **What is the total mass of the two scales?** (2800 g or 2 kg 800g) **What is the difference in mass?** (400 g)

• Discuss the Let's learn section and the paired activity. Work through the Guided practice example together.

Practise 🔲 [TWM.01]

• Workbook

Title: Measuring in kilograms and grams

Pages: 188–189

• Refer to the variation in Activity 2 from the Additional practice activities.

Apply 👥 💻 [TWM.03/04]

• Display **Slide 4**. Learners make up a problem similar to the example given and then use the diagram they have been shown to find the answer.

Review 📊

• Recap the learning of the four lessons on mass. 🗣 Ask: **How many grams are equivalent to 1kg?**

• Practise converting from kilograms to grams and vice versa.

• Use the **Weighing tool** set at 5 kg in 0·5 kg. Give learners masses to show on the scales in different ways. For example, 4 kg 500 g–can they find at least five ways to show this?

Assessment for learning

• In which units do we measure mass? Do we usually use only grams or kilograms or do we use a mixture of both?

• What is 4500 g in kilograms and grams?

• Mel has $10. She buys 3 kg of rice. Rice costs $2 per kilogram. How much money does she have left? How can a diagram help you to find the answer?

Same day intervention

Support

• Focus on converting whole kilograms to grams and multiples of 100 g to kilograms. Ask learners to find totals and differences that are based on these units, for example: the total of 600 g and 800 g, the difference between 1 kg and 700 g.

Additional practice activities

Activity 1

Learning objectives
- Estimate and measure masses in kilograms and grams
- Use instruments that measure mass

Resources
eight objects from around the classroom (per group); 1 kg bag of sugar or rice, 5 kg bag of potatoes or similar, 500 g packet of butter or similar (per group); paper and pens (per group); large sheet of paper (per group) balance scale (per group); selection of 1 kg, 500 g, 200 g and 100 g weights (per group); kitchen scales, for the variation (per group)

What to do [TWM.06]
- Learners find eight objects and estimate the masses, using the sugar, potatoes and butter as guides.
- On a large sheet of paper learners make a table with five columns and the headings: 'object', 'less than 500 g', 'between 500 g and 1 kg', 'between 1 kg and 5 kg' and 'more than 5 kg'. They write the names of their objects and their estimated masses in the appropriate sections of the table.

- They then add a sixth column headed 'Actual mass'. They place each of their objects in one pan of the balance scale and balance it with the weights. They write the actual mass of each object in the table, in kilograms or grams as appropriate, before comparing with their estimates.

Variation
Learners repeat the activity, but rather than using the balance scale and the selection of 1 kg, 500 g, 200 g and 100 g weights, they use kitchen scales to find the actual masses.

Activity 2

Learning objectives
- Estimate and measure masses in kilograms and grams
- Use instruments that measure mass

Resources
balance scale (per group); four plastic bags (per group); container of sand or similar (per group); large spoon (per group); paper and pens (per group)

What to do   [TWM.04/05/06]
- In groups, learners put different amounts of sand into each plastic bag. They order these, from lightest to heaviest. They then estimate the mass of the lightest bag. All members of the group need to agree on the most sensible estimate. They write this down on the paper.
- Learners measure the bag and write the actual mass. They write it next to their estimate and compare the two.
- They use this mass to estimate the next bag. Then they find the mass of it. They do this for the other two bags.

- Learners should consider whether their estimates improve once they have actual masses to use as a guide.

Variation
Once learners have carried out the main task, they work together to create word problems for their bags of sand that involve addition and subtraction and if appropriate, doubling and halving.

They could share these with the rest of the class, who attempt to solve them.

Geometry and Measure – Geometrical reasoning, shapes and measurements

Unit 24: Capacity and temperature

Collins International Primary Maths
Recommended Teaching and
Learning Sequence: Term 3, Week 9

Learning objectives

Code	Learning objective
3Gg.07	Estimate and measure capacity in millilitres (ml) and litres (l), and understand their relationships.
3Gg.11	Use instruments that measure [length, mass,] capacity and temperature.

Unit overview

Learners will begin by exploring the appropriate equipment and units to use when estimating and then measuring and recording different capacities. This includes whether to use millilitres and litres. Learners will develop an understanding of the relationship between the units of millilitres and litres. They will use various strategies to find equivalences between them; for example, using what they already know to generate other facts and scaling up and down. They will measure capacities in practical experiments to learn how to read intervals on a scale.

It is important that sufficient time is allowed for the Additional practice activities in this unit. The key for a rounded understanding of measurement is through repeated 'hands-on' learning and familiarity with real units of measure, rather than just the theory; for example, having an awareness of how much 1 l actually looks like.

Lesson 4 builds on previous work on temperature, focusing on using thermometers for measuring temperature.

Prerequisites for learning

Learners need to:
• understand that the capacity of a container is the maximum amount that it can hold
• know how to use non-standard or standard units to estimate and measure capacity
• understand a measuring scale as a continuous number line where intermediate points have value
• be able to read a temperature on a thermometer.

Vocabulary

capacity, litre (l), millilitre (ml), equivalent, empty, full, scale, measuring jug, measuring cylinder, interval, temperature, thermometer, degrees, Celsius, Fahrenheit

Common difficulties and remediation

As with length and mass, some learners may forget equivalences between units; for example, how many millilitres are equivalent to one litre. It is helpful to revise these facts regularly and refer to them when covering fractions of measures and mental and written calculations.

At this stage, learners' understanding of capacity should be that it is the amount of liquid that a container can hold, and that it is measured in litres and millilitres. The concept of volume as being a measure of the actual amount of liquid in a container is not introduced until Stage 6, so there is no need to mention it here.

Some learners may have difficulty estimating capacity. It is important to give them something to compare with the capacity they are estimating. Let them see, for example, the capacities of 500 ml, 1 litre and 2 litre bottles in order to make sensible estimates.

Supporting language awareness

In this unit, vocabulary includes: litres, millilitres, capacity, empty, full, scales, intervals. You may find it useful to refer to the audio glossary on Collins Connect. It is important to display the words used on cards in the classroom, with visual clues, to help learners recognise and understand what each one means. It is a good idea to write the abbreviations used for litre (l) and millilitre (ml) beside these words and explain how these abbreviations have been developed.

Promoting Thinking and Working Mathematically

During this unit, learners will have opportunities to think and work mathematically; for example, they work out other facts that they know if they are given a fact such as 1 l = 1000 ml, which involves specialising (TWM.01) and making generalisations (TWM.02). They make estimates of capacities and then test these out to find how close they are, which involves making conjectures (TWM.03) and convincing (TWM.04). They look at similarities and differences between measuring cylinders, which involves characterising (TWM.05) and classifying (TWM.06).

Success criteria

Learners can:
• estimate and measure capacity in litres and millilitres
• understand the relationship between units
• use instruments to measure capacity
• use thermometers to measure temperature.

Geometry and Measure – Geometrical reasoning, shapes and measurements

Lesson 1: **Units of capacity**

Geometry and Measure – Geometrical reasoning, shapes and measurements

Learning objectives

Code	Learning objective
3Gg.07	Estimate and measure capacity in millilitres (m*l*) and litres (*l*), and understand their relationships.
3Gg.11	Use instruments that measure [length, mass,] capacity [and temperature].

Resources

1 litre bottle (per group); selection of measuring jugs or cylinders (per group); water (per group)

Revise

Use the activity *How many cups?* from Unit 24: *Capacity and temperature* in the Revise activities to reinforce learners' understanding of non-standard units of measure.

Teach [SB] [TWM.01/02]

• **[T&T]** Display **Slide 1**. Ask: **What can you see on the slide?** Agree that there are bottles that have different capacities. Say: **These containers hold different amounts of liquid. The capacity of the largest bottle is 2 litres.** Ask: **What do you think the capacities of the other bottles might be?** Encourage learners to think in terms of fractions; for example, the second bottle might have half the capacity of the largest one, so will have a capacity of 1 litre. The third bottle might be $\frac{3}{4}$ of the size of the second, so will have a capacity of three-quarters of a litre.

• **[T&T]** Ask: **What can you tell me about the relationship between millilitres and litres?** Agree that litres are the greater of the two units for measuring capacity. Give groups a 1-litre bottle, measuring jugs or cylinders and water. Ask them to use these to find out how many millilitres there are in one litre. Establish that there are 1000 millilitres in one litre.

• **[TWM.01/02]** Ask questions such as: **How many millilitres are there in 4 litres/12 litres/6 litres? How many litres there in 3000 m*l*/4000 m*l*/6000 m*l*?** Expect learners to consider if their answer is sensible and to explain the strategy they used.

• Direct learners to the Student's Book. Discuss the Let's learn section before learners work on the paired activity. Then discuss the Guided practice example.

Practise [WB]

• Workbook

Title: Units of capacity

Pages: 190–191

• Refer to the variation in Activity 1 from the Additional practice activities.

Apply 👥 🖥 [TWM.01]

• Display **Slide 2**. Learners work with a partner to make up new equivalent facts from the fact that 1 litre = 1000 m*l*. They use strategies such as doubling, halving and finding fractions of 1000.

• They explain their strategies to their partner in preparation for sharing with the class.

Review

• Invite pairs to share some of their equivalent capacities they made in the **Apply** activity.

• Make a list of these on the board. Then ask learners to find more by adding and subtracting those that have been suggested.

• Finish the lesson by asking learners to estimate the capacities of different containers in the classroom and decide whether they would be measured in litres or millilitres.

Assessment for learning

• How many millilitres are there in one litre? What about 2 litres/5 litres/10 litres?

• How many litres are there in 5000 ml? What about 2500 ml/9500 ml/3250 ml?

• Which unit would we use to measure the capacity of a swimming pool?

• Which unit would we use to measure the capacity of a cup?

Same day intervention
Support

• Give learners whole litre measurements to convert to millilitres and multiples of 1000 to convert to litres, for example: 3 litres is the same as 3000 m*l*. Encourage them to say with you, for example: '4000 millilitres is the same as 4 litres.'

Lesson 2: **Measuring capacity**

Learning objectives

Code	Learning objective
3Gg.07	Estimate and measure capacity in millilitres (m*l*) and litres (*l*), and understand their relationships.
3Gg.11	Use instruments that measure [length, mass,] capacity [and temperature].

Resources

set of measuring spoons, optional (per class); 2 *l* bottle of water (per pair); container, smaller than the 2 *l* bottle (per pair); measuring jug (per pair); tray (per pair)

Revise

Use the activity *Units of capacity* from Unit 24: *Capacity and temperature* in the Revise activities to review and consolidate measuring capacity using the units of millilitres and litres.

Teach 🔲 💻

- **[T&T]** Display **Slide 1**. Ask **What do you think the spoons in the photograph are used for?** Agree that these are often used in cooking to measure quantities in millilitres. Discuss the different spoon sizes expressed in millilitres.

- Establish that the spoons measure amounts that are a very small part of a litre. If you have them available, you could show different spoons to show approximately how much liquid those in the photograph hold.

- **[T&T]** 🔖 Ask: **What do we mean by capacity?** Agree that the capacity of a container is the amount of liquid it holds when it's full.

- **[T&T]** Display **Slide 2**. Ask: **What equipment and units can we use for measuring capacity?** Take feedback. Agree that we use millilitres for small quantities and litres for larger quantities. We measure using measuring jugs or cylinders that have a scale on them.

- Discuss the divisions on the jug, linking the fractions with the number of millilitres. Make clear that the fractions relate to one whole litre.

- Direct learners to the Student's Book. Discuss the Let's learn section before learners work on the paired activity. Then discuss the Guided practice example.

Practise 📓 [TWM.04]

- Workbook

Title: Measuring capacity

Pages: 192–193

- Refer to the main activity in Activity 1 from the Additional practice activities.

Apply 👥 💻 [TWM.04]

- Display **Slide 3**.

- Pairs use a 2 *l* bottle of water, container, measuring jug and a tray so that the water can be contained.

- They fill the container with water from the bottle and then pour the water into the measuring jug to find the capacity of the container.

Review

- Discuss the difficulty of making estimates without having anything to compare against.

- Point to different containers that haven't been used and ask learners to tell you whether they think these would hold less than or more than a litre.

Assessment for learning

- What units do we use to measure small capacities?

- What units do we use to measure large capacities?

- What equipment might we use to measure the capacity of a bath?

Same day intervention

Support

- Work with groups that need support during the **Apply** and **Additional practice activity**. Support them in reading the scales on the measuring jugs to the nearest 100 m*l*.

Lesson 3: **Measuring in litres and millilitres**

Learning objectives

Code	Learning objective
3Gg.07	Estimate and measure capacity in millilitres (ml) and litres (l). Understand the relationship between units.
3Gg.11	Use instruments that measure [length, mass,] capacity [and temperature].

Resources

measuring jugs/cylinders (per group); two 2 l bottles of water (per pair); tray (per pair) three different-sized containers, smaller than the 2 l bottles (per pair); ruler (per learner) (for the Workbook)

Revise

Use the activity *Millilitre or litre?* from Unit 24: *Capacity and temperature* in the Revise activities to assess learners' understanding of units of measure for capacity.

Teach 📖 💻 [TWM.01/02/05]

- 👥 **[TWM.01/02]** On the board, write 7000 ml = 7 l. Ask learners to use doubling, halving, adding, multiplying, and so on, to generate as many new facts as they can. Create a mind-mapping diagram similar to the diagram on page 98 of the Student's Book (Unit 24, Lesson 1).
- **[T&T]** Display **Slide 1**. Ask: **What do you notice about the measuring cylinder on the slide? What do you think it is used for?** Establish that it has one scale showing millilitres to 1000 ml or 1 l. The labelled intervals are in divisions of 100 ml. Between each of these are unlabelled intervals of 10 ml. Agree that this can be used to measure the capacity of a container.
- **[TWM.05]** Give groups of learners measuring jugs or cylinders. **[T&T]** Ask learners to compare the measuring vessels they have with the one on the slide. Ask: **What do you notice? What is the same? What is different?** Expect them to compare the size of them and the divisions on the scales. Ask learners to describe how they would use one. Establish that they would pour, for example, water from a container into it and read where the level stops to find its capacity.
- Direct learners to the Student's Book. Discuss the Let's learn section before learners work on the paired activity. Then discuss the Guided practice example.

Practise 📒 [TWM.01]

- Workbook

Title: Measuring in litres and millilitres

Pages: 194–195

- Refer to Activity 2 from the Additional practice activities.

Apply 👥 💻

- Display **Slide 2**. Pairs use three different-sized containers, a measuring jug or cylinder, two 2 l bottles of water and a tray so that the water can be contained.
- They find the capacity of each container by pouring water into it and then that water into a measuring vessel.
- They then write statements comparing the capacities of the three containers.

Review 📊

- Recap what has been learned about capacity over the last three lessons. Ensure that you discuss what capacity is, the units used to measure it and the types of measuring instruments. Finish the lesson by using the **Capacity tool** set with the jug. Invite learners to add certain amounts of fluid to the jug using the tap. Then invite learners to take certain amounts out and to say how much is now in the jug.

Assessment for learning

- How would you find the capacity of a container?
- What do you do if the water level comes between two labelled divisions?

Same day intervention
Support

- Work with learners to help them measure the capacity of containers to the nearest 100 ml.

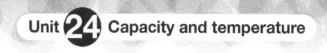

Lesson 4: **Temperature**

Learning objective

Code	Learning objective
3Gg.11	Use instruments that measure [length, mass, capacity and] temperature.

Resources

thermometer (per group); coloured pencil (per learner) (for the Workbook); paper (per learner) (for the Workbook)

Revise

Use the activity *Reading scales* from Unit 24: *Capacity and temperature* in the Revise activities to review and consolidate learners' knowledge of reading scales.

Teach 🆂🅱 💻 📊

- **[T&T]** ✋ Display **Slide 1**. Ask: **What do you notice? What is the same? What is different?** Ask learners to discuss this with their partner. After a few minutes, take feedback. Establish that the first photograph shows somewhere that is very hot (Death Valley, California, which is the hottest part of the world) and the second photograph shows somewhere that is very cold (Antarctica, which is the coldest part of the world). If possible, locate these on a map of the world.

- **[T&T]** Ask: **How do we know that these are the hottest and coldest parts of the world?** Remind the class that the word to describe how hot or cold a place or an object is, is 'temperature'. Explain that in most countries temperature is measured in degrees Celsius (°C) but some countries use degrees Fahrenheit (°F). The highest temperature recorded in Death Valley is just below 57 °C or 135 °F. Antarctica is colder than the ice in a freezer. Very cold things have temperatures close to 0 °C, and a kettle will boil water at 100 °C.

- **[T&T]** ✋ Ask: **What instrument do we use to measure temperature?** Agree thermometer. Display **Slide 2**. Ask: **Who can tell me what these are? What's the same about them? How are they different?**

- Allow a few minutes discussion, then elicit that they are all thermometers, used to measure temperature. The first is a normal thermometer, the second is a clinical thermometer, used to take a person's temperature and the third has a round scale often used in cooking. Encourage learners to see if they can read the scales. Write the temperatures (23 °C, 37 °C, 14 °C) on the board so that learners are familiar with how to show the units.

- Display **Slide 3**. Ask: **What temperatures do these thermometers show?** Give learners a few minutes to discuss. Take feedback.

- Use the **Thermometer tool** showing three thermometers. Invite learners to read different temperatures. You could make up a scenario, for example, in the morning the temperature was 100, it rose by 7 degrees in the afternoon.

The temperature dropped by eight degrees in the evening. Learners show these three temperatures on the three thermometers.

- Give opportunities for learners to take temperatures; for example, in a cupboard, outside, a cold glass of water. Write results on the board and compare.

- Direct learners to the Student's Book. Discuss the Let's learn section before learners work on the paired activity. Then discuss the Guided practice example.

Practise 🆆🅱 [TWM.04/05]

- Workbook

Title: Temperature

Pages: 196–197

- Refer to the variation in Activity 2 from the Additional practice activities.

Apply 👥 💻 [TWM.03/04]

- Display **Slide 4**. Learners are asked to imagine what it would be like to visit a very hot place, like Death Valley, and a very cold place, like Antarctica.

- They write three statements about how they might prepare for a visit.

Review 💻

- ✋ Ask: **In what units do we measure temperature? What equipment would we use?**

- Display **Slide 5** and ask learners to read the temperature on each of the thermometers. Then ask them to match each object to its temperature.

Assessment for learning

- What units do we use to measure temperature?
- What numbers do we use if the temperature falls below zero?
- Where might we be if the temperature is above 40 °C? Where else?
- Where might we be if the temperature falls below 0 °C ? Where else?

Same day intervention

Support

- Focus on reading thermometers with fully numbered scales or where a temperature is only given for numbered divisions on a partially numbered scale.

Additional practice activities

Activity 1 👥 ▲2

Learning objectives
- Estimate and measure capacities in litres and millilitres
- Use instruments that measure capacity

Resources
selection of six different-sized containers (per group); 500 ml, 1 l and 2 l bottles of water (per group); measuring vessel (per group); tray to work in (per group); poster paper and pens (per group); sticky labels (per group); A3 or A4 plain paper (per pair) (for variation)

What to do [TWM. 04/05/06]
- Learners look at the six different containers and estimate their capacities, using the different bottles of water as a guide. They should think in terms of 'less than 500 ml', 'between 500 ml and 1 l', 'between 1 l and 2 l' and 'more than 2 l'.
- Learners label each container with letters from A to F with the sticky labels.
- They make a table with five columns on some poster paper with the headings above and write the letters A to F in the first column and their estimated capacities in the appropriate sections of the table.

- They then pour water into each container to test their estimates, then consider whether their estimates were close.
- They then add a sixth column headed 'Actual capacity' and complete it by writing in each containers actual capacity.

Variation
▲2 Give pairs a sheet of A3 or A4 plain paper. They write '5 litres is the same as 5000 ml' in the middle of the paper. Together, they use this fact to make up as many other facts as they can in the time allowed. Encourage them to use doubling, halving and the four operations.

Take feedback and invite learners to explain how they made their new facts.

Activity 2 👥 ▲2

Learning objectives
- Estimate and measure capacities in litres and millilitres
- Use instruments that measure capacity and temperature

Resources
measuring vessels (per group); water in bottles (per group); large container (per group); Resource sheet 17: Thermometers (per group) (for variation)

What to do [TWM. 04/05/06]
- Prior to the activity, write a list of around eight different capacities on the board, for example: 500 ml, 1 l 200 ml, 900 ml.
- Volunteers take turns to pick an amount from the board.
- They pour what they estimate will be that amount into a large container. They then measure that amount with a measuring vessel and find out how close their estimate was.
- Each learner must have the opportunity to estimate and measure.

Variation
▲2 Provide each learner with thermometer from Resource sheet 17: Thermometers. Note that all four thermometers show different scales. Working in pairs, learners take turns to set scenarios that involve counting on and back to find temperatures. Learners make sure that their partner can use their own thermometer to work out the answer to each scenario. So, if their partner has the second thermometer from Resource sheet 17, the learner might say, for example, 'The temperature was 12 °C at 10 o'clock in the morning. It rose another 10 degrees by 2 o'clock in the afternoon. The temperature had dropped 11 degrees by 5 o'clock in the afternoon. What temperature was it at 5 o'clock in the afternoon?'

Geometry and Measure – Geometrical reasoning, shapes and measurements

Unit 25: Position, direction, movement and reflection

Collins International Primary Maths Recommended Teaching and Learning Sequence: Term 2, Week 8

Learning objectives

Code	Learning objective
3Gp.01	Interpret and create descriptions of position, direction and movement, including reference to cardinal points.
3Gp.02	Sketch the reflection of a 2D shape in a horizontal or vertical mirror line, including where the mirror line is the edge of the shape.

Unit overview

This unit focuses on position direction, movement and reflection. Learners will explore the movements described as clockwise and anticlockwise, up, down, above, below, next to, between, under, right, left, straight on. They will do this by making up oral and written instructions. Learners will create and interpret simple grids, using this vocabulary as well as the cardinal points – the four main points of the compass (north, south, east and west). They will have the opportunity to sketch reflections of 2D shapes in horizontal and vertical mirror lines, including where the mirror line is the edge of the shape. These sketches need to be on plain paper rather than grids because the concept of reflection is more important than accuracy.

Prerequisites for learning

Learners need to:
- understand what is meant by clockwise and anticlockwise
- be able to make oral descriptions of movement and turn
- be able to sketch the reflection of 2D shapes
- recognise reflection in the environment.

Vocabulary

position, on top, above, below, beside, between, underneath, inside, direction, clockwise, anticlockwise, right, left, turn, straight, compass, north, east, south, west, reflect, horizontal, vertical, mirror line of symmetry

Common difficulties and remediation

Learners may confuse clockwise and anticlockwise. Clockwise is a turn to the right and anticlockwise is a turn to the left. To help learners, you can demonstrate the links with the right and left hand. For example, ask them to use their right arm and make a sweeping gesture to the right, making a clockwise movement. Repeat, using the left arm, for an anticlockwise turn.

You can also make the link to clocks. Clockwise is the direction in which the hands on a clock move.

Supporting language awareness

Recap the words relating to position, movement and direction that learners should know from their work in Stage 2. You may find it useful to refer to the audio glossary in the digital component. You could write these words on cards and hold them up, one at a time, for learners to read out loud.

Encourage learners to enunciate each word correctly, copying how you do it. Encourage them to put the words into sentences, for example: 'The book is behind the pen.' Learners could also act out their sentence as they say it.

Introduce new vocabulary by writing new words on cards and displaying them in the classroom, along with appropriate drawings or diagrams as clues.

As in other units, ask questions such as: **What is the same? What is different? What else do you know? Can you give me another example?** Expect learners to answer in complete sentences.

Promoting Thinking and Working Mathematically

During this unit, learners will have opportunities to think and work mathematically; for example, discussing similarities and differences between different images will involve specialising (TWM.01) and generalising (TWM.02). Learners make and describe the movements in patterns that involve making conjectures (TWM.03), convincing (TWM.04), characterising (TWM.05) and classifying (TWM.06).

Success criteria

Learners can:
- explain and make descriptions of position, direction and movement
- understand the positions of N, S, W and E
- sketch reflections of 2D shapes in a mirror line.

Geometry and Measure – Position and transformation

Lesson 1: **Position**

Learning objective

Code	Learning objective
3Gp.01	Interpret and create descriptions of position [,direction and movement, including reference to cardinal points].

Resources

selection of 3D shapes: cubes, cuboids, cylinders, spheres prisms and pyramids (per pair); pencil and paper (per learner)

Revise

Use the activity *Where is it?* from Unit 25: *Position, direction, movement and reflection* in the Revise activities to assess learners' vocabulary for position.

Teach 🅂🄱 🖳

- 🄿 Display **Slide 1**. Ask: **What can you see?** Agree that the photograph shows some stacked boxes filled with different objects. Ask learners to identify the objects, for example: cuddly toys, slinky, building blocks.

- **[T&T]** Ask: **Can you describe the position of each toy?** After a few minutes, ask learners to tell you where the toys are in relation to each other. For example, the teddy bear is in the box above the building blocks, the red building block in the green box is to the right of the green building block. Highlight any positional language used by writing it on the board.

- **[T&T]** Say: **Tell your partner all the words you can think of that we could use to describe position.** Take feedback. Add any new words to the list you began on the board. Ask learners to choose a word from the list and to make up a sentence to show what it means.

- **[T&T]** Display **Slide 2**. Ask: **What do you notice about the cube?** Give learners a few minutes to discuss this. Take feedback. They should notice that its position can be described in different ways, depending on the shape they are comparing it to; for example, it is to the left of the cuboid, in front of the pyramid.

- Direct learners to the Student's Book. Discuss the Let's learn section.

- Give learners the opportunity to discuss the paired activity and to work through the Guided practice activity.

Practise 🅆🄱

- Workbook

Title: Position

Pages: 198–199

- Refer to Activity 1 from the Additional practice activities.

Apply 👥 🖳 [TWM.01]

- Display **Slide 3**. Learners use a selection of 3D shapes. They take turns to give instructions to their partner about where to place the shapes to make a display.

- They sketch their display and write some sentences to describe the positions of the shapes.

Review

- Give pairs a cube and a cuboid. Invite learners to take turns to give instructions to the class for positioning the two shapes, for example: 'Place the cube in front of the cuboid. Place the cuboid on top of the cube.' How many different positions do learners say?

- Ask learners to reflect on each other's descriptions. Were they good or could they be improved?

Assessment for learning

- What words can you tell me that would describe position?
- Can you place this cube below the pyramid?

Same day intervention

Support

- Focus on fewer words that are related, such as 'on top' and 'under', 'left' and 'right'.

Lesson 2: **Direction and movement**

Learning objective

Code	Learning objective
3Gp.01	Interpret and create descriptions of [position,] direction and movement [,including reference to cardinal points].

Resources

piece of large (2 cm) squared paper and a counter (per learner)

Revise

Use the activity *Moving around* from Unit 25: *Position, direction, movement and reflection* in the Revise activities to review direction and movement.

Teach [SB] 🖥

• [T&T] Display **Slide 1**. Ask: **What do you think these road signs mean?** After a few minutes, agree that the first sign means turn left and the second sign indicates a roundabout, and tells people the direction to go round it.

• [T&T] Ask: **What words could you use to give directions?** Agree 'left', as in the first road sign, 'right' and 'straight on'. Ask learners to find a space and walk three steps straight ahead. Tell them to turn right and walk another three steps and then turn left. Repeat this a few times, varying the instructions to cover different directions.

• Display **Slide 2**. Remind learners that they can also use 'clockwise' and 'anticlockwise' to turn. Ask: **How would we move if we turned in a clockwise direction?** Agree 'to the right'. Link this to the movement of the hands on a clock, identifying 'anticlockwise' as the opposite direction from 'clockwise'.

• Ask learners to make a one quarter turn in a clockwise direction, and then another. Ask: **What size turn have you made?** Agree a half turn. Ask them to make another quarter turn clockwise. Agree that altogether they have made a three-quarter turn. Ask them to make another quarter turn clockwise and agree that they have made a whole or full turn.

• Ask learners to make different clockwise and anticlockwise turns.

• Direct learners to the Student's Book. Give them the opportunity to complete the paired activity. Discuss the Guided practice section.

Practise 🗒

• Workbook

Title: Direction and movement

Pages: 200–201

• Refer to the variation in Activity 1 from the Additional practice activities.

Apply 👥 🖥

• Display **Slide 3**. Learners work with a partner. They each need a counter and a piece of squared paper with squares large enough to enclose the counters.

• They take turns to give instructions to their partner so that they move their counter around the paper.

• Discuss the activity. Invite learners to give instructions to the class for them to move the counter around their paper.

Review

• Recap the vocabulary of 'right', 'left', 'straight', 'clockwise' and 'anticlockwise' by asking learners to move in different directions.

• Point out the relationship between right and clockwise turns and left and anticlockwise turns.

Assessment for learning

• What words can we use to describe direction?

• Can you show me an anticlockwise turn?

• What do we mean by a 'right turn'?

Same day intervention

Support

• Focus on the language of 'left', 'right' and 'straight on' until learners are familiar with them. Then introduce 'clockwise' and 'anticlockwise'.

Geometry and Measure – Position and transformation

Lesson 3: **Compass points**

Geometry and Measure – Position and transformation

Learning objective

Code	Learning objective
3Gp.01	Interpret and create descriptions of [position,] direction and movement, including reference to cardinal points.

Resources

Resource sheet 18: 10 x 10 grid (one copy per learner and one copy per pair); pencil and paper (per pair)

Revise

Use the activity *Moving around* from Unit 25: *Position, direction, movement and reflection* in the Revise activities.

Teach 🆂🅱 🖥

- **[T&T]** 🔁 Display **Slide 1**. Ask: **What can you see on the slide? What do you think it is?** Give learners a few minutes to discuss with a partner. Take feedback. Establish that it shows a compass. Inform learners that a compass is an instrument for finding directions. It uses a magnetic needle that turns on a pivot and points to the magnetic north, which is in the Arctic.

- **[T&T]** Ask: **What do the letters N, E, S and W mean?** Give learners a few minutes to discuss this. Take feedback. Agree that the letters stand for north, east, south and west.

- Discuss who might use a compass; for example, people who fly planes and sail boats, people who trek.

- If possible, locate the position of north in the classroom and using the PPT (not PDF) version of the slide, use the Rotate function to move the compass so that north is in the correct position. Ask learners to stand and face it. If this is not possible, ask them to face the front of the classroom and imagine that they are facing north. Ask them to turn to face east, then south, then west. Repeat this a few times to help learners remember these positions.

- Next ask them to make quarter, half and three-quarter turns in clockwise and anticlockwise directions and to tell you what direction they are facing.

- Give each learner a copy of Resource sheet 18: 10 x 10 grid. Ask learners to draw an X in the top left-hand square. Then give them instructions to follow, for example: Move south one square, move west three squares, move south five squares and so on. Ensure you cover each direction.

- Give learners the opportunity to give instructions to a partner to follow.

- Direct learners to the Student's Book. Discuss the Let's learn section.

- Give learners time to work on the paired activity and then discuss the Guided practice.

Practise 🆆🅱 [TWM.03]

- Workbook

Title: Compass points

Pages: 202–203

- Refer to Activity 2 from the Additional practice activities.

Apply 👥 🖥 [TWM.05]

- Display **Slide 2**. Give each pair a copy of Resource sheet 18: 10 x 10 grid and pencil and paper.

- They draw a 'journey' from the top of the grid to the bottom.

- They then work with a partner to write the directions to get from the start to the finish of their journey.

- They must use the compass directions and write the number of squares moved.

Review 🖥

- 🔁 Ask: **What is a compass used for? What compass points have we learned?**

- Conclude the lesson by asking learners to stand up and face the real or an imaginary south (if possible, display the PPT (not PDF) version of **Slide 1**, with the compass facing north). Give them instructions that involve making quarter, half and three-quarter turns in clockwise and anticlockwise directions. Ask them to tell you which direction they are facing.

Assessment for learning

- Who would use a compass? Why?
- What does the N on a compass stand for? What about the E? W? S?

Same day intervention

Support

- Provide learners with a prepared compass that shows N, E, S and W. Ask them to identify objects in the classroom that face in each position.

Enrichment

- Ask learners to consider the half way points between the main four compass directions. Elicit that the direction between N and E is NE, the direction between N and W is NW, that between S and E is SE and that between S and W is SW. You could incorporate some of these directions in the suggested activities.

Lesson 4: **Reflections**

Learning objective

Code	Learning objective
3Gp.02	Sketch the reflection of a 2D shape in a horizontal or vertical mirror line, including where the mirror line is the edge of the shape.

Resources

plain paper (at least 2 sheets per learner); mirror (per learner); coloured pencils (per learner); ruler (per learner)

Revise

Use the activity *Reflection* from Unit 25: *Position, direction, movement and reflection* in the Revise activities to assess learners' understanding of symmetry.

Teach 🔲 🖥

- **[T&T]** Display **Slide 1**. Ask learners to describe what they see. Do they know what this building is and where in the world it can be found? Tell them that it is the Taj Mahal, which is a famous palace in India.
- 🗣 Ask: **Can you describe what is in the water in the photograph?** Agree that there is an image of the building in the water. Explain that this is a reflection. Remind learners that a reflection is an image of a shape as it would be seen in a mirror.
- **[T&T]** Display **Slide 2**. Explain that each shape has been reflected in a mirror line. Ask: **Can you tell me where the mirror line is in these reflections?** Give learners a few minutes to discuss this with their partner. Take feedback. Agree that in the first diagram the mirror line is vertical and that in the second the mirror line is horizontal. Ask: **What do you notice about these lines?** Establish that these are also lines of symmetry.
- **[T&T]** Display **Slide 3**. Ask: **What is the same about this slide and Slide 2? What is different?** Again, give learners a few minutes to discuss this and take feedback. Agree that the reflections are still vertical and horizontal. The difference is that in **Slide 3** the mirror line does not touch the shapes, like it does in **Slide 2**.
- Give each learner a piece of plain paper and a mirror. Ask them to draw a simple polygon on the left side of the paper. They put the mirror against one edge of their shape. They then use the mirror to visualise the reflection and draw it. Repeat several times.
- Direct learners to Let's learn in the Student's Book. Discuss the information.
- Give learners the opportunity to work on the paired activity, then discuss the Guided practice.

Practise 🔲 [TWM.04]

- Workbook

Title: Reflections

Pages: 204–205

- Refer to the variation in Activity 2 from the Additional practice activities.

Apply 👥 🖥

- Display **Slide 4**. Learners work with a partner. They each fold a piece of plain paper in half. The fold represents the mirror line.
- They draw a picture or pattern on one side of the fold. They then swap papers and draw the reflection of their partner's picture.
- They use a mirror to check they are correct.

Review

- 🗣 Ask: **What do we mean by reflection?**
- Discuss when reflections can be seen in everyday life, for example when looking in a mirror.
- Recap the other elements covered in this unit: position, direction, movement and the four cardinal points of a compass.

Assessment for learning

- What is a reflection?
- In which directions have we been making reflections?
- Where do we see reflections in everyday life? Where else?
- What helps make effective reflections?

Same day intervention

Support

- Focus on reflecting shapes where one edge is on the mirror line.

Geometry and Measure – Position and transformation

Additional practice activities

Activity 1

> **Learning objective**
> • Interpret and create descriptions of position, direction and movement [,including reference to cardinal points]

Resources

ten 1 cm cubes and a sheet of 1 cm squared paper OR ten interlocking cubes and a sheet of 2 cm squared paper (per pair); pencil (per pair)

What to do

• Learners play with a partner.
• Learners place ten cubes on the squared paper. They discuss where each cube is in relation to the others.
• They then draw lines from one cube to another, so that they are all the cubes are joined together in a path-like way. They should only use vertical and horizontal lines.

• Learners describe how to get along their path, using the language of 'right', 'left', 'up' and 'down' and including the number of squares moved over.

Variation

Learners vary their description of how to get along the path, using the vocabulary 'clockwise' and 'anticlockwise'. For example, 4 to the right, make a clockwise turn and move 6 squares in a straight direction.

Activity 2

> **Learning objective**
> • Interpret and create descriptions of position, direction and movement, including reference to cardinal points

Resources

Resource sheet 18: 10 x 10 grid (per learner); pencil (per learner); coloured pencils (per learner) (for the variation)

What to do

• Give each learner a copy of Resource sheet 18: 10 x 10 grid.
• They draw their own journey from the top row of squares to the bottom, using vertical and horizontal lines over the square lines of their paper.
• Once they have done this, they write the journey in steps, using compass directions and stating the number of squares crossed.

Variation

Using a copy of Resource sheet 18: 10 x 10 grid, learners draw a vertical line to join the north and south points down the centre of the grid. They then treat this as a mirror line. They create a pattern on one side and reflect it onto the other.

Learners can use another copy of Resource sheet 18: 10 x 10 grid and repeat, drawing a horizontal (east/west) mirror line across the centre of the grid.

Geometry and Measure – Position and transformation

Unit 26: Statistics

Collins International Primary Maths
Recommended Teaching and
Learning Sequence: Term 2, Week 9

Learning objectives

Code	Learning objective
3Ss.01	Conduct an investigation to answer non-statistical and statistical questions (categorical and discrete data).
3Ss.02	Record, organise and represent categorical and discrete data. Choose and explain which representation to use in a given situation: - Venn and Carroll diagrams - tally charts and frequency tables [- pictograms and bar charts].
3Ss.03	Interpret data, identifying similarities and variations, within data sets, to answer non-statistical and statistical questions and discuss conclusions.

Unit overview

Statistics is an area of mathematics that is used regularly in everyday life. It is important, as teachers, to be aware of just how often we use statistics and to bring this to learners' notice whenever appropriate. It would be helpful to have a variety of real-life charts, diagrams and graphs, as well as examples of the way data is collected, in school for learners to look at. You could create a display of examples over the duration of this unit.

In this unit, learners are introduced to the statistical cycle:

1. What's the problem? Make a plan.
2. Record, organise and represent the data.
3. Interpret the data.
4. Discuss the data and make predictions.

Data can be presented in a variety of ways, for example, as simple lists, drawings and tables, diagrams and charts, pictograms, and block and bar graphs. It is an area of mathematics that should be included in other mathematics lessons, as well as other curriculum areas, such as science and social studies.

In this unit, learners will review the different ways of collecting and recording data that they learned in Stage 2 and progress to new methods, including Venn diagrams and Carroll diagrams with two criteria.

Prerequisites for learning

Learners need to:
• carry out an investigation to answer a question
• know how to use lists and tables
• know how to use Venn and Carroll diagrams
• know how to use tally charts
• know how to use block graphs and pictograms.

Vocabulary

record, organise, represent, data, rule, criterion, criteria, Venn diagram, set, intersection, Carroll diagram, tally, tally chart, frequency table, statistical cycle

Common difficulties and remediation

When drawing conclusions from data representations, some learners may make very simple statements, for example: 'Three learners like cakes.' They need to be encouraged to think more deeply and draw more complex conclusions, for example: 'Five more learners like biscuits than liked cakes, so Jim should buy biscuits for the party.' Some learners may not understand that a Carroll diagram shows information about what satisfies a particular criterion and information about what does not satisfies this criterion. Without this understanding, they will simply be creating a table.

Supporting language awareness

You may find it useful to refer to the audio glossary on Collins Connect. You could write these words on cards and hold them up one at a time for learners to read out loud. Break up any words that learners find difficult to enunciate into parts. As in other units, ask questions, such as: **What is the same? What is different? What else do you know?** and **Can you give me another example?** Expect learners to answer in complete sentences.

Promoting Thinking and Working Mathematically

During this unit learners will have opportunities to think and work mathematically; for example, when they sort numbers according to a variety of criteria, they are characterising (TWM.05) and classifying (TWM.06). Statistics is a particularly good topic for this. They will also be discussing similarities and differences between the various ways of representing data, and this involves specialising (TWM.01), generalising, (TWM.02) conjecturing (TWM.03) and convincing (TWM.04).

Success criteria

Learners can:
• conduct an investigation to answer questions
• record, organise and represent data in Venn diagrams, Carroll diagrams, tally charts and frequency tables
• interpret data to answer questions.

Unit 26 Statistics

Statistics and Probability – Statistics

Lesson 1: **Venn diagrams**

Learning objectives

Code	Learning objective
3Ss.01	Conduct an investigation to answer non-statistical and statistical questions (categorical and discrete data).
3Ss.02	Record, organise and represent categorical and discrete data. Choose and explain which representation to use in a given situation: - Venn [and Carroll] diagrams [- tally charts and frequency tables - pictograms and bar charts].
3Ss.03	Interpret data, identifying similarities and variations, within data sets, to answer non-statistical and statistical questions and discuss conclusions.

Resources

selection of 3D shapes: sphere, cylinder, cube, cuboid, square-based pyramid, triangular prism (per group); sheet of A3 paper (per group); sheet of A3 paper (per pair)

Revise

Use the activity *Diagrams* from Unit 26: *Statistics* in the Revise activities to assess learners' understanding of Venn diagrams.

Teach 📖 🖥 [TWM.05/06]

- 🔁 Display **Slide 1**. Ask: **How many of these shapes can you name? What are their properties? Where might you see them in real life?** After a few minutes take feedback. You could use this as an assessment opportunity for work on 3D shapes covered in Unit 21: 3D shapes.

- [T&T] Ask: **How can we sort these shapes?** Remind learners that in Stage 2 they sorted objects into Venn diagrams, according to one criterion or two criteria. Explain the meaning of 'criterion' and 'criteria'. Inform learners that in this lesson they will be sorting according to two criteria.

- Demonstrate how to sort the shapes according to whether they have only flat faces or only curved surfaces. Draw two intersecting sets on the board. In one set draw the shapes with only flat faces, and in the other the name draw the shape with only a curved surface. In the overlap draw the shape with both.

- Display **Slide 2** to confirm what you have drawn on the board. Explain that this is called a universal set. It contains everything we are interested in. The two circles inside the rectangle have sorted the shapes within the universal set.

- [T&T] [TWM.05/06] Ask: **How else can we sort these shapes?** Give small groups of learners a sheet of A3 paper and a selection of 3D shapes. Ask them to sketch two intersecting sets to make a Venn diagram. Give them a few minutes to decide on the criteria that they will use and then to sort their shapes into the Venn diagram. Take feedback. Accept any ideas that work, for example 'slide' and 'roll'.

- Direct learners to the Student's Book and discuss the Venn diagrams shown there. Remind learners that the word 'multiple' refers to the product of two numbers that are multiplied together.

So, for example, multiples of 2 are numbers that are answers to 2 times tables, and multiples of 3 are numbers that are answers to 3 times tables.

- [TWM.06] Learners work on the paired activity, drawing the Venn diagram, writing the headings and numbers in each region of the diagram.

- Discuss the Guided practice example with the class.

Practise 📓 [TWM.05/06]

- Workbook

Title: Venn diagrams

Pages: 206–207

- Refer to Activity 1 from the Additional practice activities.

Apply 👥 🖥 [TWM.05/06]

- Display **Slide 3**. Learners sketch a Venn diagram showing two criteria for sorting.

- They think of criteria that they could use to sort numbers, for example odd numbers and multiples of 3. They label each set and then think of numbers to go into each region.

Review

- Ask learners to explain how Venn diagrams work. Agree that they have been using Venn diagrams with two criteria and that the intersection contains things that meet both criteria.

Assessment for learning

- How would you use a Venn diagram to sort ...?
- What does the word 'criteria' mean?

Same day intervention
Support

- Focus on one criterion Venn diagrams where anything that does not fit is placed outside the set. Then introduce the second set.

Lesson 2: **Carroll diagrams**

Learning objectives

Code	Learning objective
3Ss.01	Conduct an investigation to answer non-statistical and statistical questions (categorical and discrete data).
3Ss.02	Record, organise and represent categorical and discrete data. Choose and explain which representation to use in a given situation: - Venn and Carroll diagrams [- tally charts and frequency tables - pictograms and bar charts].
3Ss.03	Interpret data, identifying similarities and variations, within data sets, to answer non-statistical and statistical questions and discuss conclusions.

Resources

selection of 3D shapes: sphere, cylinder, cone, cube, cuboid, square-based pyramid, triangular prism (per group); sheet of A3 paper (per group); sheet of A3 paper (per pair); paper (per learner) (for the Workbook)

Revise

Use the activity *Diagrams* from Unit 26: *Statistics* in the Revise activities to review and reinforce Carroll diagrams. Focus on the variation.

Teach 📖 🖥

- **[T&T]** Display **Slide 1**. Ask: **What are the names of these animals? Can you describe them and say where they live?** Learners should be able to identify the crab, snake, fish and lion. Agree that two of them have legs and two don't and that two of them live in the sea and two don't.

- **[T&T]** Ask: **How could we sort these animals? Could we use a Venn diagram?** Take feedback. Agree that they could be sorted into a Venn diagram. Ask learners to work with a partner to explore ways to do this. After a few minutes take feedback; for example, one set for animals with legs and another for those that live in the sea. The crab would go in the middle region and the snake would be placed outside the two sets.

- Ask: **Is there another diagram that we could use?** Elicit that they could use a Carroll diagram. Establish that a Carroll diagram can be used to sort according to whether something meets or does not meet a criterion. Display **Slide 2** and discuss the way the animals have been sorted. Point out that this Carroll diagram is sorting by two criteria and make sure that learners understand how to place each animal.

- Give small groups of learners a sheet of A3 paper and a selection of 3D shapes. Ask them to create their own Carroll diagram with two criteria and to sort their shapes into it. Take feedback, inviting learners to share their criteria for their Carroll diagram.

- Direct learners to the Student's Book. Discuss the two Carroll diagrams, asking what is the same about them and what is different.

- Learners work through the paired activity, writing down four facts they can state from the second Carroll diagram.

- Discuss the Guided practice example with the class. Once again remind learners that the word 'multiple'

refers to the product of two numbers that are multiplied together. So, for example, multiples of 2 are numbers that are answers to 2 times tables, and multiples of 3 are numbers that are answers to 3 times tables.

Practise 📝 [TWM.05/07]

- Workbook

Title: Carroll diagrams

Pages: 208–209

- Refer to the variation in Activity 1 from the Additional practice activities.

Apply 👥 🖥 [TWM.03/04]

- Display **Slide 3**. Pairs draw a Carroll diagram on a sheet of A3 paper with the headings multiple of 3/not multiple of 3 and multiple of 6/not multiple of 6.

- They think of numbers to put into each section of their diagram.

- Invite learners to share their Carroll diagrams with the class. Ask the class to suggest more numbers to add to each part.

Review

- On the board, draw a Carroll diagram without headings. Have your own headings in mind and add numbers to each section. Ask the class to work out what your headings must be.

Assessment for learning

- What is a Carroll diagram?
- How would you use a Carroll diagram to sort shapes? What about numbers?
- If one of my column headings is 'wild animals' what must the other column heading be?

Same day intervention
Support

- Focus on single criterion Carroll diagrams and simple criteria such as multiple of 2/not multiple of 2.

 **Unit 26 Statistics**

Lesson 3: **The statistical cycle**

Statistics and Probability – Statistics

Learning objectives

Code	Learning objective
3Ss.01	Conduct an investigation to answer non-statistical and statistical questions (categorical and discrete data).
3Ss.02	Record, organise and represent categorical and discrete data. Choose and explain which representation to use in a given situation: [- Venn and Carroll diagrams] - tally charts [and frequency tables - pictograms and bar charts].
3Ss.03	Interpret data, identifying similarities and variations, within data sets, to answer non-statistical and statistical questions and discuss conclusions.

Resources

paper (per pair); 2 cm squared paper (per per); ruler (per pair), coloured pencil (per pair)

Revise

Use the activity *Tally charts* from Unit 26: *Statistics* in the Revise activities to assess learners' knowledge of tally charts.

Teach 🆂🅱 💻 [TWM.04]

- Begin by briefly recapping tally charts. Learners should be familiar with these from Stage 2.
- [T&T] 🖐 Display **Slide 1**. Agree that the photograph shows lions, giraffes, elephants and zebras. Discuss similarities and differences between the animals.
- [T&T] 🖐 Ask: **How could we record how many there are of each type of animal?** Establish that the numbers of each type of animal can be shown in a tally chart, then display **Slide 2**.
- Display **Slide 3** and introduce the statistical cycle. Emphasise that data is collected to answer a question or to solve a problem. In order to do this, a plan needs to be made. Discuss learners' thoughts concerning the three questions on the slide. Agree that when these decisions have been made, the data can be recorded, organised and represented.
- Ask: **In which ways do we know how to represent data?** Expect learners to respond with data representations they have considered in this unit and in Stage 2, including block graphs and pictograms.
- Referring back to the cycle, discuss how once data has been represented it can be interpreted, discussed and possibly the problem solved.
- Give learners this scenario: The PE teacher has been given some money to spend on equipment for the school. The problem is, he doesn't know what to spend it on. Ask: **What do you think he should do?** Take suggestions. If necessary, direct learners to consider asking everyone at his school to help him decide what sport he should invest in.
- Ask learners to give you six sports that are popular inside and outside of school. Make a list. Give learners one choice. Make a tally of their choices.
- Learners then reflect on the result and discuss how this might help the teacher decide.

- [TWM.04] Direct learners to the Let's learn section in the Student's Book and recap on the features of the statistical cycle.
- Learners complete the paired activity, then as a class discuss the Guided practice example.

Practise 🆆🅱

- Workbook

Title: The statistical cycle

Pages: 210–211

- Refer to Activity 2 from the Additional practice activities.

Apply 👥 💻

- Display **Slide 4**. In pairs, learners think of a question they want the class to answer. They decide on the data they need, collect the data in a tally chart and represent it in a way that they think is appropriate. Finally, they make statements and conclusions about the data they collected.

Review

- Recap the parts of the statistical cycle. Invite learners to explain what each stage involves linking them to what they did when solving the PE teacher's problem during the lesson.

Assessment for learning

- How can we use data to solve a problem?
- How does the statistical cycle help?
- What does interpret the data mean? Can you explain in a different way?

Same day intervention
Support

- When making up their own problems to solve, work with learners to help them decide what data needs collecting and how best to represent it.

Lesson 4: **Frequency tables**

Learning objectives

Code	Learning objective
3Ss.01	Conduct an investigation to answer non-statistical and statistical questions (categorical and discrete data).
3Ss.02	Record, organise and represent categorical and discrete data. Choose and explain which representation to use in a given situation: [- Venn and Carroll diagrams] - tally charts and frequency tables [- pictograms and bar charts].
3Ss.03	Interpret data, identifying similarities and variations, within data sets, to answer non-statistical and statistical questions and discuss conclusions.

Resources

coloured counters (per pair); paper (per pair); paper (per learner) (for the Workbook)

Revise

Use the activity *Tally charts* from Unit 26: *Statistics* in the Revise activities to reinforce and consolidate learners' understanding of frequency tables.

Teach ⬛ 🖥 **[TWM.03/04/05/06]**

- **[T&T]** 🔁 Display **Slide 1**. Ask: **Is it possible to know how many of each colour jelly beans there are as soon as you look at them?** Agree that there are too many to know for sure. Ask: **What could we do?** Agree learners could make a tally. Give pairs a few minutes to list the colours that they can see and make a tally of how many of each colour they think there may be.

- Tell learners that often when we make tally chart we use it to draw up a frequency table. Display **Slide 2** and referring to the table on the left inform learners that the table shows a count of the numbers of jelly beans from **Slide 1**. Tell learners that the column on the right shows the frequency. Ask: **What's the difference between tallies and frequencies?** Agree tallies are made from marks and frequencies show numbers. Ask learners to check that each tally and number match.

- Direct learners to the table on the right inform and explain that we can also show just the frequencies in a table, and that this is also called a frequency table.

- **[TWM.03/04/05/06]** Ask, pairs of learners to take a handful of coloured counters. They sort them into colours. They count how many of each colour there are and make a tally mark for each colour so creating a tally chart. Once they have done this, they add a column to make it into a frequency table.

- Ask learners to make up questions about their frequency tables. Encourage them to include 'sum', 'difference' and 'more' in their questions.

- Direct learners to the Student's Book. Discuss the Let's learn section.

- Learners work through the paired activity.

- Discuss the Guided practice example with the class.

Practise 📒

- Workbook

Title: Frequency tables

Pages: 212–213

- Refer to the variations in Activity 2 from the Additional practice activities.

Apply 👥 🖥

- Display **Slide 3**. Pairs of learners make a tally chart and frequency table to show the information that Carly collected about favourite sports in Stage 3 at her school.

- Once they have done this, they find out how many learners are in Stage 3 at her school.

- Discuss the activity. Invite learners to show their tally charts and frequency tables. Have all the class made similar charts and tables?

- Do they agree that there are 89 learners in Stage 3 at Carly's school?

Review

- Discuss the similarities and differences between tally charts and frequency tables. Which do learners think is a more efficient way to represent data?

Assessment for learning

- How would you describe a frequency table to someone who has never seen one before?

- Can you describe this in another way?

Same day intervention

Support

- Focus on numbers below 20 for the frequency table.

Enrichment

- Focus on problem solving. Ask learners to make up a question that can be answered using information collected in a frequency table. They list the data they wish to collect, collect it from the learners in their class and record it in a frequency table. Once they have, they answer their question.

Statistics and Probability – Statistics

Additional practice activities

Activity 1

Learning objectives
- Conduct an investigation to answer non-statistical and statistical questions
- Record, organise and represent data in Venn and Carroll diagrams
- Interpret data

Resources

Resource sheet 15: Venn diagram template (per pair); Resource sheet 14: Carroll diagram template (per pair); number cards 0–20 from Resource sheet 16: 0–100 number cards (per pair); coloured pen (per learner)

What to do [TWM.05/06]
- Learners think of two criteria for a Venn diagram, for example 'multiples of 2' and 'multiples of 3'. They write labels in the circles on the Venn diagram template.
- Learners take turns to pick two number cards and perform an operation with them, for example add, subtract, multiply or divide, to make a number that will fit into one of the three regions on the Venn diagram.

- Any numbers they make that will not fit need to be written outside the diagram but inside the rectangle as part of the universal set.
- As learners make each number, they explain to their partner what they did, and the partner checks to see if they are correct.

Variation

2 Learners repeat the activity but with different criteria. They draw and label a Carroll diagram to record their results.

Activity 2

Learning objectives
- Conduct an investigation to answer non-statistical and statistical questions
- Record, organise and represent data in tallies and frequency tables
- Interpret data

Resources

A3 paper (per group); pens (per group)

What to do [TWM.03/04]
- Groups think of six different animals and make a list of them.
- Each group asks the class to choose their favourite animal from their list.
- Groups record the choices in a tally chart.
- Learners make statements using the information they have presented, such as 'More people like lions than elephants' or 'The favourite animal on our tally chart is monkeys'.

Variations

2 Repeat the activity with a different scenario/category. This time learners add a column to their tally chart to make it into a frequency table.

3 Once learners have completed their frequency table, they make up a question that can be answered from this information.

Unit 27: Statistics and chance

Collins International Primary Maths
Recommended Teaching and
Learning Sequence: Term 3, Week 6

Learning objectives

Code	Learning objective
3Ss.01	Conduct an investigation to answer non-statistical and statistical questions (categorical and discrete data).
3Ss.02	Record, organise and represent categorical and discrete data. Choose and explain which representation to use in a given situation: [– Venn and Carroll diagrams – tally charts and frequency tables] – pictograms and bar charts.
3Ss.03	Interpret data, identifying similarities and variations, within data sets, to answer non-statistical and statistical questions and discuss conclusions.
3Sp.01	Use familiar language associated with chance to describe events, including 'it will happen', 'it will not happen', 'it might happen'.
3Sp.02	Conduct chance experiments, and present and describe the results.

Unit overview

Statistics is an area of mathematics used regularly in everyday life. It is important, as teachers, to be aware of just how often we use statistics and to remind learners of this whenever appropriate. It would be helpful to have a variety of real-life charts, diagrams and graphs, as well as examples of the way data is collected in school, for learners to look at.

Learners continue to apply the statistical cycle introduced in Unit 26:

1. What's the problem? Make a plan.
2. Record, organise and represent the data.
3. Interpret the data.
4. Discuss the data and make predictions.

Learners will represent data using bar charts and pictograms with symbols representing two units.

Learners will also classify and make judgements about familiar events and explain why they have made their judgements. They will use the vocabulary of 'it will happen', 'it will not happen', 'it might happen'.

Prerequisites for learning

Learners need to:
• carry out an investigation to answer a question
• be able to use block graphs and pictograms
• be able to describe familiar patterns and more random patterns.

Vocabulary

record, organise, represent, information, pictogram, bar chart, sort, data, objects, criteria, vertical axis, interval, chance, will happen, might happen, will not happen

Common difficulties and remediation

Some learners may find it difficult to read axes on graphs. It may be helpful to link axes, particularly vertical ones, to number lines. In this way, learners should be able to see that the axes on graphs are just number lines.

When drawing conclusions from graphs, some learners may make very simple statements, for example: 'Three learners like cakes.' They need to be encouraged to think more deeply and draw more complex conclusions, for example: 'Five more learners like biscuits than like cakes, so Jim should buy biscuits for the party.'

Supporting language awareness

You may find it useful to refer to the audio glossary on Collins Connect. You could write these words on cards and hold them up, one at a time, for learners to read out loud. Break any words that learners find difficult to enunciate into parts. As in other units, ask questions, such as: **What is the same? What is different? What else do you know?** and **Can you give me another example?** Expect learners to answer in complete sentences.

Promoting Thinking and Working Mathematically

During this unit, learners will have opportunities to think and work mathematically; for example, they look at similarities and differences between various ways of representing data and they create their own using mathematical criteria that they make up. This involves making conjectures (TWM.03) and convincing (TWM.04), characterising (TWM.05) and classifying (TWM.06). When exploring chance, they consider whether their choices are always, sometimes or never true, which involves specialising (TWM.01) and generalising (TWM.02).

Success criteria

Learners can:
• record, organise and represent data, using pictograms and bar charts
• interpret data to answer questions
• use familiar language associated with chance
• conduct chance experiments, and present and describe the results.

Lesson 1: **Pictograms**

Learning objectives

Code	Learning objective
3Ss.01	Conduct an investigation to answer non-statistical and statistical questions (categorical and discrete data).
3Ss.02	Record, organise and represent categorical and discrete data. Choose and explain which representation to use in a given situation: [– Venn and Carroll diagrams – tally charts and frequency tables] – pictograms [and bar charts].
3Ss.03	Interpret data, identifying similarities and variations, within data sets, to answer non-statistical and statistical questions and discuss conclusions.

Resources

several sheets of paper (per pair); ruler (per pair/learner); coloured pencils (per pair/learner); 2 cm squared paper (per learner) (for the Workbook)

Revise

Use the activity *Charts and tables* from Unit 27: *Statistics and chance* in the Revise activities to assess understanding of tally charts and frequency tables.

Teach 🆂🅱 💻 📊

- **[T&T]** 🗣 Display **Slide 1**. Agree there are a selection of fruits including apples, pomegranates, kiwi fruit, bananas, oranges, limes.
- Give learners a scenario such as: **Mr Nolan wants to open a fruit stall in the market. He wants to sell eight different types of fruit.**
- **[T&T]** Ask: **How could we help Mr Nolan decide which fruits to sell?** Establish that finding the eight most popular fruits among learners would be a good starting point.
- **[T&T]** Ask: **Which fruit do you think is the most popular fruit in the class? How could we find out?** Take time to discuss and agree that each learner could select their favourite fruit. The data could be collected using a tally chart and then converted into a frequency table. Then the data could be represented on a pictogram, where each symbol represents one learner. Collect the information using a tally chart and, as a class, make a frequency table and then a pictogram using the **Pictogram tool**. Ask appropriate questions in order to analyse and discuss the data in the pictogram. Display **Slide 2** and discuss what the class has just done in relation to the statistical cycle.
- Display and discuss **Slide 3**. Highlight that each banana represents two people. Ask: **How many people does half a banana represent?** Agree one. Ask questions such as: **How many bananas were eaten on Saturday? How many more bananas were eaten on Friday than on Tuesday? What is the total number of bananas eaten on Monday and Thursday?**
- Discuss the Let's learn section in the Student's Book, then ask learners to work through the paired activity before discussing Guided practice.

Practise 🆆🅱

- Workbook

Title: Pictograms

Pages: 214–215

- Refer to Activity 1 from the Additional practice activities.

Apply 👥 💻

- Display **Slide 4**. Give each pair of learners several sheets of paper, a ruler and coloured pencils.
- Learners carry out an investigation, collecting data on learner's preferred item from a given list. They list six items (e.g. favourite food, animal or sport). They collect the data using a frequency table, then create a pictogram to represent the data, with one symbol standing for two learners.

Review

- 🗣 Ask: **Can you tell me the different ways we have represented data so far?** Agree... tally charts, frequency tables, Venn diagrams, Carroll diagrams and pictograms.
- Ask learners to discuss the usefulness of each. Talk about when Venn and Carroll diagrams are helpful; for example, for sorting against one or two criteria. Talk about when tally charts, frequency tables and pictograms are helpful; for example, for collecting, organising and representing sets of data.

Assessment for learning

- What is a pictogram?
- How is it the same as a tally chart and a frequency table? How is it different?
- What do the symbols on a pictogram tell us?

Same day intervention
Support

- Focus on making tally charts and converting these to frequency tables.

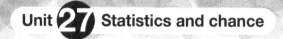

Lesson 2: Bar charts

Learning objectives

Code	Learning objective
3Ss.01	Conduct an investigation to answer non-statistical and statistical questions (categorical and discrete data).
3Ss.02	Record, organise and represent categorical and discrete data. Choose and explain which representation to use in a given situation: [– Venn and Carroll diagrams – tally charts and frequency tables – pictograms and] bar charts.
3Ss.03	Interpret data, identifying similarities and variations, within data sets, to answer non-statistical and statistical questions and discuss conclusions.

Resources

several sheets of paper (per pair); 2 cm squared paper (per pair/learner); ruler (per pair/learner); coloured pencils (per pair/learner)

Revise

Use the activity *Pictograms and bar charts* from Unit 27: *Statistics and chance* in the Revise activities.

Teach ⬛ 🖥 📊

- **[T&T]** Display **Slide 1**. Ask: **What can you see? What instruments can you name? Have you ever played any of these instruments?** Discuss how we can find out by listing the instruments, then marking a tally beside each one for each learner that has played it. We can represent the information in a frequency table or pictogram.

- **[T&T]** Say: **We know how to represent information in a tally chart, a frequency table and a pictogram.** Ask: **How else can we represent information?** Establish that information can also be represented on a bar chart.

- Make a list of instruments played by learners. You might want to include instruments that learners would like to play. Make a tally beside each and then use the **Bar charter tool** to represent the data where the scale on the vertical axis increases in ones (i.e. set Increment and Sub-divide each to 1). Explain that the vertical line and the horizontal line are the axes. The numbers up the vertical axis tell us how many of each data item there are (this is the frequency) and the data items are listed along the horizontal axis. Take care to remind learners that there should be a small gap between one bar and the next.

- **[T&T]** 🖥 Display **Slide 2**. Ask: **What does this bar chart show?** Agree it shows favourite fruits. Ask: **How is this the same as the bar graph we drew for musical instruments? How is it different?** After a few minutes, take feedback. Agree that they are both bar charts that show data. They are different because of the subject and the scales on the vertical axes.

- Give learners a problem, for example, we have decided to organise an additional after school club. What would you like it to be? Ask learners to work with a partner and choose four or five possible clubs. They then collect votes from the class and represent the information as a bar chart. They share their recommendations for the extra club with the class.

- Discuss the Let's learn section in the Student's Book, then ask learners to work through the paired activity, before discussing the Guided practice example.

Practise 📖

- Workbook

Title: Bar charts

Pages: 216–217

- Refer to the variation in Activity 1 from the Additional practice activities.

Apply 👥 🖥

- Display **Slide 3**. Give each pair of learners several sheets of paper, 1 cm or 2 cm square paper (optional), ruler and coloured pencils. Learners are given this problem: The pet shop wants to add an extra pet to sell. They want your help. Choose six pets for your class to choose from. They collect the data, represent it as a bar chart. They then decide which pet the pet shop should sell.

Review

- Discuss the similarities and differences between pictograms and bar charts, highlighting the fact that they can both be used to show data in similar ways and the use of axes on the bar chart.

Assessment for learning

- What is a bar chart? How is it the same as a pictogram? How is it different?
- What do the intervals on a bar chart tell us?

Same day intervention

Support

- Focus on bar charts with intervals of one.

Unit 27 Statistics and chance

Lesson 3: **Chance (1)**

Learning objective

Code	Learning objective
3Sp.01	Use familiar language associated with chance to describe events, including 'it will happen', 'it will not happen', 'it might happen'.

Resources

coin (per pair); mini whiteboard and pen (per learner); 1–6 dice (per pair)

Revise

Use the activity *Random or repeat* from Unit 27: *Statistics and chance* in the Revise activities to remind learners about patterns and randomness.

Teach [SB] 🖥 [TWM.01/02]

- **[T&T]** 🗫 Display **Slide 1**. Ask: **What do you think the slide shows?** Discuss the events shown under the arrows and, for each one, ask: **Do you think this can happen?** Ask learners to discuss this with their partner. After a few minutes, take feedback. Agree that the answers are 'no', 'perhaps' and 'yes'.
- Establish that a cat will never have a rhinoceros baby and that Tuesday will follow Monday.
- Ask: **Why is 'If I toss a coin it will land heads up' something that 'might happen'?** Agree that a coin has two sides: a heads side and a tails side. The chances of the coin landing heads up are the same as it landing tails up, so this chance must be 'might happen'.
- Give each pair of learners a coin and paper or a whiteboard. Ask them to toss their coin and see how it lands. They take turns to do this and record H or T, depending on how it lands. Ask: **How many times did your coin land heads up?** Take feedback and highlight that the coin landing heads up is not *always* going to happen, but neither will it *never* happen, so it *might* happen.
- **[T&T] [TWM.01/02]** Display **Slide 2**. Ask learners to discuss the statements shown and to decide if they 'will happen,' 'will not happen,' or 'might happen'. Expect them to give reasons for their answers. After a few minutes, take feedback. Invite learners to share what they think with their reasoning.
- Discuss the Let's learn section in the Student's Book, then ask learners to work through the paired activity, before discussing the Guided practice example.

Practise [WB] [TWM.04]

- Workbook

Title: Chance (1)

Pages: 218–219

- Refer to Activity 2 from the Additional practice activities.

Apply 👥 🖥 [TWM.04]

- Display **Slide 3**.
- Each learner rolls a 1–6 dice ten times to see how many odd and even numbers they get.
- They record their results.
- Learners then discuss whether or not the chance of rolling an odd number is the same as rolling an even number.

Review

- 🗫 Ask: **What do we mean by 'chance'? What are the three types of chance that we looked at during the lesson?**

Assessment for learning

- What do we mean by 'chance'?
- Can you explain it another way?
- What do we mean by 'it may happen'?
- Can you give an example of something that will not happen?
- Can you give an example of something that will happen?

Same day intervention

Support

Focus on events that might happen initially. When learners are confident with these, introduce events that will and then will not happen.

Enrichment

- Ask learners to write statements of things that will happen, will not happen and might happen and display these around the room.

Lesson 4: **Chance (2)**

Learning objective

Code	Learning objective
3Sp.02	Conduct chance experiments, and present and describe the results.

Resources

Resource sheet 19: Sack race (per pair/group); two 1–6 dice (per pair); squared paper (per group); Resource sheet 20: Snakes and ladders (per pair); Resource sheet 21: Spinners (per pair); pencil and paper clip, for the spinner (per pair)

Revise

Use the activity *Will it or won't it?* from Unit 27: *Statistics and chance* in the Revise activities to consolidate learners' understanding of chance

Teach 🆂🅱 💻 [TWM.01/02]

• **[T&T] [TWM.01/02]** 🔁 Display **Slide 1**. Ask: **What do you think this slide is about?** After a few minutes, take feedback. Tell learners that it is a race game. Explain the rules of playing. Each learner picks the number of the racer they want to be. They roll two dice and find the sum. If their sum matches the number of their racer, they move forward one space on the track. The winner is the first racer to get into the finish section. Ask: **Which number would you choose to have? Why?** Give learners a few minutes to discuss this.

• Try the game out with the whole class. Give each pair or group of three a copy of Resource sheet 19 Sack race, two 1–6 dice and give each learner a racer number. They take turns to roll the dice and add together the two numbers generated. If the sum matches their racer number, they move forward one place on the gameboard.

• **[T&T] [TWM.01/02]** Ask: **Which number will not be rolled? Why?** Agree 1 because you cannot score 1 by rolling two 1–6 dice, so that racer will never move. Ask: **What numbers might not be rolled?** Agree 12 (because learners would need to roll double 6) and 2 (because learners would need to roll double 1). Ask: **Are there any other numbers that will not win?** Agree 3 (need a 1 and a 2) and 11 (need a 5 and a 6). Ask: **Which is the best number, that definitely might win?** Agree 7 (because they could roll 1 and 6, 2 and 5 or 3 and 4). After the class have played the game, consider if what happened confirms this.

• Encourage learners to play the game again, in groups of four. This time, they each choose two racer numbers. At the end of their games, discuss the outcomes. Did number 7 win? Ask: **Why might 6 or 8?** Agree that there are several ways to make these two numbers and so their chance of winning is better than some other numbers.

• Discuss the Let's learn section in the Student's Book, then ask learners to work through the paired activity, before discussing the Guided practice example.

Practise 🆆🅱 [TWM.04]

• Workbook

Title: Chance (2)

Pages: 220–221

• Refer to the variation in Activity 2 from the Additional practice activities.

Apply 👥 💻 [TWM.03/04]

• Display **Slide 2**.

• Learners decide which spinner they would prefer to use when playing a game of snakes and ladders.

• They discuss why with their partner.

• Give pairs of learners a game board and spinners from Resource sheets 20 and 21. They play the game to find out if their choice was sensible.

Review

• Discuss the activities that learners have been involved with.

Assessment for learning

• If there are 6 red counters and 1 yellow counter in a bag, which colour are you most likely to pick? Why?

• What words can we use if something will happen?

• What about if something will not happen? Might happen?

Same day intervention
Support

• When playing the race game, ensure that less confident learners play alongside more confident learners.

Enrichment

• Give learners the opportunity to devise their own chance experiments and to provide a written description of the results.

Statistics and Probability – Statistics/Probability

Additional practice activities

Activity 1

Learning objectives
- Conduct an investigation to answer non-statistical and statistical questions
- Record, organise and represent data in pictograms and bar charts
- Interpret data

Resources
interlocking cubes (per pair); 2 cm squared paper (per pair); ruler (per pair); coloured pencils (per pair)

What to do
- Before the activity, write this scenario on the board: The ice-cream seller wants to sell two new flavours of ice cream but does not know which to pick. She is thinking of pistachio, banana, liquorice, butterscotch and pomegranate.

- Learners need to help her decide. They allocate each flavour to a colour of interlocking cubes.
- Each learner in the class picks a coloured cube. They make towers of each colour, to find the two most popular flavours.
- Learners use the information they found to make a pictogram, in which each symbol represents two learners.

Variation
Learners draw their results as a bar chart. They make comparison statements involving the data.

Activity 2

Learning objectives
- Use familiar language associated with chance to describe events
- Conduct chance experiments, and present and describe the results

Resources
two sheets of paper (per pair); two 1–6 dice (per pair)

What to do [TWM.03/04]
- Give pairs of learners two 1–6 dice and two sheets of paper.
- They write the headings: 'will not happen', 'might not happen', 'might happen' and 'will happen' on one of the sheets of paper.
- Explain that they are going to roll the two 1–6 dice and find the total of the numbers rolled.
- Before they roll the dice, they discuss and decide which of the totals, from 1 to 12, to put under each heading. For example, 1 will go under 'will not happen'. They might place 12 and 2 under 'might not happen', 7 could be placed under 'might happen' because there are more combinations of numbers to make 7 and 3, 4 and 10 possible. They write the numbers 1 to 12 underneath the

appropriate headings to show the likelihood of the totals.
- They take turns to roll the two 1–6 dice and find the total of the two numbers rolled. They write this on a different sheet of paper.
- They each roll the dice ten times and write their ten totals.
- Once each learner has rolled the dice 10 times they look at the totals rolled and decide whether they made the right decision for the numbers under the headings.
- Finally, learners discuss and write some statements about their results.

Variation
Ask pairs to devise a game similar to those explored in Lesson 4.

Statistics and Probability – Statistics/Probability

100 square

1	2	3	4	5	6	7	8	9	10
11	12	13	14	15	16	17	18	19	20
21	22	23	24	25	26	27	28	29	30
31	32	33	34	35	36	37	38	39	40
41	42	43	44	45	46	47	48	49	50
51	52	53	54	55	56	57	58	59	60
61	62	63	64	65	66	67	68	69	70
71	72	73	74	75	76	77	78	79	80
81	82	83	84	85	86	87	88	89	90
91	92	93	94	95	96	97	98	99	100

Numbers in numerals and words

one	two	three	four
five	six	seven	eight
nine	ten	eleven	twelve
thirteen	fourteen	fifteen	sixteen
seventeen	eighteen	nineteen	twenty
thirty	forty	fifty	sixty
seventy	eighty	ninety	hundred
0	1	2	3
4	5	6	7
8	9	and	and

Digit cards

0	1	2	3
4	5	6	7
8	9	0	1
2	3	4	5
6	7	8	9

Number pairs to 10 ✂ Number pairs to 10

1 + 9 = 10	1 + 9 = 10
2 + 8 = 10	2 + 8 = 10
3 + 7 = 10	3 + 7 = 10
4 + 6 = 10	4 + 6 = 10
5 + 5 = 10	5 + 5 = 10
6 + 4 = 10	6 + 4 = 10
7 + 3 = 10	7 + 3 = 10
8 + 2 = 10	8 + 2 = 10
9 + 1 = 10	9 + 1 = 10

Spots grid

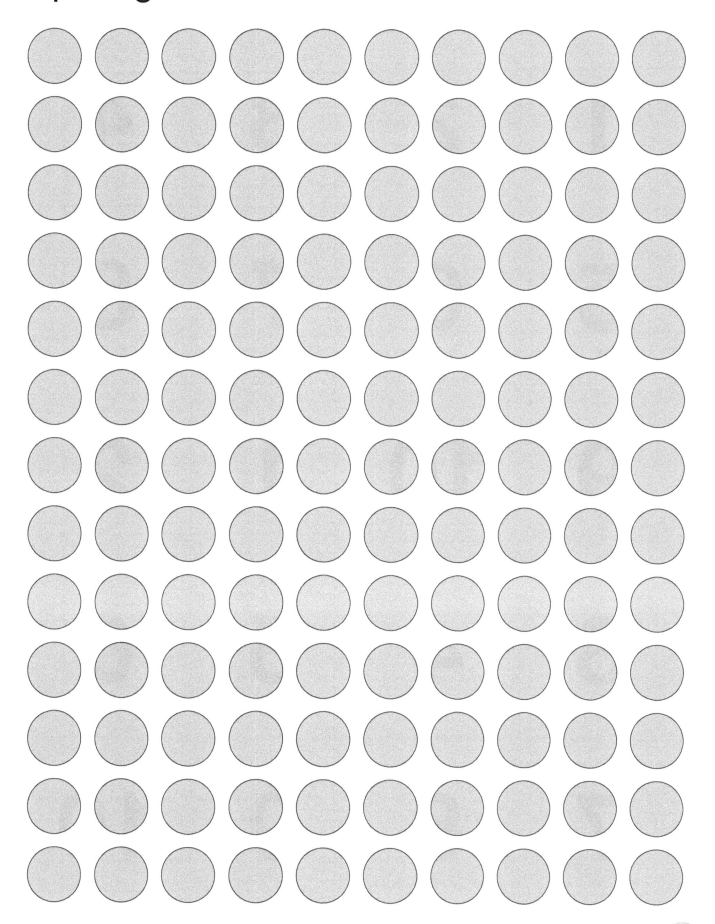

Numbers 1 to 10

1	2	3	4
5	6	7	8
9	10	1	2
3	4	5	6
7	8	9	10

Times tables grid

×	1	2	3	4	5	6	7	8	9	10
1	1	2	3	4	5	6	7	8	9	10
2	2	4	6	8	10	12	14	16	18	20
3	3	6	9	12	15	18	21	24	27	30
4	4	8	12	16	20	24	28	32	36	40
5	5	10	15	20	25	30	35	40	45	50
6	6	12	18	24	30	36	42	48	54	60
7	7	14	21	28	35	42	49	56	63	70
8	8	16	24	32	40	48	56	64	72	80
9	9	18	27	36	45	54	63	72	81	90
10	10	20	30	40	50	60	70	80	90	100

Place value counters

100	100	100	100	100	100	100	100	100
100	100	100	100	100	100	100	100	100
100	100	100	100	100	100	100	100	100
10	10	10	10	10	10	10	10	10
10	10	10	10	10	10	10	10	10
10	10	10	10	10	10	10	10	10
1	1	1	1	1	1	1	1	1
1	1	1	1	1	1	1	1	1
1	1	1	1	1	1	1	1	1

Place value chart

100s	10s	1s

100s	10s	1s

Place value abacus

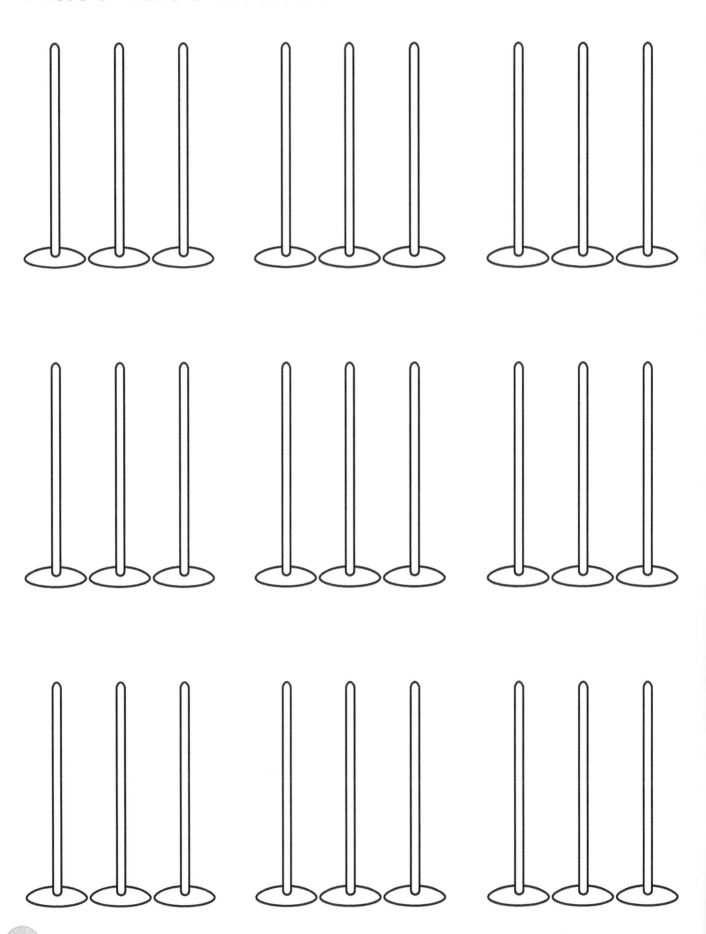

Comparison symbols

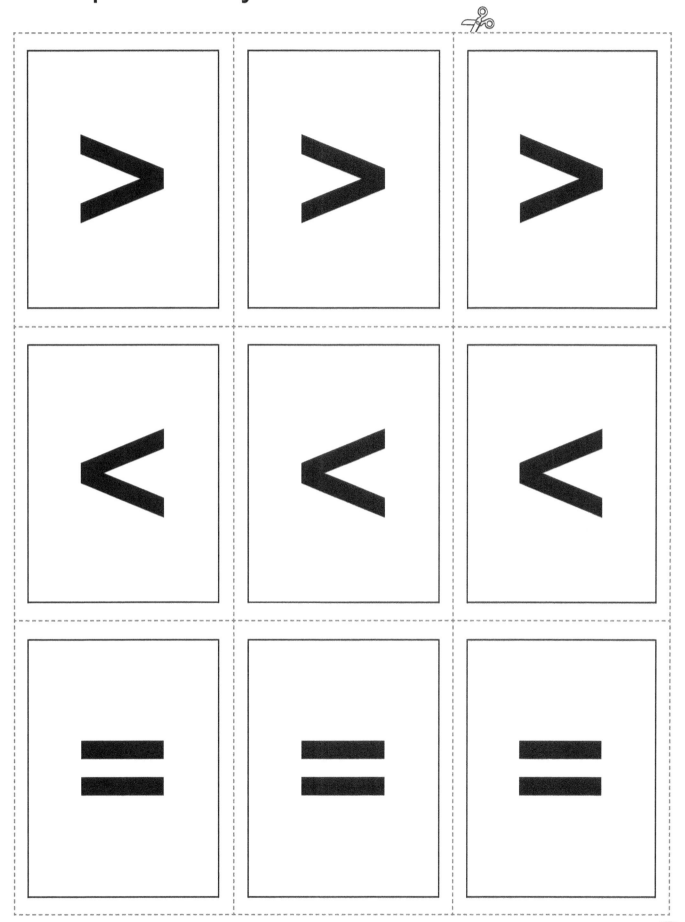

Clock faces

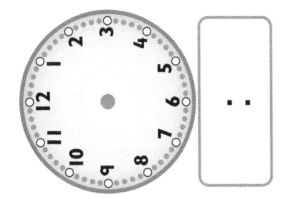
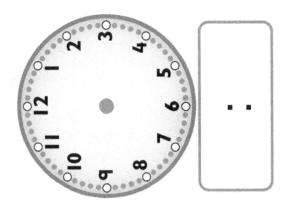
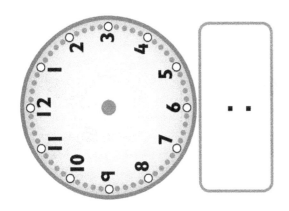
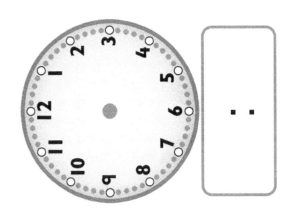
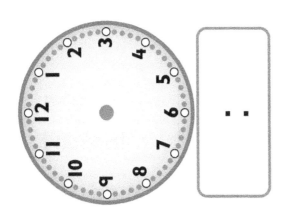
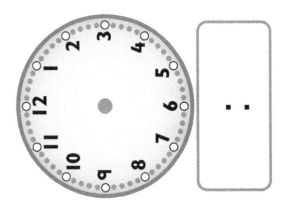

Tangram set

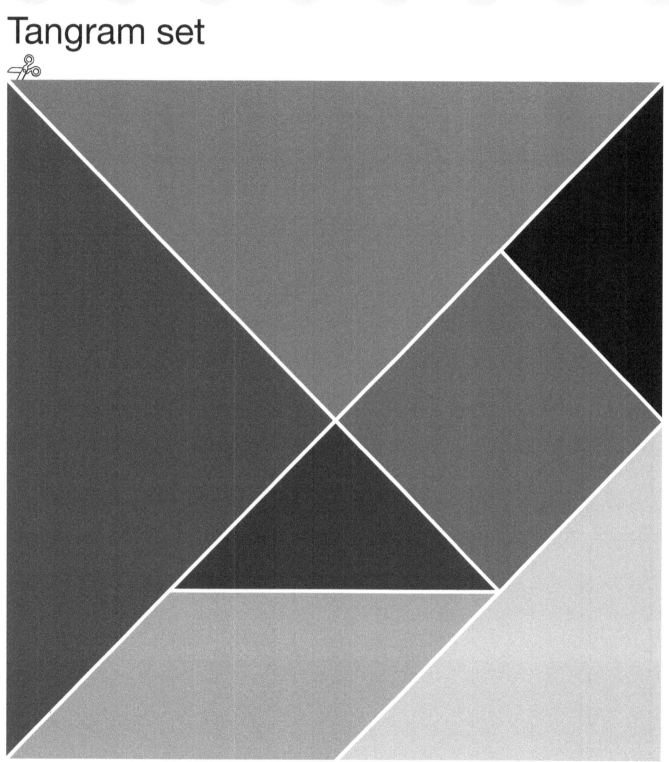

Carroll diagram template

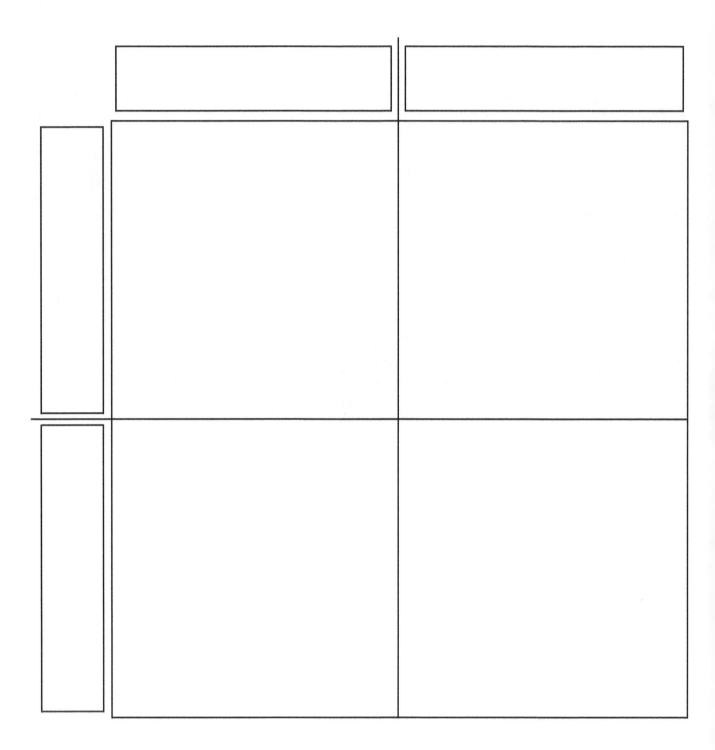

Venn diagram template

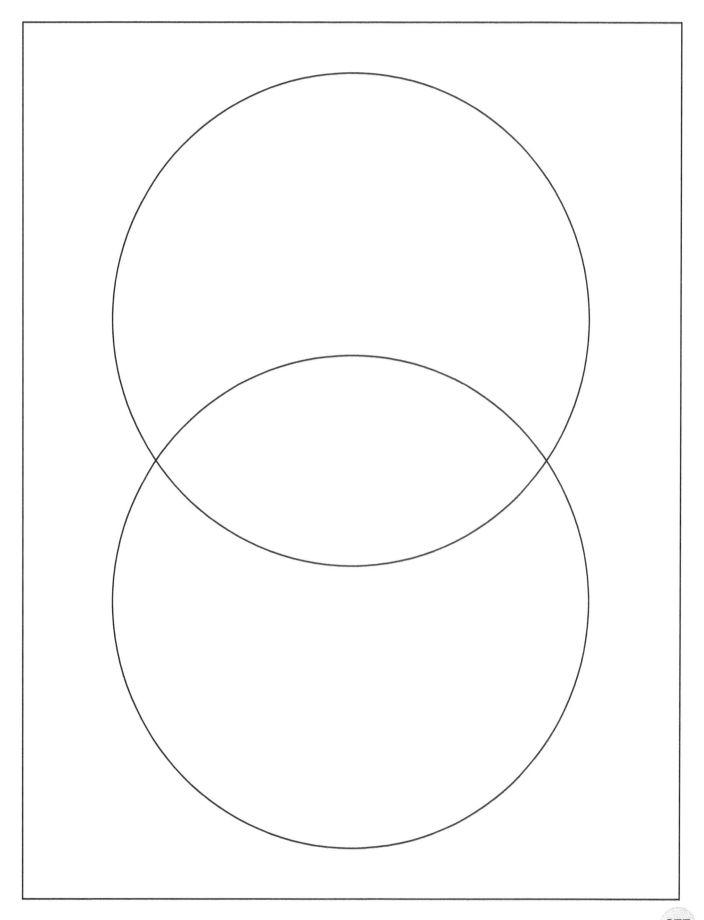

0–100 number cards

6	13	20	27
5	12	19	26
4	11	18	25
3	10	17	24
2	9	16	23
1	8	15	22
0	7	14	21

0–100 number cards

34	41	48	55
33	40	47	54
32	39	46	53
31	38	45	52
30	37	44	51
29	36	43	50
28	35	42	49

0–100 number cards

62	69	76	83
61	68	75	82
60	67	74	81
59	66	73	80
58	65	72	79
57	64	71	78
56	63	70	77

0–100 number cards

88	93	98	100
87	92	97	99
86	91	96	
85	90	95	
84	89	94	

Thermometers

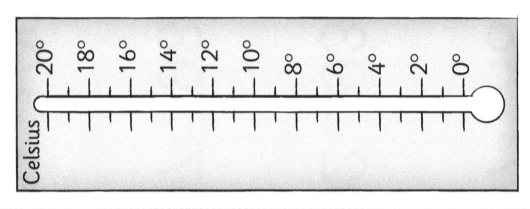

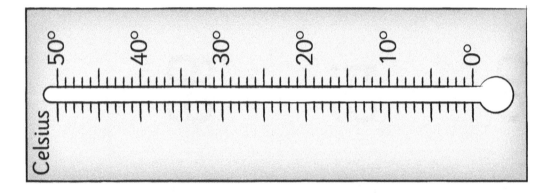

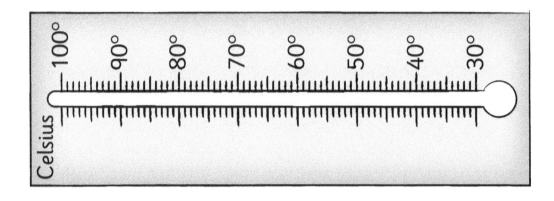

10 x 10 grid

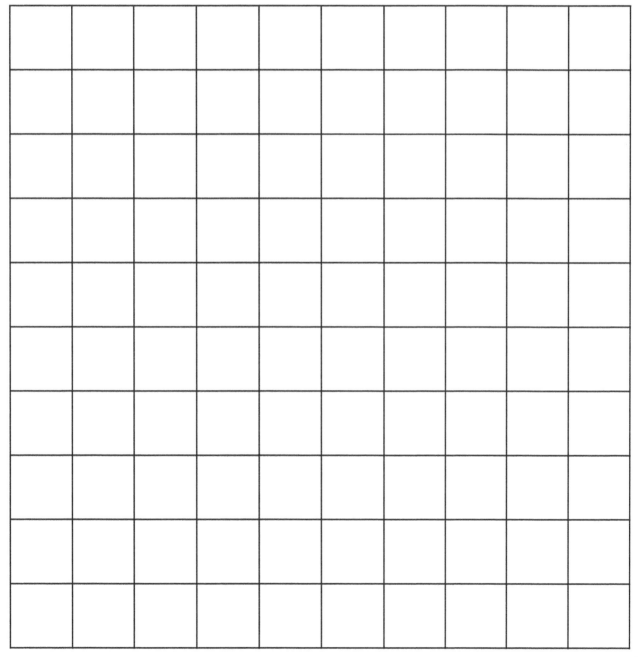

Sack race

1											
2											
3											
4											
5											
6											
7											
8											
9											
10											
11											
12											

F I N I S H

Snakes and ladders

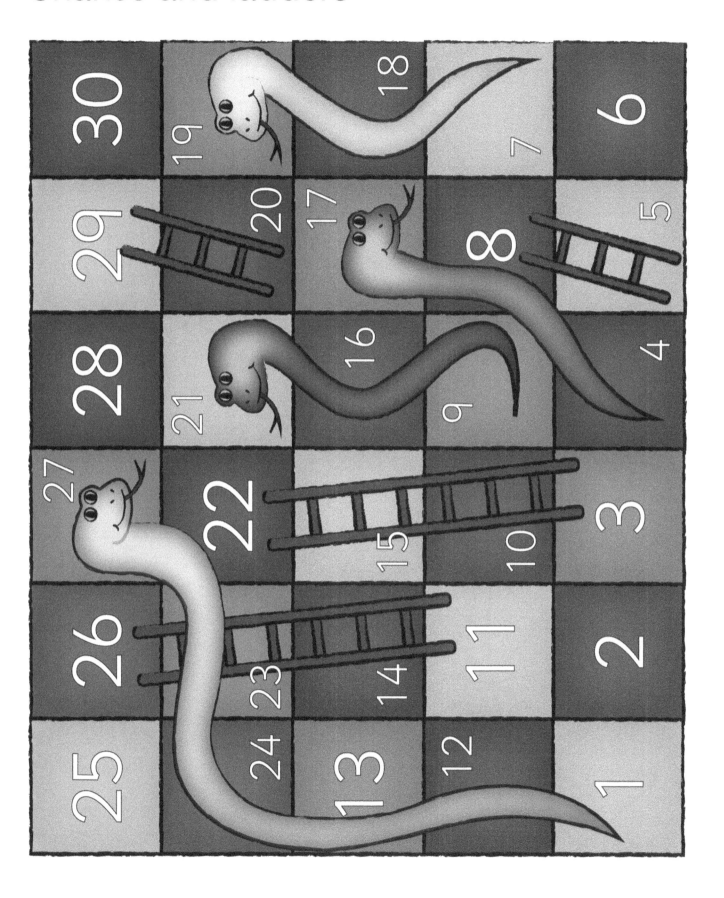

Spinners

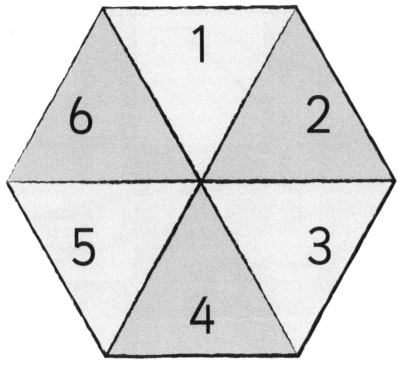

How to use the spinner

Hold the paper clip in the centre of the spinner using the pencil and gently flick the paper clip with your finger to make it spin.

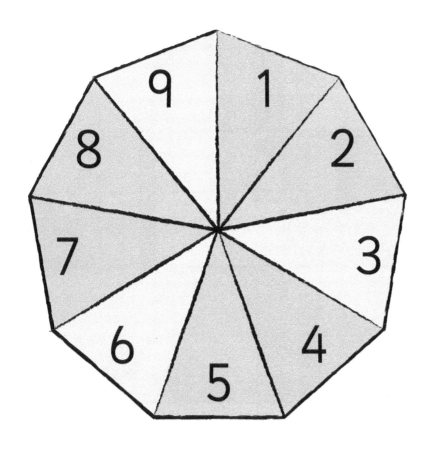

Unit 1

Lesson 1: Counting

Challenge ❶

1 a   34, 54, 64, 74
   b   38, 48, 58, 78
   c   33, 43, 73, 83

2 88, 78, 68, 58, 48, 38, 28, 18

Challenge ⚠

3 462, 452, 442, 432, 422, 412, 402, 392
   Learner's own explanation, e.g. all even numbers, 10s number changes each time

4 197, 207, 217, 227, 237, 247, 257, 267
   Learner's own explanations, The ones digit remains the same;
   After the first number the tens digits increase by one each time;
   After the first number the hundreds digit remains the same.

5 845, 745, 545, 445, 245, 145, 45

6 392

7 147, 231

8 430

Challenge ❸

9 336, 326, 316, 306, 296, 286, 276, 266
   Learner's own explanation implying that 100s digit changes.

10 849, 859, 869, 879, 889, 899, 909, 919
   Learner's own explanation implying that 100s digit changes.

11 450

Lesson 2: Even and odd numbers

Challenge ❶

1 12, 44, 68, 90

2 25, 37, 99, 21

3 14, 18, 20, 22, 26, 28

4 23, 27, 29, 31, 35, 37

Challenge ⚠

5 Odd: 85, 297, 689, 183 Not odd: 94, 368, 560, 236

6 Even numbers and odd numbers OR even numbers and not even numbers

7 No. Learner's explanation e.g. consider 1s digit only. If it is 2, 4, 6, 8, or 0 then the number is even. 574 is even.

8 Any numbers starting with 4 or 5 and ending with 2, 4, 6, 8, 0.

9 463 and 643

Challenge ❸

10 Learner's explanation, e.g. divisible by 2, multiple of 2, in 2 times table

11 Learner's explanation, e.g. not divisible by 2, not a multiple of 2, not in 2 times table

12 523, 525, 527, 529, 531

Lesson 3: More about even and odd numbers

Challenge ❶

1 Lines must divide shapes into two equal parts.

2 Lines must show two unequal groups.

3 Answers will vary – numbers correctly put into the odd and even columns.

Challenge ⚠

4 The numbers are even. Even numbers: 94, 82, 96, 118, 40, 162, 74, 200. Odd numbers: 59, 125, 41, 17, 91, 43, 89.

5 No. Learner's explanation, e.g. 1 is odd because it is one on its own. 1 can be divided into two equal parts but each part is a fraction and not a whole number.

6 Learner's own explanation, e.g. ends with 8, multiple of 2, divisible by 2.

7 Learner's own explanation, e.g. ends with 7, not a multiple of 2, not divisible by two.

Challenge ❸

8 Learner's own explanation, e.g. ends with 2, 4, 6, 8, 0, divisible by 2, multiple of 2. As above.

9 Learner's own explanation, e.g. ends with 1, 3, 5, 7, 9, not divisible by 2, not a multiple of 2. As above.

Lesson 4: Estimating

Challenge ❶

1 a   40–50   b   20–30   c   70–80

2 30–40 or 40–50

Challenge ⚠

3 a   Learner's own choice, e.g. 50–60
   b   Ensure bananas are grouped correctly.
   c   58

4 a   Learner's own choice, e.g. 40–50
   b   Ensure oranges are grouped correctly
   c   48

5 Learner's own explanation, e.g. numbers are more specific than ranges

Challenge ❸

6 Answers will vary – they should be accurate based on realistic estimates

Unit 2

Lesson 1: More about estimating

Challenge ❶

1 a 70–80 **b** 90–100 **c** 80–90

2 a Learner's own choice – a sensible estimate from 38–46

b Ensure the fir cones are grouped correctly. 42

Challenge ⚠

3 a Learner's own choice, sensible estimate given

b Ensure there are sixteen groups of 10 and two singles

c 162

4 a Learner's own choice, sensible estimate given

b Ensure there are fourteen groups of 10 and 4 left over; 144

5 It is not a good range because it is below her estimate. A good range would be 600–700 or 680–690

Challenge ❸

6 Answers will vary, sensible estimate given for children in school.

7 Answers will vary.

Lesson 2: Counting on and back

Challenge ❶

1 a Numbers joined in this order: 2, 4, 6, 8, 10, 12, 14

b even

2 a Numbers joined in this order: 5, 10, 15, 20, 25, 30, 35

b both

3 a 8, 20, 24, 32, 40, 44 **b** even

Challenge ⚠

4 a

1	2✓	3✓	4✓✓	5	6✓✓✓
7	8✓✓	9✓	10✓	11	12✓✓✓✓
13	14✓	15✓	16✓✓	17	18✓✓✓
19	20✓✓	21✓	22✓	23	24✓✓✓✓
25	26✓	27✓	28✓✓	29	30✓✓✓
31	32✓✓	33✓	34✓	35	36✓✓✓✓
37	38✓	39✓	40✓✓		

b All the numbers are in the same columns. All the numbers are even.

c The numbers are not always in the same columns. The numbers are a mixture of odd and even.

5 a 39, 45

b 48, 36

c 30, 36, 60

Challenge ❸

6 a

1	2	3	4	5	6	7	8	9	10
11	12		13	14	15			16	17	18		19	20		
21		22	23	24				25		26		27		28			29	30			
31	32			33		34		35	36				37	38		39		40			
41	42			43	44			45			46		47	48				49	50		

b 24, 30, 36, 40 and 48

c Learner's reason, for example: For four lines, the numbers must be even and in the 2 times, 3 times, 4 times and 5 times tables.

Lesson 3: Making sequences with numbers

Challenge ❶

1 a 28, 38, 48, 68, 78, 88, 98

2 Add 10 onto previous number

3 89, 94, 99, 109, 114, 119

4 Add 5 onto previous number

Challenge ⚠

5 235.

6 a 58, 56, 54, 50, 48, 46

b Subtract 2 from the previous number

7 a 315, 325, 330, 340, 345, 350

b Add 5 onto the previous number

8 a 590, 600, 610, 630, 640

b Add 10 onto the previous number

9 Check that learner's sequence follows the rule: add 10

Challenge ❸

10 Yes. Starting number 745, step 1 = 747, step 2 = 749, step 3 = 751, step 4 = 753, step 5 = 755

11 He has counted forwards instead of back.

Lesson 4: Making patterns with numbers

Challenge ❶

1 Repeatedly adding 2s, to give totals of 6, 8, 10, 12

2 Drawing of 7 lots of 2 squares.

Challenge ⚠

3 4th: six triangles: 3 + 3 + 3 + 3 + 3 + 3 = 18 sides; 5th: seven triangles: 3 + 3 + 3 + 3 + 3 + 3 + 3 = 21 sides

4 2 lots of 4 squares: 4 + 4 = 8; 3 lots of 4 squares: 4 + 4 + 4 =12; 4 lots of 4 squares: 4 + 4 + 4 + 4 = 16

5 Learner's choice: could be four lots of 5 squares or 4 pentagons, 5 + 5 + 5 + 5, three lots of 5 squares or 3 pentagons, 5 + 5 + 5, two lots of 5 squares or 2 pentagons, 5 + 5, one lot of 5 squares or one pentagon, 5.

6 a Learner's choice: could be 1 lot of 3 triangles or one triangle, increasing to 4 lots of 3 triangles or four triangles

b Add 3 onto the previous number

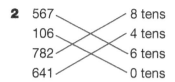

Challenge ❸

7 6 + 6 + 6 + 6 + 6 + 6 + 6 + 6 = 48

8 There are two ways. 9 groups of one square or 1 group of nine squares.

Unit 3

Lesson 1: Numerals and words (A)
Challenge ❶

1 **a** eighteen = 18

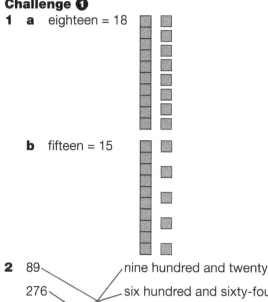

  b fifteen = 15

2
89 ———— nine hundred and twenty
276 ———— six hundred and sixty-four
664 ———— eighty nine
709 ———— two hundred and seventy-six
920 ———— seven hundred and nine

Challenge ⚠

3 **a** five hundred and forty-eight, eight hundred and forty-five

  b three hundred and sixty-seven, seven hundred and sixty-three

4 **a** Circle ate, fourty
    eight hundred and forty-nine

  b Circle nin, fivety, free
    nine hundred and fifty-three

5 **a** 356 **b** 54 **c** 899 **d** 105

Challenge ❸

6 358, 351, 853, 851, 153, 158, 378, 371, 873, 871, 173, 178

7 135, 137, 139, 153, 157, 159, 173, 175, 179, 193, 195, 197

Lesson 2: Numerals and words (B)
Challenge ❶

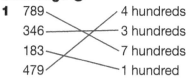

1
789 ———— 4 hundreds
346 ———— 3 hundreds
183 ———— 7 hundreds
479 ———— 1 hundred

2
567 ———— 8 tens
106 ———— 4 tens
782 ———— 6 tens
641 ———— 0 tens

Challenge ⚠

3 **a** 263 **b** two hundred and sixty-three
4 **a** 435 **b** four hundred and thirty-five
5 ◐○○○○○○○○●●●●
three hundred and fifty-four
6 **a** two hundred and fifty-four
  b six hundred and seven
  c four hundred and ninety-eight
  d one hundred and sixty

Challenge ❸

7 567, five hundred and sixty-seven
576, five hundred and seventy-six
657, six hundred and fifty-seven
675, six hundred and seventy-five
756, seven hundred and fifty-six
765, seven hundred and sixty-five

8 567, five hundred and sixty-seven
657, six hundred and fifty-seven
765, seven hundred and sixty-five
675, six hundred and seventy-five
4 (possibilities)

Lesson 3: Reading and writing even numbers to 1000
Challenge ❶

1 nine hundred and fourteen
four hundred and eight
seven hundred and twelve

2 three hundred and forty-four
four hundred and forty
five hundred and eight

Challenge ⚠

3 **a** 456 = four hundred and fifty-six,
  b 142 = one hundred and forty-two,
  c 740 = seven hundred and forty,
  d 108 = one hundred and eight,
    214 = two hundred and fourteen

4 216, 508, 342, 618, 794

5 438 = four hundred and thirty-eight, 348 = three hundred and forty-eight, 384 = three hundred and eighty-four, 834 = eight hundred and thirty-four

Challenge ❸
6   Yes, at least six: 614 = six hundred and fourteen, 814 = eight hundred and fourteen, 314 = three hundred and fourteen, 698 = six hundred and ninety-eight, 398 = three hundred and ninety-eight, 396 = three hundred and ninety-six, 896 = eight hundred and ninety-six

7   Grace is incorrect. Learner's explanation, e.g. 3 could be the number of units of a hundred or ten, but she could have 2, 4, 6, 8 or 0 as the units of one. Example: 354, 730.

Lesson 4: Reading and writing odd numbers to 1000

Challenge ❶
1   one hundred and forty-three
two hundred and sixty-seven
seven hundred and five
four hundred and nine
2   three hundred and nineteen
one hundred and nine
nine hundred and seventeen

Challenge ⚠
3   a   two hundred and thirty-one
    b   four hundred and fifty-three
    c   five hundred and seventy-nine
    d   one hundred and seventeen
    e   eight hundred and five
4   a   513   b   307   c   845
    d   911   e   209
5   765, 567, 657, 675

Challenge ❸
6   Yes, at least six: 917 = nine hundred and seventeen, 965 = nine hundred and sixty-five, 517 = five hundred and seventeen, 569 = five hundred and sixty-nine, 417 = four hundred and seventeen, 465 = four hundred and sixty-five, 469 = four hundred and sixty-nine

7   Manu is incorrect. Learner's explanation, e.g. 8 could be the number of units of a hundred or ten, he could have 1, 3, 5, 7 or 9 as the units of one. Example: 871, 485.

Unit 4

Lesson 1: Commutativity

Challenge ❶
1
$$10 + 40 \qquad 13 + 14$$
$$24 + 3 \qquad 30 + 90$$
$$50 + 28 \qquad 40 + 10$$
$$8 + 7 \qquad 7 + 4$$
$$14 + 13 \qquad 3 + 24$$
$$90 + 30 \qquad 55 + 42$$
$$42 + 55 \qquad 28 + 50$$
$$4 + 7 \qquad 7 + 8$$

2   a $\underline{30 - 10 = 20}$ $10 - 20 = 30$
    b $\underline{5 - 2 = 3}$ $2 - 5 = 3$
    c $20 - 57 = 37$ $\underline{57 - 20 = 37}$
    d $20 - 50 = 30$ $\underline{50 - 20 = 30}$
    e $\underline{76 - 42 = 34}$ $42 - 76 = 34$

Challenge ⚠
3   Learner's own examples
4   Learner's own examples
5   a   Learner's own explanation, for example: It doesn't matter because in an addition the numbers can be added in any order.
    b   Answers will vary.
6   Learner's own explanation, which needs to include the fact that subtraction cannot be done in any order.

Challenge ❸
7   a $50 + 60 + 8 + 9$, $50 + 60 + 9 + 8$, $50 + 9 + 60 + 8$, $50 + 9 + 8 + 60$, $50 + 8 + 60 + 9$, $50 + 8 + 9 + 60$, $60 + 50 + 8 + 9$, $60 + 50 + 9 + 8$, $60 + 9 + 50 + 8$, $60 + 9 + 8 + 50$, $60 + 8 + 50 + 9$, $60 + 8 + 9 + 50$, $8 + 60 + 50 + 9$, $8 + 60 + 9 + 50$, $8 + 50 + 60 + 9$, $8 + 50 + 9 + 60$, $8 + 9 + 50 + 60$, $8 + 9 + 60 + 50$, $9 + 60 + 50 + 8$, $9 + 60 + 8 + 50$, $9 + 50 + 60 + 8$, $9 + 50 + 8 + 60$, $9 + 8 + 50 + 60$, $9 + 8 + 60 + 50$.
    b   Learner's own explanation, which must include the fact that it doesn't matter in which order numbers are added.

Lesson 2: Complements of 100 and multiples of 100

Challenge ❶
1   a   60   b   30   c   40   d   10
2   a   These pairs coloured: 5 and 95, 15 and 85, 25 and 75, 35 and 65, 45 and 55.
    b   Learner's examples, e.g. $5 + 15 = 20$; $15 + 5 = 20$; $20 - 15 = 5$; $20 - 5 = 15$

Challenge ⚠
3   a $61 + 39 = 100$, $39 + 61 = 100$, $100 - 61 = 39$, $100 - 39 = 61$
    b $46 + 54 = 100$, $54 + 46 = 100$, $100 - 46 = 54$, $100 - 54 = 46$
    c $300 + 400 = 700$, $400 + 300 = 700$, $700 - 300 = 400$, $700 - 400 = 300$
4   a   Yes. Answers may vary.
5   a   24   b   49   c   200
    d   300   e   400   f   900
    g   82   h   8
6   No. $57 + 43 = 100$

Challenge ❸
7   a   400 and 600
    b   900 and 100
    c   300 and 700, or 100 and 900
    d   200 and 800

Lesson 3: Addition and subtraction of 2-digit numbers

Challenge ❶

1   **a**   64   **b**   81   **c**   77

     d   55   **e**   46   **f**   74

Challenge ⚠

2   **a**   Estimate between 80 and 90, sum 89

     b   Estimate between 80 and 90, sum 83

     c   Estimate between 90 and 100, sum 98

     d   Estimate between 150 and 160, sum 155

3   **a**   Estimate between 20 and 30, difference 26

     b   Estimate between 50 and 60, difference 54

     c   Estimate between 20 and 30, difference 25

     d   Estimate between 40 and 50, difference 48

Challenge ❸

4   **a** $25 + 36 = 61$ or $36 + 25 = 61$ or $26 + 35 = 61$ or $35 + 26 = 61$

     b $52 + 63 = 115$ or $63 + 52 = 115$ or $62 + 53 = 115$ or $53 + 65 = 115$

5   **a** $35 - 26 = 9$ **b** $65 - 23 = 42$

Lesson 4: Unknowns!

Challenge ❶

1   Examples:

     a $6 + 4 = 10$ **b** $10 = 2 + 8$

     c $3 + 2 = 5$ **d** $4 + 4 = 8$

     e $3 + 6 = 9$ **f** $5 + 2 = 7$

2   **a**   4   **b**   7   **c**   5

     d   9   **e**   3   **f**   8

Challenge ⚠

3   **a**   6   **b**   10   **c**   12

     d   5   **e**   13   **f**   30

4   Learner's own work. Ensure problems fit with solutions.

Challenge ❸

5   Learner's own work. Ensure problems fit with solutions.

6   Learner's own strategies, for example:

     a $50 - 5 = 45$, 45 is unknown number

     b $50 - 14 = 36$, 36 is unknown number

Unit 5

Lesson 1: Associative property

Challenge ❶

1   3 + 4 + 7   1 + 9 + 5

     6 + 7 + 4   3 + 7 + 4

     5 + 1 + 9   7 + 3 + 8

     8 + 7 + 3   6 + 4 + 7

2   Learner's own explanation, which must include that 3 and 7 equal 10 and 10 and 4 equal 14.

Challenge ⚠

3   Learner's own examples

4   Learner's own examples

5   Yes Jack is correct. Leaners' own explanations, to include Jack added 8 and 7 to give 15 and then added 3 to give a total of 18.

     Learners' own explanation, which must talk about making 10 with 7 and 3 and then adding 8.

6   Learner's own explanation.

Challenge ❸

7   **a**   Example: $70 + 30 = 100$, $100 + 90 = 190$

     b   Learner's own explanation.

8   **a**   No. The total is 180.

     b   Learner's own explanations, e.g. Felix could make 100 by adding 80 and 20 and then adding 80.

Lesson 2: Addition and subtraction of multiples of 10

Challenge ❶

1   **a**   250   **b**   350   **c**   280   **d**   380

     Learner's explanation

2   **a**   220   **b**   230   **c**   150   **d**   140

Challenge ⚠

3   **a**   420   **b**   810   **c**   820

     d   810   **e**   610   **f**   920

     Learner's explanation

4   **a**   370   **b**   190   **c**   580

     d   480   **e**   130   **f**   270

     Learner's explanation

Challenge ❸

5   **a**   120   **b**   210   **c**   210   **d**   230

     e   420   **f**   140   **g**   210   **h**   110

     Learner's explanation

6   No, the answer should be 880. Learner's explanation to include making the total 10 times greater/multiplying the total by 10.

7   **a**   470 and 340. Learner's explanation

     b   590 and 260. Learner's explanation

Lesson 3: Addition and subtraction with 3-digit numbers

Challenge ❶

1   **a** $345 = 300 + 40 + 5$

     b $276 = 200 + 70 + 6$

     c $387 = 300 + 80 + 7$

     d $164 = 100 + 60 + 4$

     e $253 = 200 + 50 + 3$

     f $418 = 400 + 10 + 8$

2   **a**   249   **b**   381   **c**   470

     d   216   **e**   712   **f**   478

Challenge △
3   a   437   b   695   c   546
    d   615   e   571   f   613
    Learner's own explanation
4   a   537 + 7   and   358 + 7
    b   562 – 70
    Learner's own explanation

Challenge ❸
5   393
6   434

Lesson 4: More unknowns!

Challenge ❶
1   a   8 + 2 = 10, 10 – 2 = 8, 10 – 8 = 2
    b   8 + 12 = 20, 20 – 8 = 12, 20 – 12 = 8
    c   800 + 200 = 1000, 1000 – 200 = 800,
        1000 – 800 = 200
    d   9 + 1 = 10, 10 – 1 = 9, 10 – 9 = 1
    e   9 + 11 = 20, 20 – 9 = 11, 20 – 11 = 9
    f   90 + 10 = 100, 100 – 90 = 10, 100 – 10 = 90
2   a   8   b   80   c   6
    d   40   e   1   f   100

Challenge △
3   a   50   b   40   c   10   d   100
    e   13   f   14   g   20   h   50
    i   1   j   50
4   Learner's own work. Ensure correct solutions are
    provided.
5   a   Sunita number is 179.
    b   Bobbie's number is 329.

Challenge ❸
6   26. Learner's idea for story
7   85. Learner's own explanation

Unit 6

Lesson 1: Add 3-digit numbers and
          tens (A)

Challenge ❶
1   a   146 + 53   b   234 + 33
    c   224 + 51   d   143 + 32
2   Accept estimates, not actual totals, around these
    amounts:
    a   170   b   290   c   480
    d   500   e   570   f   670

Challenge △
3   248
4   a   294   b   377   c   487
    d   587   e   766   f   686
    g   Learner's own explanation of their strategy

5   a   237   b   364
        +   42   +   35
        ―――――――   ―――――――
                9   9
               70   90
        +   200   +   300
        ―――――――   ―――――――
              279   399

6   a   897   b   367
7   369; mental calculation strategy might include
    345 + 20 + 4 = 369. Ensure learners show the
    other two methods.

Challenge ❸
8   a   751 + 43 or 753 + 41
    b   Learner's own method, e.g. 751 + 40 + 3 = 794
        or 753 + 40 + 1 = 794

Lesson 2: Add 3-digit numbers and
          tens (B)

Challenge ❶
1   a   70   b   60   c   80
    d   90   e   40   f   30
2   a   72 + 10 – 1 = 81   b   26 + 20 – 1 = 45
    c   67 + 30 – 1 = 96   d   124 + 40 – 1 = 163
    e   236 + 50 – 1 = 285   f   348 + 30 – 1 = 377

Challenge △
3   254
4   a   265   b   388   c   453
    d   587   e   683   f   789
    g   Learner's own explanation
    h   Learner's own explanation of their strategy
5   a   376   b   294   c   691
    d   519   e   593   f   893
    Ensure that learners set out the calculations
    correctly.
6   576. Ensure learners use a mental calculation
    strategy of their choice and show the expanded
    and formal written methods.

Challenge ❸
7   a   429 + 58   b   445 + 46   c   493, 456 + 37

Lesson 3: Add two 3-digit numbers (A)

Challenge ❶
1   a   50 + 3 = 53 or 43 + 10 = 53
    b   60 + 5 = 65 or 55 + 10 = 65
    c   50 + 32 = 82 or 42 + 40 = 82
    d   60 + 21 = 81 or 51 + 30 = 81
2   a   241   b   355   c   264   d   372

Challenge △
3   Accept estimates, not actual totals, around these
    amounts:
    a   446   b   340   c   500   d   480

4 Ensure calculations are set out correctly.
  a 574   **b** 784   **c** 781
  d 574   **e** 492   **f** 687
5 **a** 237 + 125 = 362   **b** 265 + 118 = 383
  c 328 + 235 = 563   **d** 278 + 117 = 395

Challenge 3
6 Steph: 234 + 128 = \$362
Robbie: 246 + 128 = \$374
Abraham: 355 + 128 = \$483
Casey: 324 + 128 = \$452

Lesson 4: Add two 3-digit numbers (B)

Challenge 1
1 Accept estimates, not actual totals, around these amounts:
  a 620   **b** 770   **c** 660   **d** 920
2 Ensure the expanded method is correctly recorded.
  a 338   **b** 325   **c** 336   **d** 445

Challenge 2
3 **a** 655   **b** 437   **c** 724
  d 619   **e** 519   **f** 618
4 Learner's own work. Ensure the problem fits with the calculation and is answered in a complete sentence. (628)
5 \$9.59

Challenge 3
6 **a** 436   **b** 688

Unit 7

Lesson 1: Subtract 3-digit numbers and tens (A)

Challenge 1
1 **a** 112   **b** 132   **c** 121   **d** 114
2 Accept estimates, not actual differences, around these amounts:
  a 150   **b** 220   **c** 350
  d 250   **e** 130   **f** 200

Challenge 2
3 343
4 **a** 124   **b** 124   **c** 225
  d 222   **e** 311   **f** 311
  g Learner's own explanation of their strategy
5 **a** 314   **b** 136   **c** 224   **d** 441
6 **a** 812   **b** 322   **c** 232   **d** 557

Challenge 3
7 Estimate: 530, answer: 526

Lesson 2: Subtract 3-digit numbers and tens (B)

Challenge 1
1 **a** 70   **b** 50   **c** 20
  d 30   **e** 60   **f** 30
2 **a** 32 − 10 + 1 = 23   **b** 25 − 10 + 1 = 16
  c 132 − 10 + 1 = 123   **d** 153 − 10 + 1 = 144

Challenge 2
3 138
4 **a** 146   **b** 248   **c** 115
  d 225   **e** 336   **f** 412
  g Learner's own explanation of their strategy
5 **a** 118   **b** 205
Ensure that learners set out the calculations correctly.
6 **a** 217   **b** 305
7 127. Ensure learners use a mental calculation strategy of their choice and show the expanded and formal written methods.

Challenge 3
8 **a** 296 − 48
Learner's own explanation
  b 384 − 75
Learner's own explanation
  c 126, 182 − 56

Lesson 3: Subtract two 3-digit numbers (A)

Challenge 1
1 **a** 9   **b** 17   **c** 9   **d** 7
2 **a** 18   **b** 125   **c** 215   **d** 204

Challenge 2
3 Accept estimates, not actual differences, around these amounts:
  a 120   **b** 240   **c** 240   **d** 250
4 Ensure calculations are set out correctly.
  a 19   **b** 128   **c** 224
5 **a** 365 − 119 = 246   **b** 461 − 325 = 136
  c 341 − 215 = 126   **d** 582 − 357 = 225

Challenge 3
6 Bertie: 335 − 137 = \$218
Farhar: 486 − 137 = \$349
Adnan: 546 − 137 = \$409
Sabrina: 464 − 137 = \$327

Lesson 4: Subtract two 3-digit numbers (B)

Challenge 1
1 Accept estimates, not actual differences, around these amounts:
  a 40   **b** 170   **c** 300   **d** 180
2 Ensure the expanded method is correctly recorded.
  a 83   **b** 262   **c** 276   **d** 262

Challenge A

3 a 74 b 171 c 172

   d 265 e 152 f 258

4 $65.

5 Learner's own work (392).

Challenge B

6 Samir spent $3.94. He has $4.81 left.

7 183

Unit 8

Lesson 1: Multiplication and division

Challenge 1

1 a $10 \times 2 = 20$ b $5 \times 4 = 20$

   c $8 \times 2 = 16$

2 a $20 \div 10 = 2$ b $20 \div 5 = 4$

   c $16 \div 8 = 2$

Challenge A

3 a 3 rows of 5 circles b 5 rows of 2 circles

   c 6 rows of 2 circles d 4 rows of 5 circles

4 a $40 \div 5 = 8, 40 \div 8 = 5$

   b $10 \div 2 = 5, 10 \div 5 = 2$

   c $30 \div 5 = 6, 30 \div 6 = 5$

   d $50 \div 10 = 5, 50 \div 5 = 10$

5 a 8, 14, 12, 20, 18, 6, 16

   b 10, 5, 9, 6, 8, 7, 3

   c 4, 10, 9, 5, 2, 3, 8

Challenge B

6 a $5 \times 4 = 20, 4 \times 5 = 20$

   b $5 \times 9 = 45, 9 \times 5 = 45$

   c $10 \times 3 = 30, 3 \times 10 = 30$

   d $5 \times 6 = 30, 6 \times 5 = 30$

Lesson 2: Checking multiplication and division

Challenge 1

1 a $5 \div 5 = 1$ b $10 \div 5 = 2$

   c $15 \div 5 = 3$ d $20 \div 5 = 4$

   e $12 \div 2 = 6$ f $16 \div 2 = 8$

   g $30 \div 5 = 6$ h $60 \div 10 = 6$

2 a ⚫⚫ ⚫⚫ ⚫⚫ b ⚫⚫ ⚫⚫ ⚫⚫ ⚫⚫ c ⚫⚫ ⚫⚫ ⚫⚫ ⚫⚫ ⚫⚫ d ⚫⚫ ⚫⚫ ⚫⚫ ⚫⚫ ⚫⚫ ⚫⚫

Challenge A

3 a $2 \times 10 = 20$ b $3 \times 10 = 30$

   c $4 \times 10 = 40$ d $5 \times 10 = 50$

   e $6 \times 10 = 60$ f $7 \times 10 = 70$

   g $8 \times 10 = 80$ h $9 \times 10 = 90$

4 a Written explanation to say that he could divide 20 by 4.

   b Written explanation to say that she could multiply 5 and 8.

5 Ensure 4 rows of 10. $10 \times 4 = 40, 4 \times 10 = 40, 40 \div 4 = 10, 40 \div 10 = 4$

6 a $70 \div 10 = 7$ b $2 \times 8 = 16$

   c $5 \times 9 = 45$ d $45 \div 5 = 9$

   e $30 \div 10 = 3$ f $5 \times 8 = 40$

   g $15 \div 5 = 3$ h $10 \times 5 = 50$

Challenge B

Reza is incorrect. Learner's explanation, Reza should divide 30 by 5.

Lesson 3: Commutativity

Challenge 1

1 a $2 \times 3 = 6, 3 \times 2 = 6$

   b $2 \times 4 = 8, 4 \times 2 = 8$

   c $2 \times 5 = 10, 5 \times 2 = 10$

   d $2 \times 6 = 12, 6 \times 2 = 12$

   e $2 \times 7 = 14, 7 \times 2 = 14$

2 a $2 \times 5 = 10$ b $2 \times 6 = 12$

   c $2 \times 7 = 14$ d $2 \times 8 = 16$

Challenge A

3 Learner's explanation. Ensure it shows that it doesn't matter which way around you multiply numbers. The answer will always be the same.

4 Learner's explanation. Ensure it indicates that Pierre could work out 9×2.

5 $2 \times 10 = 20, 10 \times 2 = 20, 5 \times 4 = 20, 4 \times 5 = 20$
   Accept $1 \times 20 = 20, 20 \times 1 = 20$

6 $5 \times 8 = 40, 8 \times 5 = 40, 10 \times 4 = 40, 4 \times 10 = 40$
   Accept $2 \times 20 = 40, 20 \times 2 = 40$

Challenge B

7 a $2 \times 2 = 4, 5 \times 5 = 25, 10 \times 10 = 100$

   b Learner's explanation. Ensure answer states that there is only one fact, where other arrays have two.

   c Learner's choice. Ensure that there is only one fact possible.

Lesson 4: Using place value to multiply

Challenge 1

1 a $10 + 2$ b $10 + 8$ c $10 + 4$

   d $10 + 5$ e $10 + 6$ f $10 + 9$

2 Ensure correctly recorded.

   a 26 b 60 c 30 d 70

Challenge A

3 a 38 b 85 c 32 d 70

4 a 72 b 48 c 51 d 76

5 80. Learner's own explanation.

Challenge B

6 Learner's own work

Unit 9

Lesson 1: 5 and 10 times tables

Challenge ❶

1
a	10	b	5	c	3	d	20
e	5	f	50	g	10	h	8
i	90	j	10				

2
a $40 \div 10 = 4$ b $30 \div 5 = 6$
c $20 \div 10 = 2$ d $35 \div 5 = 7$
e $50 \div 10 = 5$ f $45 \div 5 = 9$

Challenge ⚠

3 $5 \times 2 = 10 \times 1$
$5 \times 4 = 10 \times 2$
$5 \times 8 = 10 \times 4$
$5 \times 6 = 10 \times 3$
$5 \times 10 = 10 \times 5$
Learners explanation, for example: The numbers are doubled or halved.

4 a 80 b 90 c 100 d 110

5 Learner's explanation

6 47. Learner's explanation, for example: It is the only one that isn't in the 5 times table.

Challenge ❸

7 240. Learner's explanation, for example: Manni can halve 48 and multiply by 10.

Lesson 2: 2, 4 and 8 times tables

Challenge ❶

1
a	16	b	8	c	10	d	3
e	10	f	6	g	9	h	14
i	2						

2 $2 \times 7, 4 \times 7$
$2 \times 4, 4 \times 4$
$2 \times 6, 4 \times 6$
$2 \times 8, 4 \times 8$
$2 \times 9, 4 \times 9$

Challenge ⚠

3
a	32	b	12	c	10	d	2
e	36	f	7	g	4	h	4
i	24						

4
| a | 32 | b | 48 | c | 72 |
| d | 56 | e | 24 | f | 64 |

5
a	4	b	2	c	10	d	6
e	5	f	7	g	3	h	1
i	9	j	8				

Challenge ❸

6 Learner's explanation. Accept any that give a product of 200.

Lesson 3: Multiples

Challenge ❶

1 2, 4, 6, 8, 10, 12, 14, 16, 18, 20

2 5, 10, 15, 20, 25, 30, 35, 40, 45, 50

3 10, 20, 30, 40, 50, 60, 70, 80, 90, 100

Challenge ⚠

4 a 15 because it is the only number that is not a multiple of 2.
b 18 because it is the only number that is not a multiple of 5.
c 75 because it is the only number that is not a multiple of 10.
Accept any other ideas that are correct, for example 75 because it is the only odd number.

5 Learner's choices, for example: 10, 20, 30, 40. Learner's own explanation.

6 Hamid is incorrect. Learner's own explanation, for example: Nothing that can be multiplied by 10 will give an answer of 2.

7 Kati is correct. Learner's own explanation and examples, for example: 20 is in the 2, 5, and 10 times tables so must be a multiple of all three, $10 \times 2 = 20$, $4 \times 5 = 20$, $2 \times 10 = 20$.

8 Learner's own work. All answers must meet the criteria.

Challenge ❸

9 Learner's own work, for example: A multiple is the product of two numbers, 10 is a multiple of 2 because $5 \times 2 = 10$.

Lesson 4: Counting in steps

Challenge ❶

1
| a | 2 | b | 4 | c | 5 | d | 8 |

2
| a | 4×2 | b | 2×5 | c | 8×1 | d | 3×4 |

Challenge ⚠

3
| a | T | b | F | c | F | d | T |

4 Mollie is correct. Learner's own explanation.

5 14. It is the only number that doesn't appear in a step count of 4.

6 24, 16. Learner's explanation.

Challenge ❸

7 a $5 \times 2 = 10$, there are five groups of 2
b $5 \times 4 = 20$, $3 + 1 = 4$, there are five groups of 4
c $7 \times 5 = 35$, $2 + 3 = 5$, there are seven groups of 5
d $6 \times 8 = 48$, $6 + 2 = 8$, there are six groups of 8
e $4 \times 10 = 40$, $7 + 3 = 10$, there are four groups of 10

Unit 10

Lesson 1: 3 and 6 times tables

Challenge ❶

1 **a** 24 **b** 9 **c** 9 **d** 2
   e 15 **f** 4 **g** 6 **h** 21 **i** 10

2 $3 \times 5, 6 \times 5$
   $3 \times 4, 6 \times 4$
   $3 \times 6, 6 \times 6$
   $3 \times 8, 6 \times 8$
   $3 \times 9, 6 \times 9$

Challenge ⚠

3 **a** 48 **b** 18 **c** 4 **d** 5 **e** 54
   f 10 **g** 2 **h** 6 **i** 36 **j** 7
   k 7 **l** 42

4 **a** = **b** > **c** < **d** > **e** = **f** >

5 No, they are the same. Learner's explanation,
   for example: $3 \times 6 = 18$, twice is the same as
   doubling, so 36, which is the same as 6×6.

6 Learner's reasons, for example: The products are
   the same, the numbers for the two facts are the
   same, the order is different.

7 Possible answers: $6 \times 6, 4 \times 9, 9 \times 4$, accept 12×3

Challenge ❸

8 Always true. Expect learners to give examples to
   prove this.

Lesson 2: 9 times table

Challenge ❶

1 27, 45, 54, 81

2 **a** 9 **b** 36 **c** 63 **d** 72
   e 90 **f** 18

Challenge ⚠

3 $9 \times 2 = 18, 9 \times 4 = 36, 9 \times 6 = 54, 9 \times 7 = 63$,
   $9 \times 5 = 45, 9 \times 10 = 90, 9 \times 3 = 27, 9 \times 8 = 72$

4 $72 \div 9 = 8, 63 \div 9 = 7, 81 \div 9 = 9, 27 \div 9 = 3$,
   $18 \div 9 = 2, 36 \div 9 = 4, 54 \div 9 = 6, 9 \div 9 = 1$

5 Learner's own explanation, for example:
   $3 \times 3 \times 8 = 72, 3 \times 3 \times 3 \times 3 = 81$

6 35, because it is the only number that is not a
   multiple of 9.

Challenge ❸

7 **a** T **b** T **c** F **d** T
   e F **f** T **g** F **h** F
   Learner's own explanations, e.g. c is the same
   multiplication

Lesson 3: Multiplication and division facts (1)

Challenge ❶

1 Learner's own examples

2 **a** $9 \times 7 = 63, 7 \times 9 = 63, 63 \div 9 = 7, 63 \div 7 = 9$
   b $6 \times 8 = 48, 8 \times 6 = 48, 48 \div 6 = 8, 48 \div 8 = 6$
   c $7 \times 3 = 21, 3 \times 7 = 21, 21 \div 3 = 7, 21 \div 7 = 3$
   d $7 \times 8 = 56, 8 \times 7 = 56, 56 \div 8 = 7, 56 \div 7 = 8$

Challenge ⚠

3 **a** 24 **b** 56 **c** 70 **d** 24
   e 32 **f** 50 **g** 27 **h** 35
   i 54

4 **a** 9 **b** 9 **c** 3 **d** 9
   e 5 **f** 4 **g** 8 **h** 6
   i 7

5 Tabula is not correct – there are five: 1×16,
   $16 \times 1, 2 \times 8, 8 \times 2$ and 4×4

6 Learner's examples, for example: $1 \times 24, 24 \times 1$,
   $2 \times 12, 12 \times 2, 8 \times 3, 3 \times 8, 6 \times 4, 4 \times 6$

Challenge ❸

7 Learner's choice, for example
   a $3 \times 4 \times 2 = 24, 3 \times 5 \times 2 = 30$
   b $6 \times 5 \times 2 = 60, 5 \times 5 \times 2 = 50$

Lesson 4: Multiplication and division facts (2)

Challenge ❶

1 Learner's examples, e.g. $62 \times 3 = 6$

2 Learner's examples

Challenge ⚠

3 $36, 4 \times 9 = 36, 9 \times 4 = 36, 36 \div 4 = 9, 36 \div 9 = 4$

4 **a** 25 because it is not in the 3 times table.
   b 12 because it is not in the 9 times table.
   c 32 because it is not in the 6 times table, or
         60 because it is not in the 4 times table.

5 5 and 6, 3 and 10

6 No. Learner's own explanation, for example:
   Yakoob is partly correct – 5 is a quotient of
   numbers that end with 0 or 5.

7 48, 27 or 3, 63

Challenge ❸

8 Georgia is correct. $9 \times 8 = 72$ and half of 72 is 36,
   which is a multiple of 4.

Unit 11

Lesson 1: Multiplying by repeated addition

Challenge ❶

1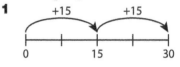

2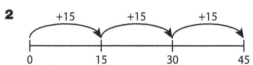

3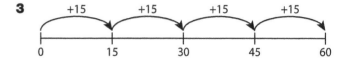

Challenge △
4 a Double
  b Double and double again
5 Humaira is correct. Learner's explanation and example, e.g. 5 is half of 10 so we can multiply the number by 10 and halve it.
6 Learner's ways, for example: Double 32 to give 64 and double again to give 128. 32 + 32 + 32 + 32 = 128
  Learner's own explanation.
7 Arsalan has used doubling but only doubled once. He should have doubled twice. The correct product is 92.

Challenge ❸
8 Learner's explanation, for example: She doubled 40 and 3 but didn't add them together.

Lesson 2: Multiplying with arrays
Challenge ❶
1 a 23   b 2   c 23×2 d 46
2 a 23   b 3   c 23×3 d 69
3 a 23   b 4   c 23×4 d 92

Challenge △
4 The array should show three rows of 3 tens counters and 2 ones counters.
  Product = 96
5 33×3
6 Learner's explanation, for example: Seth halves 32 to give 16 and multiplies by 10 to give 160.
7 $24 \times 5 = 120$, $42 \times 5 = 210$, $45 \times 2 = 90$, $54 \times 2 = 108$, $25 \times 4 = 100$, $52 \times 4 = 208$

Challenge ❸
8 a ✓   b ✗   c ✓   d ✗
  Learner's own explanations. Products should be
  b 256   d 470

Lesson 3: Multiplying by the grid method
Challenge ❶
1 a 32   b 2   c 32×2 d 64
2 a 32×2
  b Learner's explanation, for example: They both show the same calculation.
  c Learner's explanation, for example: The representations are different – one is counters, the other is a grid.

Challenge △
3

×	20	1
5	100	5

Product = 100 + 5 = 105

4 a

×	20	4
2	40	8

Product = 40 + 8 = 48

b

×	20	4
3	60	12

Product = 60 + 12 = 72

c

×	20	4
4	80	16

Product = 80 + 16 = 96

d

×	20	4
5	100	20

Product = 100 + 20 = 120

5 Yes. Learner's explanation, for example: 15 is 1 more than 14, $1 \times 3 = 3$, so 15×3 must be 3 more than 14×3.

6

×	30	6
5	150	30

Half of 36×10: $18 \times 10 = 180$

Challenge ❸
7 80, 25. Product = 425
8 Zoltan is incorrect. Learner's explanation, for example: 35×5 is 35 more than 35×4. The product is 175.

Lesson 4: Multiplying by partitioning
Challenge ❶
1 a 50 + 4   b 80 + 9
  c 30 + 2   d 20 + 8
2 a 20 + 16 = 36   b 40 + 16 = 56
  c 60 + 16 = 76   d 80 + 16 = 96

Challenge △
3 Ensure learners have used the partitioning model learned in the lesson.
  a 105   b 328
4 Jessica can do this. Ensure learner's diagram shows $8 \times 3 = 24$, $8 \times 3 = 24$, $8 \times 3 = 24$ and $4 \times 3 = 12$. 24 + 24 + 24 + 12 = 84
5 Abraham is correct. Learner's explanation. Ensure check is the partitioning strategy learned in this lesson.

Challenge ❸
6 a Doubling
  b Doubling and doubling again
  c Multiplying by 10 and halving or vice versa

Unit 12

Lesson 1: Dividing using known facts
Challenge ❶
1 a 8   b 8   c 8   d 8   e 9
  f 9   g 6   h 9   i 6

2 **a** $15 \div 3 = 5$ **b** $30 \div 5 = 6$
    c $20 \div 4 = 5$ **d** $18 \div 2 = 9$

Challenge ⚠

3 **a** halve   **b** double and divide by 10.
4 Learners need to show evidence that they have halved and halved again.
    a 6   **b** 11   **c** 12   **d** 14
    e 22   **f** 18
5 Attiqa is incorrect. Learner's explanation, for example: Attiqa should have doubled, not halved, and divided by 10 to give a quotient of 16.
6 Learner's explanation, for example: Bernie has used halving but only halved once. He should have halved twice. The correct quotient is 21.

Challenge 🔳

7 Learner's explanation, for example: Tooba halved 40 and 8 but didn't add them together.

Lesson 2: Dividing by partitioning

Challenge ❶

1 Learner's examples to include 40 and 6, 30 and 16, 20 and 26, 10 and 36
2 **a** 35   **b** 45   **c** 55
    d 65   **e** 75   **f** 95

Challenge ⚠

3 40 and 16
    30 and 26
    20 and 36
    10 and 46

4 **a** 30 and 6, $10 + 2 = 12$
    b 30 and 12, $10 + 4 = 14$
    c 30 and 21, $10 + 7 = 17$
    d 30 and 27, $10 + 9 = 19$
5 Learner's explanation, for example: Partition 92 into 80 and 12. I know that $2 \times 4 = 8$ so $20 \times 4 = 80$, that means 80 divided by 4 is 20, 12 divided by 4 is 3, so the quotient is $20 + 3$, which is 23.

Challenge 🔳

6 Learner's choices, for example: 52, 56, 60. Any number that is a number multiplied by 4.

Lesson 3: Dividing with arrays

Challenge ❶

1 **a** 23   **b** 32   **c** The digits
    d The order of the digits
2 2 groups of tens and 1 group of ones; 21

Challenge ⚠

3 Step 1: partition 48 into 40 and 8
    Step 2: make a group of 3 tens
    Step 3: exchange the ten for 10 ones
    Step 4: make 6 groups of 3 ones

4 Daisy is incorrect. Learner's explanation, for example: Daisy has treated 48 as 4 and 8. 4 tens are 40 and, when divided by 2, the quotient is 20 not 2.
5 Always true. Learner's explanation, for example: Any number that ends with 5 is a multiple of 5 and can be divided by 5. Learner's own example.
6 24. Learner's explanation, for example: Partitioning 96 into 80 and 16 and halving each part twice.

Challenge 🔳

7 Learner's ideas and explanations.

Lesson 4: Division with remainders

Challenge ❶

1 **a** Ensure 1 group of 3 tens and 2 groups of 3 ones, 12
    b 12 above 36
2 **a** Ensure 3 groups of 2 tens and 3 groups of 2 ones, 33
    b 33 above 66

Challenge ⚠

3 **a** 21 remainder 2   **b** 22 remainder 1
    c 22 remainder 2   **d** 24 remainder 1
    e 24 remainder 2
4 **a** 14 remainder 1   **b** 13 remainder 3
    c 25 remainder 1   **d** 37 remainder 1
    e 24 remainder 1   **f** 16 remainder 3
5 Learner's explanation, for example: Adisa can make 2 groups of 4 tens, he has 1 left, which he should have exchanged for 10 ones, giving 14 ones. 14 divided by 4 is 3 remainder 2. Quotient should be 23 remainder 2.

Challenge 🔳

6 If dividing by 2: 29 groups, 1 remaining
    If dividing by 3: 19 groups, 2 remaining
    If dividing by 4: 14 groups, 3 remaining
    If dividing by 5: 11 groups, 4 remaining

Unit 13

Lesson 1: Writing money

Challenge ❶

1 100
2 60c or $0.60
3 Ensure learner's choice totals $1.12. There are many possibilities including: 10c, 1c, 1c and 5c, 5c, 1c, 1c

Challenge ⚠

4 $12.35
5 $6.32
6 **a** $3.50   **b** $9.25   **c** $10.06   **d** $21.11

Challenge 3

7 **a–c** All have many possibilities. Ensure each totals the correct amount.

8 **a** $24.25 and $13.05 **b** $11.20

Lesson 2: Finding totals

Challenge 1

1 **a** $1.70

   b Total drawn: one dollar note, one 50c and two 10c coins.

2 **a** $2.40

   b Total drawn: two dollar notes and four 10c coins.

Challenge 2

3 $25.50

4 **a** $2.65

   b Total drawn: two dollar notes, 50c, 10c and 5c coins.

Challenge 3

5 **a** $3.20

   c Total drawn: three dollar notes, two 10c coins

Lesson 3: Finding change

Challenge 1

1 **a** $1.20

2 **a** 20c

Challenge 2

3 **a** $7.00 **b** $2.50 **c** $1.50 **d** $33.00

Challenge 3

4 $0.85 or 85c

5 No. Learners explanation, she needs 40c

Lesson 4: Solving problems with money

Challenge 1

1 **a** $9.00 **b** $2.00

Challenge 2

2 **a** $21.00 **b** $33.00 **c** $51.00

Challenge 3

3 **a** $37.50 **b** $25.00 **c** $40.00

Unit 14

Lesson 1: Understanding place value (A)

Challenge 1

1 **a** 132 **b** 256

2 Base 10 equipment drawn correctly.

Challenge 2

3 **a** 435 **b** 354 **c** 271

   d 148 **e** 354 **f** 148

4 **a** 6, 4, 7 **b** 3, 5, 7 **c** 9, 4, 9 **d** 7, 3, 3

5 No. Learner's explanation, for example: Amina needs to talk about how the digits are placed in different positions, they are multiplied to give their values and then the values are added together.

Challenge 3

6 **a** 124 **b** 642

   c 612, 614, 216, 214, 412, 416

   d 421 or 426

Lesson 2: Understanding place value (B)

Challenge 1

1 **a** 152 **b** 317 **c** 640

   d 204 **e** 845 **f** 422

2 **a** 300 40 8 **b** 200 70 3

Challenge 2

3 **a** 303 (The odd number)

   b 212 (The number with 2 ones)

   c 324 (The greatest number)

   d 144 (The number with 4 tens)

   e 124 (The smallest number)

4 **a** 500 + 30 + 2 **b** 200 + 40 + 5

   c 700 + 0 + 1 **d** 300 + 80 + 6

   e 300 + 50 + 9 **f** 400 + 60 + 0

   g 701

5 No. 275, 527, 752

Challenge 3

6 **a** 900 **b** 108

   c 405, 414, 423, 432, 441, 450

   d 900, 711, 522, 333, 144

Lesson 3: Regrouping

Challenge 1

1 **a** 389 **b** 354 **c** 821

   d 309 **e** 426 **f** 613

2 **a** 14 tens, 5 ones **b** 24 tens, 5 ones

   c 34 tens, 5 ones **d** 44 tens 5 ones

Challenge 2

3 **a** 425 **b** 638 **c** 810 **d** 107

   e 280 **f** 594 **g** 359

4 453

5 Yes. Learner's explanation, for example: He has kept the hundreds the same and regrouped 4 tens for 40 ones, giving 46 ones.

6 **a** 1 hundred, 2 tens, 5 ones

   b 12 tens, 5 ones

   c 11 tens, 15 ones

7 Learner's own examples, with Base 10 equipment drawn correctly

Challenge 3

8 Learner's own examples, giving correct ways of regrouping 246

Workbook Answers

Lesson 4: Comparing numbers

Challenge 1
1. a less   b greater
2. a ✓   b ✗   c ✓
   d ✗   e ✓   f ✓

Challenge 2
3. a <   b >   c >   d <
   e <   f >   g <   h >
   i >   j >
4. a 312 > 302   b 708 < 870   c 544 > 459
5. Learner's own examples. Ensure they are accurate.
6. Learner's own examples. Ensure they are within the parameters.

Challenge 3
7. a 652   b 129   c 843   d 520
   e 259, 295, 529, 592
   f 381, 813, 831

Unit 15

Lesson 1: Ordering numbers

Challenge 1
1. a 49 and 18   b 35 and 15
   c 99 and 93   d 77 and 50
   e 60 and 49
2. a 146, 346, 746   b 821, 621, 321

Challenge 2
3. a 21, 39, 47, 78   b 13, 30, 52, 94
   c 23, 26, 59, 75   d 348, 384, 438, 834
   e 196, 619, 691, 916
   f 257, 275, 527, 725
4. a 853. 832, 835, 849, 852, 853
   b 437. 409, 410, 421, 437, 439
   c 571. 517, 538, 565, 569, 571
5. Adil is correct. Learner's explanation, for example: To find the greatest number, you need to find the hundreds first as this is going to give the highest value, then the tens and then the ones.

Challenge 3
6. Various possibilities, for example:
   a 512, 554, 573, 586, 590
   b 231, 264, 275, 282, 290
   c 719, 727, 735, 740, 768
7. No. Learner's explanation of how to order, particularly focusing on the digits in the hundreds.

Lesson 2: Multiplying by 10

Challenge 1
1. $8 \times 10 = 80$, $12 \times 10 = 120$, $4 \times 10 = 40$, $15 \times 10 = 150$, $10 \times 10 = 100$, $16 \times 10 = 160$

2. a 84, 840   b 96, 960
   c 50, 500   d 13, 130

Challenge 2
3. a 640   b 910   c 170
   d 280   e 350   f 140
4. 470, 880, 75, 42, 190, 20, 320
5. Always true. Learner's explanation
6. Learner's explanation of how to multiply by 10

Challenge 3
7. 7, 3, 2, 1. Other possible calculations: 37×10, 12×10, 23×10, 32×10, 17×10, 71×10 27×10, 72×10, 13×10, 31×10

Lesson 3: Rounding to the nearest 10

Challenge 1
1. a 68, 71, 69, 73   b 31, 29, 25
   c 87, 85, 93, 91
2. a 230   b 220   c 260
   d 280   e 290   f 290

Challenge 2
3. a 380   b 590   c 390   d 880
   e 850   f 740
4. No. Learner's explanation, for example: The number has 5 in the ones position and so should be rounded up to 350.
5. No, it should be 230 + 130 = 360
6. Learner's explanation, for example: Numbers with 1, 2, 3 or 4 in the ones position all round down and numbers with 5, 6, 7, 8 or 9 in the ones position all round up.

Challenge 3
7. Duke's estimate. Learner's explanation, for example: 456 is rounded to 460 and 315 is rounded to 320, 460 + 320 = 780

Lesson 4: Rounding numbers to the nearest 100

Challenge 1
1. a 500 b 200 c 300 d 800 e 700
   f 700 g 300 h 900 i 900

Challenge 2
2. a Learner's own examples of numbers that round down.
   b Learner's own examples of numbers that round up.
3. a 900   b 500   c 500   d 200
   e 700   f 300
4. a 300 + 200 = 500   b 600 + 400 = 1000
   c 200 + 200 = 400   d 500 + 300 = 800
   e 700 + 400 = 1100   f 400 + 500 = 900
5. a 340, 300   b 760, 800
   c 900, 900   d 680, 700
   e 150, 100   f 320, 300

6 The total is 581, so Flo's estimate is better. Learner's explanation, for example: Rounding to 10 will give a closer estimate than rounding to 100.

Challenge ❸

7 Keisha's number could be, for example: 491, 482, 473, 464, 455

Learner's own puzzle

Unit 16

Lesson 1: Equal parts: quantities

Challenge ❶

1 a 5   **b** 7   **c** 10   **d** 6

2 16→8, 22→11, 6→3, 18→9

Challenge ⚠

3 Each part labelled

  a $\frac{1}{2}$ **b** $\frac{1}{5}$ **c** $\frac{1}{4}$ **d** $\frac{1}{10}$

4 Each part labelled

  a 8   **b** 4   **c** 3   **d** 3

5 a $\frac{1}{4}$ **b** $\frac{1}{2}$ or $\frac{3}{6}$ **c** $\frac{1}{3}$ **d** $\frac{2}{5}$

  e $\frac{2}{3}$ **f** $\frac{1}{5}$

6 $\frac{7}{10}$ is the odd one out because it is the only one that is not a unit fraction.

Challenge ❸

7 25. Learner's explanation, for example: $\frac{1}{5}$ is 5, so $\frac{5}{5}$ will be 25.

Lesson 2: Equal parts: shapes

Challenge ❶

1 Ensure correct parts are coloured to match the given fractions.

2 The two fractions that show half are $\frac{2}{4}$ and $\frac{5}{10}$.

3 a $\frac{2}{5}$ **b** $\frac{3}{5}$ **c** $\frac{3}{4}$ **d** $\frac{7}{10}$

Challenge ⚠

4 a $\frac{3}{4}$ **b** $\frac{2}{4}$ or $\frac{1}{2}$ **c** $\frac{4}{10}$ or $\frac{2}{5}$

  d $\frac{2}{4}$ or $\frac{1}{2}$ **e** $\frac{1}{4}$ **f** one whole

  g $\frac{3}{4}$ **h** $\frac{5}{10}$ or $\frac{1}{2}$

5 Any two of **b**, **d** and **h**

6 Learner's own examples with fractions correctly given

Challenge ❸

7 Yes. Learner's explanation, for example: Both halves of the square are made from two equal parts. Therefore, all parts are equal.

Lesson 3: Same fraction, different whole

Challenge ❶

1 Ensure each circle is shaded correctly.

Challenge ⚠

2 All shapes show $\frac{1}{5}$. Learner's explanation, for example: The fraction is part of the whole shape so it doesn't matter what the shape is or where it is positioned.

3 Learner's example. Ensure each rectangle shows $\frac{1}{4}$.

4 Kaede is correct. Learner's explanation, for example: In his two shapes the square is larger than the rectangle. $\frac{1}{4}$ of the square is greater than $\frac{1}{2}$ of the rectangle.

5 Learner's own explanation that they are all quarters.

Challenge ❸

6 Learner's own explanation that more than half is shaded.

Lesson 4: Fractions of quantities and sets

Challenge ❶

1 a 10   **b** 6   **c** 9   **d** 11

Challenge ⚠

2 melon: $\frac{3}{4}$, pineapple: $\frac{2}{5}$, orange: $\frac{5}{6}$, pear: 3, 2, banana: 5, 3, watermelon: 10, 4

3 $100

$100			
Charity	$20	$20	$20	$20

4 40

40		
Boys	10	10	10

Challenge ❸

5 Learner's examples, following the example given in the Workbook

Unit 17

Lesson 1: Same value, different appearance

Challenge ❶

1 Any 2 parts of each shape should be coloured.

2 Any 4 parts of each shape should be coloured.

Challenge △

3 **a** 2 parts of the shape shaded, $\frac{2}{4}$

b 4 parts of the shape shaded, $\frac{4}{8}$

c 6 parts of the shape shaded, $\frac{6}{12}$

4 **a** 1 part of the shape shaded, $\frac{1}{4}$

b 2 parts of the shape shaded, $\frac{2}{8}$

c 3 parts of the shape shaded, $\frac{3}{12}$

5 **a** Ben should eat 5,

b Caila should eat 2

c Deepak should eat 4

6 b, f circled

Challenge 3

7 $\frac{3}{6}, \frac{2}{4}, \frac{50}{100}, \frac{4}{8}, \frac{6}{12}, \frac{10}{20}$. Learner's explanation, for example: The numerator is half the denominator.

Lesson 2: Equivalent fractions

Challenge ❶

1 $\frac{1}{10} = \frac{2}{10}, \frac{1}{4} = \frac{2}{8}, \frac{1}{5} = \frac{2}{10}, \frac{1}{2} = \frac{2}{4}$. Learner's own explanation.

Challenge △

2 b, c, d circled.

3 No. Learner's explanation, for example: $\frac{1}{4}$ is less than $\frac{1}{2}$.

4 Yes. Learner's explanation, for example: $\frac{6}{8}$ has been shaded, $\frac{6}{8}$ is equivalent to $\frac{3}{4}$.

5 **a** $\frac{2}{8}, \frac{3}{12}$ **b** $\frac{2}{10}, \frac{3}{15}$

Challenge 3

6 $\frac{1}{10}, \frac{3}{30}, \frac{4}{40}$. Learner's own explanation.

7 Learner's own explanation that it is sometimes true.

Lesson 3: Comparing fractions

Challenge ❶

1 **a** $\frac{1}{2}$ **b** $\frac{1}{8}$ **c** $\frac{1}{4}$

Challenge △

2 **a** > **b** < **c** < **d** > **e** =
f > **g** < **h** = **i** > **j** <

3 No. Learner's explanation, for example: $\frac{1}{2}$ is a whole divided into two equal parts, $\frac{1}{5}$ is divided into five, so $\frac{1}{2}$ is bigger than $\frac{1}{5}$.

4 No. Learner's explanation, for example: $\frac{1}{2}$ is greater than $\frac{1}{4}$, $\frac{2}{4}$ is the same as $\frac{1}{2}$, $\frac{3}{4}$ more than $\frac{2}{4}$.

5 Learner's own examples of fractions between $\frac{1}{2}$ and $\frac{1}{10}$

Challenge 3

6 One tenth. 2 bars to show eighths and sixths. Sixth. Learner's explanation.

Lesson 4: Ordering fractions

Challenge ❶

1 **a** $\frac{1}{10}, \frac{1}{4}, \frac{1}{2}$ **b** $\frac{1}{5}, \frac{3}{5}, \frac{4}{5}$

2 $\frac{1}{4}, \frac{2}{4}, \frac{3}{4}$

Challenge △

3 $\frac{1}{10}, \frac{2}{10}, \frac{3}{10}, \frac{5}{10}, \frac{6}{10}, \frac{7}{10}$

4 **a** $\frac{7}{10}$ **b** $\frac{5}{10}$ or $\frac{1}{2}$

c, d Learner's own fractions, for example: $\frac{3}{10}, \frac{1}{10}$

5 Sometimes, learner's explanation, for example: If the denominators are the same, we look at the numerators.

6 Sometimes, learner's explanation, for example: If the denominators are different, we look at the denominator to order them.

7 She has ordered them from smallest to greatest not greatest to smallest.

Challenge 3

8 Learner's own examples correctly placed

Unit 18

Lesson 1: Fractions and division (A)

Challenge ❶

1 **a** $\frac{1}{2}$ **b** 1 **c** $1\frac{1}{2}$ **d** 2

Challenge △

2 $\frac{1}{2}$. Learner's explanation

3 $\frac{3}{4}$. Learner's explanation

4 $\frac{2}{3}$. Learner's explanation

5 $\frac{3}{4}$. Learner's explanation

Challenge 3

6 Yes. Learner's explanation

Lesson 2: Fractions and division (B)

Challenge ❶

1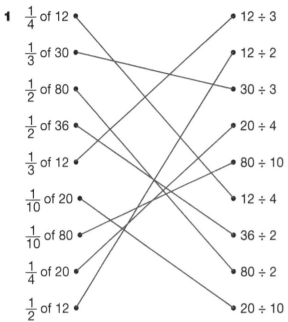

$\frac{1}{4}$ of 12
$\frac{1}{3}$ of 30
$\frac{1}{2}$ of 80
$\frac{1}{2}$ of 36
$\frac{1}{3}$ of 12
$\frac{1}{10}$ of 20
$\frac{1}{10}$ of 80
$\frac{1}{4}$ of 20
$\frac{1}{2}$ of 12

12 ÷ 3
12 ÷ 2
30 ÷ 3
20 ÷ 4
80 ÷ 10
12 ÷ 4
36 ÷ 2
80 ÷ 2
20 ÷ 10

Challenge ⚠

2

Number of rice piles	Fraction needed	Division	Answer
20	$\frac{1}{2}$	20 ÷ 2	10
16	$\frac{1}{2}$	16 ÷ 2	8
24	$\frac{1}{4}$	24 ÷ 4	6
12	$\frac{1}{3}$	12 ÷ 3	4
30	$\frac{1}{10}$	30 ÷ 10	3
18	$\frac{1}{3}$	18 ÷ 3	6
12	$\frac{1}{4}$	12 ÷ 4	3
26	$\frac{1}{2}$	26 ÷ 2	13

3 a $\frac{1}{4}$ of 20 b $\frac{1}{2}$ of 14 c $\frac{1}{5}$ of 15

 d $\frac{1}{4}$ of 24 e $\frac{1}{3}$ of 30 f $\frac{1}{2}$ of 32

 g $\frac{1}{5}$ of 25 h $\frac{1}{3}$ of 18

4 No. Learner's explanation, for example: Shay should divide by the denominator and multiply by the numerator.

Challenge ❸

5 $\frac{1}{10}$ of 40 = 40 ÷ 10 = 4, $\frac{1}{10}$ of 20 = 20 ÷ 10 = 2, $\frac{1}{2}$ of 40 = 40 ÷ 2 = 20, $\frac{1}{2}$ of 20 = 20 ÷ 2 = 10, $\frac{1}{2}$ of 24 = 24 ÷ 2 = 12, $\frac{1}{2}$ of 12 = 12 ÷ 2 = 6, $\frac{1}{3}$ of 24 = 24 ÷ 3 = 8, $\frac{1}{3}$ of 12 = 12 ÷ 3 = 4, $\frac{1}{4}$ of 40 = 40 ÷ 4 = 10, $\frac{1}{4}$ of 20 = 20 ÷ 4 = 5, $\frac{1}{4}$ of 24 = 24 ÷ 4 = 6, $\frac{1}{4}$ of 12 = 12 ÷ 4 = 3.

Lesson 3: Adding and subtracting fractions (A)

Challenge ❶

1 a $\frac{2}{3}$ b $\frac{3}{5}$ c $\frac{4}{4}$ or 1

 d $\frac{3}{8}$ e $\frac{4}{5}$ f $\frac{5}{8}$

2 a $\frac{2}{4}$ or $\frac{1}{2}$ b $\frac{1}{4}$ c $\frac{6}{8}$ or $\frac{3}{4}$

 d $\frac{1}{3}$ e $\frac{4}{8}$, $\frac{2}{4}$ or $\frac{1}{2}$ f $\frac{1}{5}$

Challenge ⚠

3 No. Learner's explanation, for example: She doesn't need to add the denominators. The answer should be $\frac{2}{4}$.

4 Yes. Learner's explanation, for example: The denominators show the number of parts, so will be the same. He just needs to subtract the numerator.

5 $\frac{5}{8}$. $\frac{3}{8}$. Learner's explanation, for example: We add the two fractions to see what has been eaten and then subtract that from the whole to find what is left.

6 $\frac{2}{5}$. Learner's explanation, for example: $\frac{5}{5} - \frac{3}{5} = \frac{2}{5}$

Challenge ❸

7 Zac eats $\frac{1}{4}$. Learner's diagram, ensure it shows one half and two quarters. Learner's own explanation.

Lesson 4: Adding and subtracting fractions (B)

Challenge ❶

1 a $\frac{2}{6}$ b $\frac{3}{7}$ c $\frac{4}{9}$

 d $\frac{3}{10}$ e $\frac{5}{7}$ f $\frac{7}{9}$

2 a $\frac{2}{6}$ b $\frac{1}{7}$ c $\frac{5}{9}$

 d $\frac{3}{10}$ e $\frac{2}{7}$ f $\frac{1}{10}$

Challenge A

3   Yes. Learner's explanation, for example: The whole is 6 parts. If she eats 2, there will be 4 left because $2 + 4 = 6$ so $\frac{2}{6}$ and $\frac{4}{6} = \frac{6}{6}$.

4   Yes. Learner's explanation, for example: Tenths are when 1 whole is divided into 10 parts. 5 of those are equivalent to one half.

5   $\frac{8}{10}, \frac{2}{10}$. Learner's explanation, for example: Add the numerators to make $\frac{8}{10}$. The difference between $\frac{8}{10}$ and $\frac{10}{10}$ is $\frac{2}{10}$.

Challenge 3

6   $\frac{3}{10}, \frac{7}{10}$. Learner's explanation, for example: $\frac{1}{5}$ is equivalent to $\frac{2}{10}$; $\frac{2}{10}$ and $\frac{1}{10}$ is $\frac{3}{10}$; $\frac{3}{10}$ and $\frac{7}{10}$ is the whole, so $\frac{7}{10}$ is left.

7   $\frac{2}{10}$. Learner's diagram and explanation. Ensure the two match.

Unit 19

Lesson 1: Units of time

Challenge 1

1   a   digital clock, stopwatch, analogue clock
    b   Learner's ideas, for example: digital clock – car dashboard; stopwatch – sports event; analogue clock – classroom wall

2   Learner's ideas, for example: brush teeth, drink a carton of juice, read a page of a book

Challenge A

3   No. Learner's ideas and explanation, for example: If there is a lot of text on each page he probably wouldn't be able to read fast enough.

4   No. Learner's ideas and explanation, for example: Most people eat breakfast in a few minutes.

5   Yes. Learner's explanation, for example: There are 7 days in a week.

6   Sunlight in a day – hours,
    Clap your hands – seconds,
    Grow a tree – years,
    Listen to a song – minutes,
    Spring – months

Challenge 3

7   Thursday 2nd to Tuesday 7th – days,
    1st March, 2021 to 1st July 2021 – months,
    7th February to 21st February – weeks,
    2019 to 2022 – years

8   a   2   b   14   c   28   d   4

Lesson 2: Telling the time (A)

Challenge 1

1   Ensure clocks show correct times.

2   a   10:00   b   6:00   c   8:00

Challenge A (TWM.03, TWM.04)

3   a   1:35   b   7:20   c   2:55   d   9:10

4   Ensure clocks show correct times.

5   No. Learner's explanation, for example: 5 is the hour number and 10 is the minutes.

Challenge 3

6   a   35 minutes past 6, 6:35 and 25 minutes to 7
    b   55 minutes past 7, 7:55, 5 minutes to 8
    c   45 minutes past 1, 1:45, 15 minutes to 2

Lesson 3: Telling the time (B)

Challenge 1

1   Ensure clocks show correct times.

2   a   3:26   b   7:44   c   6:53   d   11:08

Challenge A

3   Ensure clocks show correct times.

4   Sasha. Learner's explanation, for example: The hour hand has passed the 2 so it must be 10 to 3.

5   a   1st clock – 7:12, 2nd clock – 2:36, 3rd clock – 3:46, 4th clock – 9:16.
    b   No. Learner's explanation, for example: The minutes are all different, so we can see which ones match.

Challenge 3

6   The time could be 10 minutes to 7, 11 minutes to 7. Learner's own explanation that it could be any time around 6:45.

Lesson 4: Timetables

Challenge 1

1   a   11:00   b   Science
    c   10:30   d   1 hour
    e   10:00   f   3:00

Challenge A

2   Coach A: 8.00; Coach B: 8.45; Coach C: 9.00; Coach D: 9.45.

3   No. Learner's explanation, for example: 2:30 is only 30 minutes

4   Longest time: George
    Shortest time: Ruby

Challenge 3

5   a   Train 1: 1 hour. Train 2: 2 hours. Train 3: 2 hours. Train 4: 2 hours
    b   Trains 2, 3 and 4   c   Train 1

Unit 20

Lesson 1: 2D shapes

Challenge ❶

1 triangle, regular hexagon, regular pentagon, regular octagon

rectangle, irregular pentagon, irregular hexagon, semi-circle

2 octagon, triangle, hexagon, semi-circle, rectangle

Challenge ⚠

3 Answers will vary.

4 Learner's choice. Any one for each question part from:

a square: 4 sides, 4 vertices; pentagon: 5 sides, 5 vertices

b triangle: 3 sides, 3 vertices; heptagon: 7 sides 7 vertices; octagon: 8 sides 8 vertices; quadrilateral: 4 sides 4 vertices

5 Learner's choice, for example: square because the others are triangles, regular triangle because the others have right angles

Challenge ❸

6 a A line drawn to meet the criteria.

b Right angles crossed.

c Because it has 6 sides.

7 Learner's description, for example: a vertex is a corner, where two sides meet.

Lesson 2: Sorting 2D shapes

Challenge ❶

1 Check learners' shapes match the names. Shapes can be regular or irregular.

2 Any headings that are correct, for example: polygons and non-polygons/not polygons, all straight sides and some curved sides

Challenge ⚠

3 Ensure semi-circle drawn is correct. Learner's description, for example: A semi-circle is a non-polygon. It has one curved and one straight side.

4 a Octagon, triangle, pentagon, hexagon

b The preceding shapes with unequal sides and vertices

5 Ensure learners have drawn any four regular shapes and any four irregular shapes.

Challenge ❸

6 Ensure learners have drawn a hexagon with two right angles.

7 Learner's work. Ensure the table headings match with the drawings.

Lesson 3: Symmetry

Challenge ❶

1 Most shapes have several ways to show symmetry. Accept answers that divide them into two equal parts.

2

Challenge ⚠

3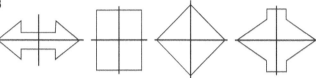

4 Learner's explanation, for example: If a shape has two identical halves it shows symmetry.

5 Each shape should have at least three lines.

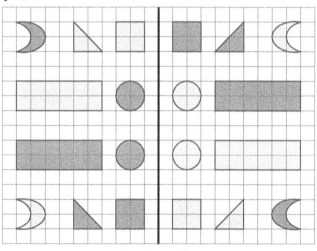

6 The third shape. It is the only one that is not symmetrical.

Challenge ❸

7

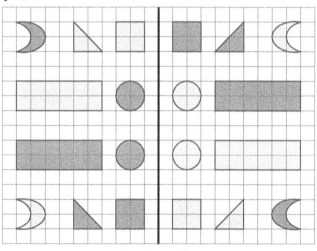

line of symmetry

8 Learner's own pattern. Check that it is symmetrical.

Lesson 4: Angles

Challenge ❶

1 1st, 4th, 5th and 7th ticked

Challenge ⚠
2   TV – 4 right angles, scissors – 2 right angles, clock – 1 right angle, filing cabinet – at least 6 right angles marked
3   **a**   quarter   **b**   two quarters or half
4   D. Learner's explanation, for example: It is the only one that is not a right angle.
5   4th and 5th angles circled

Challenge 3
6   Learner's own pattern
7   Learner's own response

Unit 21

Lesson 1: Identifying 3D shapes
Challenge ❶
1

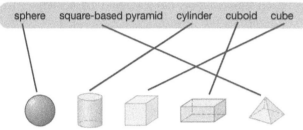

2   1 square face, 4 triangular faces, 8 edges, 5 vertices

Challenge ⚠
3

	square faces	no square faces
12 edges	cube	cuboid
not 12 edges	square-based pyramid	triangular-based pyramid sphere cylinder

4   Cylinder because it is the only shape that has a curved surface.
    Cube because it is the only shape with all faces identical.
    Accept other solutions if they are correct.
5   Triangular-based pyramid or tetrahedron

Challenge 3
6   Cube, cuboid, square-based pyramid

Lesson 2: Prisms
Challenge ❶
1   Triangular prism (circled), sphere, triangular-based pyramid, rectangular prism or cuboid (circled)
2   Prisms and not prisms or prisms and other shapes

Challenge ⚠
3

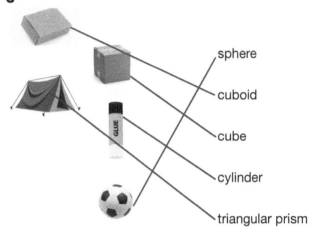

sphere
cuboid
cube
cylinder
triangular prism

4   **a**   1st, 3rd and 6th shapes ticked
    b   6, 8, 12, yes
    c   Learner's explanation, for example: A cube has 6 square faces. A cuboid has either 2 square faces and 4 rectangular faces or 6 rectangular faces.
5   Ensure that the drawing looks like a cuboid.

Challenge 3
6   Learner's ideas, for example: All 3D shapes, all have faces, all are prisms, different shaped ends, different number of edges, different number of vertices

Lesson 3: Pyramids
Challenge ❶
1   1st and 5th shapes ticked
2   Learner's explanation, for example: Pyramids have triangular faces. The sides meet at a single vertex.

Challenge ⚠
3   Sometimes. Learner's explanation, for example: A triangular-based pyramid is made from triangles. Other pyramids are made from triangles and one other shape.
4   False. Learner' explanation, for example: Triangular pyramids have 4 faces, triangular prisms have 5 faces.
5   Yes. Learner's explanation, for example: A hexagon has 6 sides. Each side has a face coming from it, that makes a total of 7.

Challenge 3
6   Triangular prism, two different pyramids

Lesson 4: 3D shapes in real life
Challenge ❶
1   Learner's ideas, e.g. tissue box
2   Learner's ideas, e.g. football
3   Learner's ideas, e.g. cuboid - box; cylinder - glue stick

Challenge ⚠
4 2 spheres, 2 triangular prisms, 1 triangular-based pyramid, 1 square-based pyramid, 2 cubes, 2 cuboids, 2 cylinders.
5 Learner's ideas, e.g. pencil and cylinder
6 Box circled. Learner's explanation, for example: It is the only object with all flat faces.

Challenge 3
7 Learner's own ideas, e.g. cubes and cuboids

Unit 22

Lesson 1: Units of length
Challenge 1
1 400 cm. Learner's explanation
2 2000 m. Learner's explanation

Challenge ⚠
3 5 cm, 25 cm, 12 m, 41 m, 3 km, 10 km
4 10
5 4
6 Learner's examples, e.g. 50 m = 5000 cm, 1 m = 100 cm
7 Learner's examples and explanations

Challenge 3
8 a 3300 m   b 330 000 cm
9 Learner's own examples and strategies. Ensure they are correct.

Lesson 2: Measuring lines
Challenge 1
1 5 cm, 10 cm, 4 cm, 1 cm, 12 cm
2 Ensure line is 11 cm long.
3 Ensure line is 8 cm long.

Challenge ⚠
4 a Ensure line is 3 cm long.
   b Ensure line is 12 cm long.
   c Ensure line is 15 cm long.
5 5 cm, 9 cm, 12 cm, 8 cm, 2 cm
6 No. Learner's explanation, for example: You need to place the 0 mark on the ruler at the beginning of the line, then read the length where the line ends.

Challenge 3
7 Learner's work

Lesson 3: Perimeter
Challenge 1
1 a 30 cm   b 20 cm
2 Learner's explanation

Challenge ⚠
3 a 18 cm   b 22 cm   c 20 cm
4 40 cm, 20 cm, 20 cm, 24 cm, 18 cm, 18 cm
5 a 1 cm × 9 cm, 2 cm × 8 cm, 3 cm × 7 cm, 4 cm × 6 cm
   b Learner's description, for example: The width increases by 1 cm and the length decreases by 1 cm.

Challenge 3
6 a 4 cm × 2 cm   b 8 cm × 4 cm
   c 10 cm × 5 cm

Lesson 4: Area
Challenge 1
1 a 6   b 12   c 18   d 20
2 Learner's explanation, for example: I counted the squares.

Challenge ⚠
3 a Ensure that a 4 × 4 grid is drawn
   b 16 square units
   c Learner's choice
   d 16 square units
4 Learner's activity - ensure all rectangles have an area of 12 square units.
5 Perimeters depend on shapes drawn.

Challenge 3
6 a 12 square units
   b 12 square units
   c 12 square units.
   Learner's explanation, for example: Different shapes can have the same area.

Unit 23

Lesson 1: Units of mass
Challenge 1
1 a 1000   b 3000   c 7000
2 a 2   b 9   c 4   d 6

Challenge ⚠
3 a 50 kg, 2 kg, 800 g, $\frac{1}{2}$ kg, 100 g, 50 g
   b 10
   c 100
4 a 800 g   b 700 g
5 8 kg = 8000 g, 5000 g = 5 kg, 500 g = $\frac{1}{2}$ kg, $\frac{1}{4}$ kg = 250 g

Challenge ❸

6 7750 g

7 Many possibilities. Learner's examples

8 No. Learner's explanation, for example: If you add three zeros to 1.5 kg, you get 1.5000 g, which is incorrect.

Lesson 2: Measuring in kilograms

Challenge ❶

1 a 2nd, 4th, 5th and 6th images ticked

   b Learner's own explanation, for example: I know what a kilogram bag of rice feels like and these would be heavier.

2 Learner's choices

Challenge ⚠

3 Learner's examples - ensure that estimations are accurate

4 Based on learner's examples.

5 a 900 g b 250 g c 650 g

6 Pointers at each mass shown

Challenge ❸

7 Learner's definition

8 1 kg 850 g

Lesson 3: Measuring in grams

Challenge ❶

1 a 3rd, 4th items ticked. Accept 1st and 5th.

   b Grams

2 Learner's choices, e.g. notebook

Challenge ⚠

3 Learner's examples, e.g. 500 g and 200 g

4 a 900 g b 250 g c 650 g

5 Pointers at each mass shown.

Challenge ❸

6 a 600 g b 100 g c 900 g

Lesson 4: Measuring in kilograms and grams

Challenge ❶

1 1 kg 500 g; ensure learner's pointer is in the 1500 g position.

2 1 kg 750 g; ensure learner's pointer is in the 1750 g position.

Challenge ⚠

3 3600 g

4 Yes. Learner's own correct examples

5 Many possibilities. Learner's choice, e.g. 4 × 1 kg

Challenge ❸

6 3 kg 200 g

7 Yuko 600 g, Samuel 1800 g or 1 kg 800 g

Unit 24

Lesson 1: Units of capacity

Challenge ❶

1 a 4000 b 8000 c 7000 d 3000

2 a 6 b 7 c 5 d 9

Challenge ⚠

3 a 3 l, 1 l, 900 ml, $\frac{1}{2}$ l, 400 ml, $\frac{1}{4}$ l

   b 10

   c 100

4 a 5 b 10 c 10

5 No. Capacity is the amount of liquid a container will hold when it's full.

6 15 litres. Learner's explanation, for example: It's the only capacity in litres.

Challenge ❸

7 100 ml + 200 ml + 400 ml = 700 ml

8 Several possibilities, learner's answers should be totals possible using two of the given amounts

Lesson 2: Measuring capacity

Challenge ❶

1 a 1st, 4th, 5th pictures ticked b millilitres

2 a 3rd, 4th and 6th pictures ticked

   b Learner's explanation, for example: These are too big to be measured in millilitres.

Challenge ⚠

3 300 ml. Learner's explanation

4 Sometimes true. Results depend on containers used.

5 They both have the same, 1500 ml.

Challenge ❸

6 a 1 l b 250 ml c 100 ml

7 a First 1 l and 250 ml crossed out

   b 30 ml and 500 ml crossed out

Lesson 3: Measuring in litres and millilitres

Challenge ❶

1 a 300 ml b 750 ml

   c 700 ml d 500 ml

2 Ensure learner's shading is correct.

Challenge ⚠

3 Learner's answers. Ensure they are between the parameters.

4 6 l 500 ml

5 5 l 800 ml, 7500 ml, 8 l 100 ml, 4200 ml

Challenge ❸

6 Learner's diagram. Check it shows 1500 ml or 1 l 500 ml.

Lesson 4: Temperature

Challenge ❶
1 a 18°C   b 4°C   c 13°C
  d 70°C   e 55°C

Challenge ⚠
2 Ensure learner's shading shows correct temperatures.
3 64°F. Learner's explanation, for example: It is the only temperature with the units in Fahrenheit.

Challenge ❸
4 Ensure that learners have drawn a thermometer with the divisions and labels. The temperature should show 8°C.
5 Learner's choice and explanation

Unit 25

Lesson 1: Position

Challenge ❶
1 The cube is on top of the cylinder.
2 Learner's drawing. Make sure the triangle is to the right of the square.

Challenge ⚠
3 a The sphere is on top of the cube
  b The sphere is behind the cube.
4 The cube is on top of the cylinder, the cube is under the pyramid, the cube is between the pyramid and the cylinder
5 Learner's drawing. Make sure the pentagon is to the right of the square and the circle is to the right of the pentagon.
6 a The sphere is to the left of/next to the cylinder.
  b The cylinder is under the pyramid/to the right of the sphere.
  c The pyramid is on top of the cylinder.
7 a sphere, cylinder, cube
  b blue, yellow, red

Challenge ❸
8 Learner's ideas, with an accurate description of the shape's position

Lesson 2: Direction and movement

Challenge ❶
1 2nd, 4th and 7th ticked
2 5th and 7th ticked

Challenge ⚠
3 a Half a turn
  b The same as where you started

4
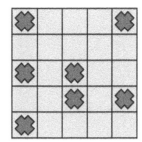

5 Learner's drawing showing a left turn
6 Learner's drawing showing a right turn
7 Yes. Learner's explanation

Challenge ❸
8 Learner's ideas - accurate sentences to describe their crosses
9 Learner's ideas, accurate sentences to describe movement

Lesson 3: Compass points

Challenge ❶
1 a South   b West
  c North   d East
2 West and north

Challenge ⚠
3 a East   b South   c South
4 Step 2 east for 3 squares
   Step 3 south for one square
   Step 4 west for 2 squares
   Step 5 south for 2 squares
5 Step 2 east for 4 squares
   Step 3 north for 2 squares
   Step 4 west for one square
   Step 5 north for 2 squares
6 West

Challenge ❸
7 a Learner's ideas, for example: Same number of squares covered
  b North is in different directions
  c Step 2 west for 3 squares, Step 3 south for 2 squares, Step 4 east for 4 squares, Step 5 north for 2 squares, Step 6 east for one square
  d Step 2 east for 3 squares, Step 3 north for 2 squares, Step 4 west for 4 squares, Step 5 south for 2 squares, Step 6 west for one square

Lesson 4: Reflections

Challenge ❶
1

2

Challenge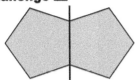

3

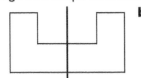

Learner's explanation, for example: The pattern is symmetrical.

4 No. Learner's explanation, for example: It is an image of a shape as it would look like in a mirror.

5 a  **b**

6 a 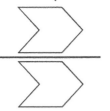 **b**

7 Learner's own thoughts

Challenge 🄸

8

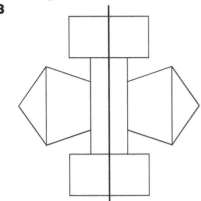

Unit 26

Lesson 1: Venn diagrams

Challenge ❶

1 Learner's explanation, for example: Children and cats both have noses. Children have two legs but cats do not. Cats are furry but children are not.

Challenge 🄰

2 6, 12, 18

3 a Curved side: circle. Straight side: triangle, square, pentagon. Intersection: semicircle.

   b Learner's statements. Ensure they are correct.

Challenge 🄸

4 First circle: multiples of 3, second circle: multiples of 4

5 Learner's exercise, answers may vary and should have a logical answer for the fruit Mr Plum should choose based on their class's most popular fruit.

Lesson 2: Carroll diagrams

Challenge ❶

1 4 sides or fewer: circle, semicircle, triangle, square, rectangle. Not 4 sides or fewer: pentagon, hexagon

Challenge 🄰

2 a 49, 65 **b** 12, 24

3 a

	Even number	Not even number
Multiples of 3	6, 36	27, 45
Not multiples of 3	8, 16	19, 35

   b Check learner's additional numbers.

   c Check learner's own statements.

4 Learner's explanation, for example: It is impossible to have an odd multiple of 2 and a number that is not odd and not a multiple of 2

Challenge 🄸

5 Multiples of 8/not multiples of 8 and multiples of 7/not multiples of 7.

6 Check learner's additional numbers.

7 Learner's exercise, answers may vary but should be correctly entered into the Carroll diagram.

Lesson 3: The statistical cycle

Challenge ❶

1 a blue 20, red 18, yellow 13, brown 8

   b 59

Challenge 🄰

2 a Kiwi fruit, mango and papaya

   b pomegranate

   c 45

3 No. Learner's explanation, for example: It's unlikely each person only bought one piece of fruit.

4 Learner's statements, accurately describing the tally chart.

Challenge 🄸

5 Learner's tally charts

   Learner's statements, accurately describing their results

Lesson 4: Frequency tables

Challenge ❶

1 a 18 **b** 5 **c** 21

2 Learner's table

Challenge 🄰

3 a Learner's table

   b 4 **c** 43 **d** 91

4 No. Learner's explanation

Challenge 🄸

5 Learner's table

   Learner's statements, accurately describing the results.

Unit 27

Lesson 1: Pictograms

Challenge ❶

1 Ensure learner's pictogram is accurate and labelled.

2 **a** Mango **b** 5 **c** 18

Challenge ⚠

3 **a** 4 **b** 16 **c** 4 **d** 55

4 Ensure learner's pictogram is accurate and labelled.

5 Learner's own statements, correctly describing their pictogram

Challenge ❸

6 Learner's own work - table correctly filled out and corresponding pictogram correctly drawn, which a logical choice for the sport the PE teacher should choose.

Lesson 2: Bar charts

Challenge ❶

1 Ensure learner's bar chart is correct and labelled.

2 Learner's own information, e.g. the most number of people like football

Challenge ⚠

3 **a** 4 **b** 18 **c** 30

4 **a** Ensure learner's bar chart is correct and labelled.

b Learner's statements correctly describing the data

Challenge ❸

5 Learner's explanation, e.g. they show numerical data

6 Learner's explanation, e.g. a pictogram uses images and a bar chart uses bars

7 **a** Ensure learner's pictogram is accurate and labelled.

b Ensure learner's bar chart is correct and labelled

c Learner's statements correctly describing the data

Lesson 3: Chance (1)

Challenge ❶

1 **a** and **d** ticked

Challenge ⚠

2 **a** Might happen (depends on learner)

b will not happen

c might happen (depends on learner)

d will happen

e might happen

f will not happen

3 **a** will not happen

b might happen

c might happen (depends on learner)

d might happen (depends on learner)

e will not happen

4 Learner's explanations, giving logical explanations for why they will not meet a dinosaur and might each some rice.

5 Yes. Learner's explanation, for example: Maisie is correct about what chance is.

Challenge ❸

6 Learner's own statements, correctly written according to the chance phrases.

Lesson 4: Chance (2)

Challenge ❶

1 **a** might happen **b** might happen

c might happen **d** will not happen

e Learner's explanation, for example: There is no 7, so it is not possible.

Challenge ⚠

2 Yes. Learner's explanation, for example: There are more grey cubes than white.

3 **a** No. Learner's explanation, for example: I might also spin a 3, 5 or 6.

b No. Learner's explanation, for example: It is possible that I might spin an odd number.

c No. Learner's explanation, for example: There is a 3, so it is possible that I might spin a 3.

d Yes. Learner's explanation, for example: The spinner has these numbers so I might spin one.

e Learner's own statements

4 Yes. Learner's explanation, for example: There is no yellow cube, so it is not possible.

Challenge ❸

5 No. Learner's explanation, for example: There are only two options for tossing s coin, but 20 options for numbers 1 to 20. Samir is more likely to get heads.

6 No. Learner's explanation, for example: There are six options. There is one chance of spinning 6 and 5 chances of spinning another number. So Lauren is likely to spin numbers other than 6 more times than a 6.

Stage 3 Record-keeping

Class: _____ **Year:** _____

KEY

A: Exceeding expectations in this sub-strand	B: Meeting expectations in this sub-strand	C: Below expectations in this sub-strand

Strand: **Number** Sub-strand: **Counting and sequences**		
Code	**Learning objectives**	
3Nc.01	Estimate the number of objects or people (up to 1000).	
3Nc.02	Count on and count back in steps of constant size: 1-digit numbers, tens or hundreds, starting from any number (from 0 to 1000).	
3Nc.03	Use knowledge of even and odd numbers up to 10 to recognise and sort numbers.	
3Nc.04	Recognise the use of an object to represent an unknown quantity in addition and subtraction calculations.	
3Nc.05	Recognise and extend linear sequences, and describe the term-to-term rule.	
3Nc.06	Extend spatial patterns formed from adding and subtracting a constant.	
A	B	C

Class: _____ **Year:** _____

Strand: **Number**		
Sub-strand: **Integers and powers**		
Code	**Learning objectives**	
3Ni.01	Recite, read and write number names and whole numbers (from 0 to 1000).	
3Ni.02	Understand the commutative and associative properties of addition, and use these to simplify calculations.	
3Ni.03	Recognise complements of 100 and complements of multiples of 10 or 100 (up to 1000).	
3Ni.04	Estimate, add and subtract whole numbers with up to three digits (regrouping of ones or tens).	
3Ni.05	Understand and explain the relationship between multiplication and division.	
3Ni.06	Understand and explain the commutative and distributive properties of multiplication, and use these to simplify calculations.	
3Ni.07	Know 1, 2, 3, 4, 5, 6, 8, 9 and 10 times tables.	
3Ni.08	Estimate and multiply whole numbers up to 100 by 2, 3, 4 and 5.	
3Ni.09	Estimate and divide whole numbers up to 100 by 2, 3, 4 and 5.	
3Ni.10	Recognise multiples of 2, 5 and 10 (up to 1000).	
A	B	C

Strand: **Number**		
Sub-strand: **Money**		
Code	**Learning objective**	
3Nm.01	Interpret money notation for currencies that use a decimal point.	
3Nm.02	Add and subtract amounts of money to give change.	
A	B	C

Class: _____ **Year:** _____

Strand: **Number**		
Sub-strand: **Place value, ordering and rounding**		
Code	**Learning objectives**	
3Np.01	Understand and explain that the value of each digit is determined by its position in that number (up to 3-digit numbers).	
3Np.02	Use knowledge of place value to multiply whole numbers by 10.	
3Np.03	Compose, decompose and regroup 3-digit numbers, using hundreds, tens and ones.	
3Np.04	Understand the relative size of quantities to compare and order 3-digit positive numbers, using the symbols =, > and <.	
3Np.05	Round 3-digit numbers to the nearest 10 or 100.	
A	B	C

Strand: **Number**		
Sub-strand: **Fractions, decimals, percentages, ratio and proportion**		
Code	**Learning objectives**	
3Nf.01	Understand and explain that fractions are several equal parts of an object or shape and all the parts, taken together, equal one whole.	
3Nf.02	Understand that the relationship between the whole and the parts depends on the relative size of each, regardless of their shape or orientation.	
3Nf.03	Understand and explain that fractions can describe equal parts of a quantity or set of objects.	
3Nf.04	Understand that a fraction can be represented as a division of the numerator by the denominator (half, quarter and three-quarters).	
3Nf.05	Understand that fractions (half, quarter, three-quarters, third and tenth) can act as operators.	
3Nf.06	Recognise that two fractions can have an equivalent value (halves, quarters, fifths and tenths).	
3Nf.07	Estimate, add and subtract fractions with the same denominator (within one whole).	
3Nf.08	Use knowledge of equivalence to compare and order unit fractions and fractions with the same denominator, using the symbols =, > and <.	
A	B	C

Class: _____ **Year:** _____

Strand: **Geometry and Measure** Sub-strand: **Time**		
Code	**Learning objectives**	
3Gt.01	Choose the appropriate unit of time for familiar activities.	
3Gt.02	Read and record time accurately in digital notation (12-hour) and on analogue clocks.	
3Gt.03	Interpret and use the information in timetables (12-hour clock).	
3Gt.04	Understand the difference between a time and a time interval. Find time intervals between the same units in days, weeks, months and years.	
A	B	C

Strand: **Geometry and Measure** Sub-strand: **Geometrical reasoning, shapes and measurements**		
Code	**Learning objectives**	
3Gg.01	Identify, describe, classify, name and sketch 2D shapes by their properties. Differentiate between regular and irregular polygons.	
3Gg.02	Estimate and measure lengths in centimetres (cm), metres (m) and kilometres (km). Understand the relationship between units.	
3Gg.03	Understand that perimeter is the total distance around a 2D shape and can be calculated by adding lengths, and area is how much space a 2D shape occupies within its boundary.	
3Gg.04	Draw lines, rectangles and squares. Estimate, measure and calculate the perimeter of a shape, using appropriate metric units and area on a square grid.	
3Gg.05	Identify, describe, sort, name and sketch 3D shapes by their properties.	
3Gg.06	Estimate and measure the mass of objects in grams (g) and kilograms (kg). Understand the relationship between units.	
3Gg.07	Estimate and measure capacity in millilitres (ml) and litres (l), and understand their relationships.	
3Gg.08	Recognise pictures, drawings and diagrams of 3D shapes.	
3Gg.09	Identify both horizontal and vertical lines of symmetry on 2D shapes and patterns.	
3Gg.10	Compare angles with a right angle. Recognise that a straight line is equivalent to two right angles or a half turn.	
3Gg.11	Use instruments that measure length, mass, capacity and temperature.	
A	B	C

Class: _____ **Year:** _____

Strand: **Geometry and Measure** Sub-strand: **Position and transformation**		
Code	**Learning objective**	
3Gp.01	Interpret and create descriptions of position, direction and movement, including reference to cardinal points.	
3Gp.02	Sketch the reflection of a 2D shape in a horizontal or vertical mirror line, including where the mirror line is the edge of the shape.	
A	B	C

Strand: **Statistics and Probability** Sub-strand: **Statistics**		
Code	**Learning objectives**	
3Ss.01	Conduct an investigation to answer non-statistical and statistical questions (categorical and discrete data).	
3Ss.02	Record, organise and represent categorical and discrete data. Choose and explain which representation to use in a given situation: - Venn and Carroll diagrams - tally charts and frequency tables - pictograms and bar charts.	
3Ss.03	Interpret data, identifying similarities and variations, within data sets, to answer non-statistical and statistical questions and discuss conclusions.	
A	B	C

Strand: **Statistics and Probability** Sub-strand: **Probability**		
Code	**Learning objectives**	
3Sp.01	Use familiar language associated with chance to describe events, including 'it will happen', 'it will not happen', 'it might happen'.	
3Sp.02	Conduct chance experiments, and present and describe the results.	
A	B	C

Class: _____ **Year:** _____

Thinking and Working Mathematically		317
Code	**Characteristics**	
TWM.01	Specialising Choosing *an example* and checking to see if it satisfies or does not satisfy specific mathematical criteria.	
TWM.02	Generalising Recognising an underlying pattern by identifying *many* examples that satisfy the same mathematical criteria.	
TWM.03	Conjecturing Forming mathematical questions or ideas.	
TWM.04	Convincing Presenting evidence to *justify or challenge* a mathematical idea or solution.	
TWM.05	Characterising Identifying and describing the mathematical properties of an object.	
TWM.06	Classifying Organising objects into groups according to their mathematical properties.	
TWM.07	Critiquing Comparing and evaluating mathematical ideas, representations or solutions to identify advantages and disadvantages.	
TWM.08	Improving Refining mathematical ideas or representations to develop a more effective approach or solution.	

A	B	C

Cambridge Global Perspectives™

Below are some examples of lessons in *Collins International Primary Maths Stage 3* which could be used to develop the Global Perspectives skills. The notes in *italics* suggest how the maths activity can be made more relevant to Global Perspectives.

Please note that the examples below link specifically to the learning objectives in the Cambridge Global Perspectives curriculum framework for Stage 3. However, skills development in a wider sense is embedded throughout this course and teachers are encouraged to promote research, analysis, evaluation, reflection, collaboration and communication as general best practice. For example, the pair work and group activities suggested throughout this Teacher's Guide offer opportunities to develop skills in communication, collaboration and reflection which build towards the specific Global Perspectives learning objectives.

Cambridge Global Perspectives	Learning Objectives for Stage 3	Collins International Primary Maths Stage 3
RESEARCH	Recording findings • Select, organise and record information from sources and findings from research in simple charts or diagrams	• Unit 20, Additional practice activity 1: Resource sheet 14 *After completion, discuss the criteria learners selected and how well they were able to organise the information on the sheet.* • Unit 21, Additional practice activity 1 *After completion, discuss the criteria learners selected and how well they were able to organise the information on the Carroll diagram.* • Unit 26, Lessons 1–4 *At the end of the Unit, encourage learners to compare the different ways of recording findings and reflect on which they find the easiest to use, how useful they are for different kinds of information, etc.* • Unit 27, Lessons 1–2 *Encourage learners to compare the different ways of recording findings in this and the previous Unit, and reflect on which they find the easiest to use, how useful they are for different kinds of information, etc.*
ANALYSIS	Interpreting data • Draw simple conclusions from graphical or numerical data	• Unit 26, Lessons 1–4 *In each lesson, ask questions to focus on the information the diagrams and tables present and encourage learners to draw conclusions.* • Unit 27, Lessons 1–2 *In each lesson, ask questions to focus on the information the pictograms and charts present and encourage learners to draw conclusions.*